MW01118937

ELSEVIER'S
SUGAR
DICTIONARY

ELSEVIER'S SUGAR DICTIONARY

in six languages

English/American, French, Spanish, Dutch,
German and Latin

compiled by

L.Y. CHABALLE

*Conference interpreter, technical translator,
Mons, Belgium*

ELSEVIER
AMSTERDAM -- OXFORD -- NEW YORK -- TOKYO 1984

ELSEVIER SCIENCE PUBLISHERS B.V.
Molenwerf 1
P.O. Box 211, 1000 AE Amsterdam, The Netherlands

Distributors for the United States and Canada:

ELSEVIER SCIENCE PUBLISHING COMPANY INC.
52, Vanderbilt Avenue
New York, N.Y. 10017

ISBN 0-444-42376-1

© Elsevier Science Publishers B.V., 1984

Printed in The Netherlands

Electronic data processing:
Büro für Satztechnik W. Meyer KG
Weissensberg, W. Germany

PREFACE

"So eine Arbeit wird eigentlich nie fertig.
Man muss sie für fertig halten, wenn man
nach Zeit und Umständen das Möglichste getan
hat." (Goethe)

Sugar has been an important factor in human nutrition for many centuries and is likely to remain one of the mainstays of our diet. The cultivation of sugar cane and sugar beet has had to be adapted to modern methods. In the manufacturing process, there has been a constant search for improvement in yields and reliability, and in the quality of sugar products. The European Economic Community has left its mark on the marketing of sugar.

The aim of this new Elsevier's Sugar Dictionary is to facilitate communication between all those involved, directly or indirectly, in the manufacture or marketing of cane sugar and beet sugar in their various forms: sugar beet and sugar cane growers, manufacturers of insecticides and agricultural equipment, sugar technologists, sugar plant designers and manufacturers, sugar manufacturers, sugar brokers, sugar research institutes, translators and interpreters.

The approach is descriptive. The author has examined numerous reliable scientific and technical documents (textbooks, handbooks, reference works, periodicals, glossaries, catalogues, brochures, etc.) and presents the reader with the terms and expressions thus found. Different companies, technologists and manufacturers may use different terms to describe almost similar processes. It should also be borne in mind that the terminology is constantly changing. Consequently, the author did not feel it was his responsibility to express a value judgement and to create standards.

This dictionary includes 2741 specific terms (+ synonyms and variants). All English/American entries are listed according to their English spelling, with their equivalents in French, German, Spanish, Dutch and Latin (weeds and pests). The dictionary also includes an alphabetical list for each language, so that corresponding translations may be readily found.

Certain terms are followed by a short definition in English and/or a qualifying label indicating the field in which they are used.

Special attention has been devoted to such areas as: sugar beet and sugar cane growing, agricultural machines, diseases, weeds and pests, beet and cane sugar manufacture (processes and machines), types of sugar, economic terms, international sugar organizations, EEC terminology of sugar regulations.

The choice of terms to be included was subjective and therefore contestable. The author would welcome criticisms, comments and suggestions from users, so that these may be taken into consideration in the next edition.

ACKNOWLEDGEMENT

The author would like to thank Dr. R.A. McGinnis, Editor of "Beet-Sugar Technology", published by the Beet Sugar Development Foundation, P.O. Box 1546, Fort Collins, Colo. 80522 (USA) and Mr. J.H. Fisher, Secretary-Manager of the Beet Sugar Development Foundation, for having given their permission for certain definitions from that work to be used in this dictionary. There can be no doubt that these short definitions will help the user of this dictionary to gain a better understanding of the complex and fascinating world of sugar manufacture.

The author would also like to thank Dr. P. Koronowski of the Biologische Bundesanstalt für Land- und Forstwirtschaft, Berlin (FRG), who provided German equivalents for terms relating to sugar cane diseases.

Finally, this dictionary would not have seen the light of day without the effective cooperation and judicious advice of the following institutes, organizations, engineering and design offices and manufacturers:
Alfa-Laval, Thermal Engineering, Lund (Sweden)
ABR Engineering, Brussels (Belgium)
BMA, Braunschweigische Maschinenbauanstalt, Braunschweig (FRG)
Bopack, Deurne (Belgium)
Buckau-Wolf, Grevenbroich (FRG)
Claas, Harsewinkel (FRG)
Commission of the European Communities, Brussels (Belgium)
CSM, Centrale Suikermaatschappij, Amsterdam (The Netherlands)
DDS, De Danske Sukkerfabrikker, Copenhagen (Denmark)
Dia-Prosim, Vitry-sur-Seine (France)
Dr. Wolfgang Kernchen, Seelze (FRG)
Faculté des Sciences Agronomiques de l'Etat, Gembloux (Belgium)
Fives-Cail Babcock, Paris (France)
Fulton Iron Works Co., St. Louis, Missouri (USA)
IIRB, Institut International de Recherches Betteravières, Brussels (Belgium)
Institut Belge pour l'Amélioration de la Betterave, Tienen (Belgium)
Institut für Landwirtschaftliche Technologie und Zuckerindustrie an der
    Technischen Universität Braunschweig, Braunschweig (FRG)
Institut für Zuckerrübenforschung, Göttingen (FRG)
IPRO, Industrieprojekt GmbH, Braunschweig (FRG)
J&L Honiron Engineering Company Inc., Jeanerette, Louisiana (USA)
Nederlandse Bietenfederatie, Kaagdorp (The Netherlands)
Peter Zeilfelder KG, Helmstedt (FRG)
Putsch Maschinenfabrik, Hagen (FRG)
Salzgitter Maschinen und Anlagen, Salzgitter (FRG)
Stork-Werkspoor Sugar, Hengelo (The Netherlands)

Süddeutsche Zucker AG, Mannheim (FRG)
Suikerstichting Nederland, Amsterdam (The Netherlands)
Syndicat Général des Producteurs de Sucre et de Rhum des Antilles Françaises, Paris (France)
Tate + Lyle, Croydon (UK)
Technische Universität Berlin, Abt. Zuckertechnologie, Berlin (FRG)
The Western States Machine Company, Hamilton, Ohio (USA)
Tiense Suikerraffinaderij, Tienen (Belgium)
Union Syndicale des Producteurs de Sucre et de Rhum de l'Ile de la Réunion, Paris (France)
Van Aarsen Machinefabriek B.V., Panheel (The Netherlands)
Verein der Zuckerindustrie, Bonn (FRG)

BIBLIOGRAPHY

Alexander, A.G., Sugarcane Physiology, Elsevier Scientific Publishing Company, Amsterdam, 1973

ATAC, La Habana, Cuba

Baxa, J., Bruhns, G., Zucker im Leben der Völker, Verlag Dr. A. Bartens, Berlin, 1967

Binkley, W.W., Composition of Cane Juice and Cane Final Molasses, Sugar Research Foundation, Inc., New York, 1953

Birch, G.G., Parker, K.J., Sugar: Science and Technology, Applied Science Publishers Ltd, London, 1979

Bruhns, G., e.a., Zuckertechniker-Taschenbuch, Verlag Dr. A. Bartens, Berlin, 1956

Cecil, F., Zuckerhausrechnungen und Saftschema-Modelle, Verlag Dr. A. Bartens, Berlin, 1957

Confédération Générale des Planteurs de Betteraves, L'Economie Betteravière 1982, SEDA, Paris, 1982

Dankowski, K., e.a., Zuckerwirtschaftliches Taschenbuch 1980/81, Verlag Dr. A. Bartens, Berlin, 1981

Dubourg, J., Sucrerie de Betteraves, Librairie J.-B. Baillière et Fils, Paris, 1952

Dunning, A., Byford, W., Pests, Diseases and Disorders of the Sugar Beet, Deleplanque & Cie, Maisons-Laffitte, 1982

Estord, G., L'Industrie du Sucre de Cannes, Dunod, Paris, 1958

Fauconnier, R., Bassereau, D., La Canne à Sucre, G.-P. Maisonneuve & Larose, Paris, 1970

Hickson, J.L., Sucrochemistry, American Chemical Society, Washington, 1977

Honig, P., Principles of Sugar Technology, Elsevier Publishing Company, Amsterdam, 1953—1963

Honig, P., Principios de Tecnología Azucarera, Compañía Editorial Continental, Mexico, 1969

Hugot, E., La Sucrerie de Cannes, Dunod, Paris, 1970

Hugot, E., Handbook of Cane Sugar Engineering, Elsevier Publishing Company, Amsterdam, 1972

Hugot, E., Manual para Ingenieros Azucareros, Compañía Editorial Continental, Mexico, 1963

Humbert, R.P., The Growing of Sugar Cane, Elsevier Publishing Company, Amsterdam, 1968

IIRB, Institut International de Recherches Betteravières, Sugar-Beet Glossary, Elsevier Publishing Company, 1967

International Sugar Journal, High Wycombe, England

Jenkins, G.H., Introduction to Cane Sugar Technology, Elsevier Publishing Company, Amsterdam, 1966

La Sucrerie Belge, Tienen, Belgium

Lejealle, F., e.a., Ennemis et Maladies de la Betterave Sucrière, Deleplanque & Cie, Maisons-Laffitte, 1982

Lopez, A.S., e.a., Los herbicidas y las malas hierbas en el cultivo de la remolacha azucarera, Estación Experimental de Aula dei Zaragoza, 1972

Lorenz, H., Grundlagen der Zuckerrübenproduktion, VEB Deutscher Landwirtschaftsverlag, Berlin, 1980

Lüdecke, H., Winner, C., Farbtafelatlas der Krankheiten und Schädigungen der Zuckerrübe, DLG-Verlag, Frankfurt am Main, 1966

Lüdecke, H., Zuckerrübenbau, Verlag Paul Parey, Berlin, 1961

Lyle, O., Technology for Sugar Refinery Workers, Chapman & Hall Ltd, London, 1957

Martin, J.P., e.a., Sugar-Cane Diseases of the World, Elsevier Publishing Company, Amsterdam, 1961—1964

McGinnis, R.A., Beet-Sugar Technology, Beet Sugar Development Foundation, Fort Collins, Colorado, 1982

Meade, G.P., Chen, J.C.P., Cane Sugar Handbook, John Wiley & Sons, New York, 1977

Müller, C.A., Glossary of Sugar Technology, Elsevier Publishing Company, Amsterdam, 1970

Murry, C.R., Holt, J.E., The Mechanics of Crushing Sugar Cane, Elsevier Publishing Company, Amsterdam, 1967

Pancoast, H.M., Junk, W.R., Handbook of Sugars, AVI Publishing Company, Inc., Westport, Connecticut, 1980

Paturau, J.M., By-products of the Cane Sugar Industry, Elsevier Publishing Company, Amsterdam, 1969

Payne, J.H., Sugar Cane Factory Analytical Control, Elsevier Publishing Company, Amsterdam, 1968

Prinsen Geerligs, H.C., Handboek ten Dienste van de Suikerriet-Cultuur en de Rietsuiker-Fabricage op Java, J.H. De Bussy, Amsterdam, 1915

Quillard, Ch., La Sucrerie de Betteraves, Librairie J.-B. Baillière et Fils, Paris, 1932

Saillard, E., Betterave et Sucrerie de Betteraves, Librairie J.-B. Baillière et Fils, Paris, 1913

Schäufele, W.R., Schädlinge und Krankheiten der Zuckerrübe, Verlag Th. Mann, Gelsenkirchen, 1982

Schneider, F., Sugar Analysis, ICUMSA, Peterborough, 1979

Schneider, F., e.a., Technologie des Zuckers, Verlag M. & H. Schaper, Hannover, 1968

Smit, M.J., De Beetwortelsuiker-Industrie, Servire, Den Haag, 1953

Société Sucrière d'Etudes et de Conseils, Manuel de Sucrerie, S.S.E.C., Tienen, 1979

Spencer, G.L., Meade, G.P., Manual del Azúcar de Caña, Montaner y Simon, Barcelona, 1967

Sucrerie Française, Paris, France

Sugar Journal, New Orleans, USA

Sugar Y Azucar, Fort Lee, N.J., USA

Suikerbietblad, Tienen, Belgium

Sümmermann, K.-H., Arbeitsabläufe in der Zuckerrübenernte, Verlag Eugen Ulmer, Stuttgart, s.d.

Taiwan Sugar, Taipei, Taiwan

Tödt, F., Betriebskontrolle und Messwesen in der Rübenzuckerindustrie, Naturwissenschaftlicher Verlag vorm. Gebrüder Borntraeger, Berlin, 1949

Urban, D., e.a., Die Zuckerherstellung, VEB Fachbuchverlag, Leipzig, 1980

Van Steyvoort, L., Ziekten en Plagen van de Suikerbiet, Deleplanque & Cie, Maisons-Laffitte, 1982

Vavrinecz, G., Atlas der Zuckerkristalle, Atlas of Sugar Crystals, Verlag Dr. A. Bartens, Berlin, 1965

Verein der Zuckerindustrie, Die Zuckermarktordnung nach dem Zweiten Weltkrieg, Verein der Zuckerindustrie, Bonn, 1975

Villarías Moradillo, J.L., Plagas y Enfermedades de la Remolacha Azucarera, Deleplanque & Cie, Maisons-Laffitte, 1982

Vukov, K., Physics and Chemistry of Sugar-Beet in Sugar Manufacture, Elsevier Scientific Publishing Company, Amsterdam, 1977

Vukov, K., Physik und Chemie der Zuckerrübe als Grundlage der Verarbeitungsverfahren, Akadémiai Kiadó, Budapest, 1972

Williams, J.R., e.a., Pests of Sugar Cane, Elsevier Publishing Company, Amsterdam, 1969

Winner, C., Zuckerrübenbau, DLG-Verlag, Frankfurt (Main), 1981

Zuckerindustrie, Berlin, FRG

RELATED DICTIONARIES

As indicated in the preface, the author has used only specific terms. The following dictionaries may prove useful complements to Elsevier's Sugar Dictionary:

Dorian, A.F., Dorian's Dictionary of Science and Technology, Elsevier Scientific Publishing Company, 1978—1981

Dorian, A.F., Elsevier's Dictionary of Chemistry, Elsevier Scientific Publishing Company, Amsterdam, 1983

Haensch, G., Haberkamp de Antón, G., Dictionary of Agriculture, Elsevier Scientific Publishing Company, Amsterdam, 1975

Haensch, G., Haberkamp de Antón, G., Dictionary of Biology, Elsevier Scientific Publishing Company, Amsterdam, 1981

Macura, P., Elsevier's Dictionary of Botany, Elsevier Scientific Publishing Company, Amsterdam, 1979

Meinck, F., Möhle, K., Dictionary of Water and Sewage Engineering, Elsevier Scientific Publishing Company, Amsterdam, 1977

Merino-Rodriguez, M., Lexicon of Plant Pests and Diseases, Elsevier Scientific Publishing Company, Amsterdam, 1966

Morton, I.D., Morton, C., Elsevier's Dictionary of Food Science and Technology, Elsevier Scientific Publishing Company, Amsterdam, 1977

Williams, G., Elsevier's Dictionary of Weeds of Western Europe, Elsevier Scientific Publishing Company, 1982

GUIDE TO THE USE OF THIS DICTIONARY

This dictionary is divided into two sections:

Basic Table: The first section, called the "Basic Table", lists the English/ American entries in alphabetical order according to their English spelling. Each entry is followed by its equivalents in the other languages. The author has chosen strict alphabetical order, i.e., hyphens or spaces between the words forming one expression have no influence on the alphabetical arrangement. This also applies to punctuation marks such as parentheses and commas. A synonym is preceded by a semi-colon. Words (or parts of words) printed in parentheses may be neglected without altering the meaning of the phrase.

Indexes: French, German, Spanish, Dutch and Latin words are listed alphabetically in five separate indexes, which together form the second section of the dictionary. Each entry carries a number which refers to the relevant entry in the basic table. As in the basic table, strict alphabetical order has been used in the indexes. The tilde replaces the first word in the line.

Abbreviations and other letter symbols are listed in both the original and abridged form.

ABBREVIATIONS

agr	agriculture
eco	economic term
EEC	European Economic Community
f	feminine
m	masculine
n	neuter
sb	sugar beet
sc	sugar cane
US	term used in the United States
v	verb
f	Français
d	Deutsch
e	Español
n	Nederlands
l	Latin

Basic table

A

1 AB massecuite
(One third of syrup and two thirds of A
molasses. Four-massecuite system)
f masse *f* cuite AB
d AB-Füllmasse *f*; AB-Kochmasse *f*
e masa *f* cocida AB
n AB-masse cuite *f*; AB-vulmassa *f*; masse
cuite *f* AB

2 absolute juice
(All the dissolved solids in the cane plus
all the water; cane minus fiber)
f jus *m* absolu
d absoluter Saft *m*
e guarapo *m* absoluto; jugo *m* absoluto
n absoluut sap *n*

3 absolute pressure
f pression *f* absolue
d absoluter Druck *m*; Absolutdruck *m*
e presión *f* absoluta
n absolute druk *m*

4 absolute viscosity
f viscosité *f* absolue
d absolute Viskosität *f*
e viscosidad *f* absoluta
n absolute viscositeit *f*

* **acanthus bristle thistle** → **1778**

5 accounted losses; determined losses;
known losses
(Quantities of sucrose determined or
estimated by measurements and analyses
to be escaping the process through
known channels of loss)
f pertes *fpl* connues; pertes *fpl*
déterminées
d bestimmte Verluste *mpl*; bekannte
Verluste *mpl*
e pérdidas *fpl* determinadas
n bepaalde verliezen *npl*; bekende
verliezen *npl*

6 acetone
(By-product)
f acétone *m*
d Aceton *n*; Azeton *n*
e acetona *f*
n aceton *nm*

7 acidification *(of the juice)*
f acidification *f*
d Ansäuerung *f*; Säuerung *f*

e acidificación *f*
n aanzuring *f*; verzuring *f*; zuur maken *n*

8 acidify *v*
f acidifier
d ansäuern; säuern; sauer machen
e acidificar
n aanzuren; verzuren; zuur maken

9 acidity of juice; acidity in juice
f acidité *f* du jus
d Acidität *f* des Saftes; Saftsäure *f*
e acidez *f* del jugo; acidez *f* del guarapo
n aciditeit *f* van het sap; sapzuurte *f*

10 acid juice
f jus *m* acide
d saurer Saft *m*
e jugo *m* ácido; guarapo *m* ácido
n zuur sap *n*

11 acid soil injury
(sb; disease)
f pied *m* noir d'acidité; dégât *m* d'acidité
d Bodensäure-Schäden *mpl*
e daños *mpl* por acidez del suelo
n door zure grond veroorzaakte schade *f*

12 acid sulphitation
f sulfitation *f* acide
d saure Sulfitation *f*
e sulfitación *f* ácida
n zure sulfitatie *f*

13 ACP preferential sugar
(EEC)
f sucre *m* préférentiel ACP
d AKP-Präferenzzucker *m*
e azúcar *m* preferencial ACP
n preferentiële suiker *m* uit ACS-landen

14 ACP-sugar
(African, Carribean and Pacific States
(EEC))
f sucre *m* ACP
d AKP-Zucker *m*
e azúcar *m* ACP
n ACS-suiker *m*

* **A-C system** → **2567**

15 activated carbon; active carbon
f charbon *m* actif
d Aktivkohle *f*
e carbón *m* activado
n actieve kool *f*

16 actuating feeler; actuating finder *(US)*
(agr)

f tâteur *m* mobile
d beweglicher Taster *m*
e palpador *m* móvil
n beweegbare taster *m*

17 Adant cube process
f procédé *m* Adant
d Adantverfahren *n*; Gußwürfelverfahren *n*
e procedimiento *m* de Adant
n Adantproces *n*

* **addition, lime ~ → 1469**

18 adsorbent
f agent *m* adsorbant
d Adsorbens *n*; Adsorptionsmittel *n*
e adsorbente *m*
n adsorptiemiddel *n*

19 adsorption
f adsorption *f*
d Adsorption *f*
e adsorción *f*
n adsorptie *f*

20 advance fixing
(EEC)
f fixation *f* à l'avance; préfixation *f*
d Vorausfestsetzung *f*
e fijación *f* anticipada
n voorfixatie *f*; prefixatie *f*

21 advance fixing certificate
(EEC)
f certificat *m* de préfixation
d Vorausfestsetzungsbescheinigung *f*
e certificado *m* de fijación anticipada
n prefixatiecertificaat *n*

* **advance payment → 1731**

22 Advisory Committee on Sugar
(EEC)
f Comité *m* Consultatif du Sucre
d Beratender Ausschuß *m* für Zucker
e Comité *m* Consultivo del Azúcar
n Adviescomité *n* voor Suiker

23 aerial flume
f caniveau *m* aérien
d oberirdische Schwemmrinne *f*; Hochrinne *f*
e canal *m* de flotación aéreo; canal *m* de
flotación sobre el suelo; saetín *m* aéreo;
saetín *m* sobre el suelo
n bovengrondse zwemgoot *f*

* **affinability → 29**

**24 affinated sugar; affination sugar; washed
raw sugar; affined sugar; affinade**
(The processed raw sugar after
affination)
f sucre *m* affiné
d affinierter Zucker *m*; Affinade *f*
e azúcar *m* afinado; afinado *m*
n geaffineerde suiker *m*

25 affination; raw sugar washing
(Process for purifying an impure
crystalline sugar by mixing it with syrup
to form a magma, then spinning the
magma with or without washing)
f affinage *m*
d Affination *f*
e afinación *f*
n affinage *f*

26 affination centrifugal
f turbine *f* d'affinage
d Affinationszentrifuge *f*
e centrífuga *f* de afinación; centrífuga *f*
para azúcar refinado
n affinagecentrifuge *f*

27 affination magma
(Magma produced by mechanically
mixing crystals with syrup)
f magma *m* d'affinage
d Affinationsmaischmasse *f*
e magma *m* de afinación
n affinagemagma *n*

28 affination massecuite
f masse *f* cuite d'affinage
d Affinationsfüllmasse *f*
e masa *f* de afinación; masa *f* cocida de
afinación
n affinage-masse cuite *f*; affinage-
vulmassa *f*

29 affination properties *(of sugar)*;
affinability
f affinabilité *f*
d Affinierbarkeit *f*
e afinabilidad *f*; calidades *fpl* de afinación
n affineerbaarheid *f*

* **affination sugar → 24**

30 affination syrup; raw syrup
(The syrup produced by and used in the
affination process)
f égout *m* d'affinage
d Affinationsablauf *m*
e jarabe *m* de afinación
n affinagestroop *m*

31 affination yield
f rendement *m* d'affinage
d Affinationsausbeute *f*
e rendimiento *m* de afinación
n affinagerendement *n*;
affinageopbrengst *f*

32 affine *v*; **wash** *v (US)*
f affiner
d affinieren
e afinar
n affineren

* **affined sugar** → 24

33 affinity for water *(of bagasse)*
f affinité *f* pour l'eau
d Wasseraffinität *f*
e afinidad *f* por el agua
n affiniteit *f* voor water

34 after-crystallization
f cristallisation *f* ultérieure
d Nachkristallisation *f*
e cristalización *f* subsiguiente
n nakristallisatie *f*

35 after evaporator; concentrator
f post-évaporateur *m*
d Nachverdampfer *m*
e post-evaporador *m*
n naverdamper *m*

36 afterfilter
f filtre *m* de sécurité; filtre *m* de finition;
filtre *m* de postfiltration
d Nachfilter *n*
e postfiltro *m*
n nafilter *n*

37 afterproduct
(Describes any material obtained after
the white-sugar boiling, including remelt
sugars, molasses, etc)
f arrière-produit *m*; bas-produit *m*
d Nachprodukt *n*; NP *n*
e tercer producto *m*; producto *m* de bajo
grado
n naprodukt *n*

* **afterproduct massecuite** → 1509

**38 afterproduct sugar; low-grade sugar; low
raw sugar**
f sucre *m* de bas-produit; sucre *m*
d'arrière-produit; sucre *m* arrière-
produit; sucre *m* inférieur
d Nachproduktzucker *m*; drittes Produkt *n*
e azúcar *m* de la última templa; azúcar *m*

de segundo producto; azúcar *m* de baja
calidad
n naproduktsuiker *m*

39 after-reaction vessel *(2nd carbonatation)*
f bac *m* de post-réaction
d Nachreaktionsgefäß *n*
e depósito *m* de reacción posterior
n nareactievat *n*

* **after worker** → 1507

40 agglomerate *v*
f s'agglomérer
d agglomerieren; zusammenballen;
zusammenbacken
e aglomerarse
n agglomereren; samenklonteren;
samenbakken

41 agitating pan; tank with stirrer
f bac *m* à agitation; bac *m* mélangeur; bac
m agitateur
d Rührkessel *m*; Rührpfanne *f*
e recipiente *m* de agitación; tanque *m* de
agitación
n roerbak *m*

* **agitation** → 2288

* **agitator** → 2287

42 air-cooled crystallizer
f malaxeur *m* à refroidissement par air
d Luftkühlmaische *f*
e cristalizador *m* enfriado por aire
n kristallisoir *m* met luchtkoeling

43 air-heater
f réchauffair *m*; réchauffeur *m* d'air
d Lufterhitzer *m*
e calentador *m* de aire
n luchtverhitter *m*

44 air-heater with effective circulation
f réchauffair *m* à circulation méthodique
d Lufterhitzer *m* mit systematischem
Umlauf
e calentador *m* de aire con circulación
sistemática
n luchtverhitter *m* met systematische
circulatie

* **airing** → 2615

* **airlift beet pump** → 1540

45 air pump
f pompe *f* à air

d Luftpumpe *f*
e bomba *f* de aire
n luchtpomp *f*

* **Aleppo grass → 1366**

46 alfalfa looper *(US)*
(sb; pest)
l Autographa californica

47 alfalfa snout beetle *(US)***; lovage weevil**
(sb; pest)
f otiorrhynque *m* de la livèche; charançon *m* de la livèche; bécare *m*
d Liebstöckelrüßler *m*; Luzernerüßler *m*; brauner Lappenrüßler *m*
e gorgojo *m* del ligústico
l Otiorrhynchus ligustici; Brachyrhinus ligustici

48 alfalfa webworm *(US)*
(sb; pest)
l Loxostege commixtalis

* **alfilaria → 586**

* **alfileria → 586**

* **alfilerilla → 586**

49 alkali mallow *(US)*
(sb; weed)
l Sida hederacea

50 alkaline juice
f jus *m* alcalin
d alkalischer Saft *m*
e jugo *m* alcalino; guarapo *m* alcalino
n alkalisch sap *n*

51 alkaline sulphitation
(Sulphitation after liming)
f sulfitation *f* alcaline
d alkalische Sulfitation *f*
e sulfitación *f* alcalina
n alkalische sulfitatie *f*

52 alkalinity
(In the product streams of a beet-sugar factory, the result of a titration with standardized acid solution to a phenolphthalein endpoint or equivalent pH, expressed as g CaO per 100 ml. In water analysis the endpoint must be specified and the result is usually expressed as ppm or gpg $CaCO_3$)
f alcalinité *f*
d Alkalität *f*

e alcalinidad *f*
n alkaliteit *f*

53 alkalinization; alkalization
f alcalinisation *f*
d Alkalisierung *f*
e alcalinización *f*; alcalización *f*
n alkalisatie *f*

54 alkali salt
f sel *m* alcalin
d Alkalisalz *n*
e sal *f* alcalina
n alkalizout *n*

* **alkalization → 53**

55 alkalizing agent
f alcalinisateur *m*
d Alkalitätslieferant *m*
e alcalinizador *m*
n alkalisch reagens *n*

* **alkanet → 566**

* **allgood → 1173**

* **allied crane fly → 571**

56 allseed; many-seeded goosefoot
(sb; weed)
f chénopode *m* polysperme
d vielsamiger Gänsefuß *m*
e cenizo *m* de muchas semillas
n korrelganzevoet *m*
l Chenopodium polyspermum

57 alternaria leaf spot
(sb; disease)
f alternariose *f*; alternaria *f*
d Alternaria-Blattbräune *f*
e alternaria *f*
n alternariaziekte *f*

58 A massecuite; first massecuite; high-grade massecuite
f masse *f* cuite A; masse *f* cuite premier jet; masse *f* cuite de premier jet; m c *f* I
d A-Füllmasse *f*; Erstprodukt-Füllmasse *f*; A-Kochmasse *f*; Weißzuckerfüllmasse *f*
e masa *f* cocida A; masa *f* cocida primera; masa *f* cocida de primera; masa *f* cocida de alta calidad; masa *f* cocida de alta graduación
n A-masse cuite *f*; A-vulmassa *f*; masse cuite *f* A; masse cuite *f* eerste produkt; hooggradige masse cuite *f*; hooggradige vulmassa *f*

* **American armyworm** → **86**

* **American licorice** *(US)* → **2695**

59 amide saponification
f saponification f des amides
d Amidverseifung f
e saponificación f de las amidas;
  saponificación f amida
n verzeping f van de amiden

60 amino acid
f acide m aminé
d Aminosäure f
e aminoácido m
n aminozuur m

61 amino nitrogen
f azote m aminé
d Aminostickstoff m
e nitrógeno m amino
n aminostikstof f

62 amino nitrogen content *(of thick juice)*
f teneur f en azote aminé
d Aminostickstoffgehalt m
e contenido m de nitrógeno amino
n aminostikstofgehalte n

63 A molasses; first molasses; A syrup
f égout m A
d A-Ablauf m
e melaza f A; melaza f de primera; miel
  f A; miel f primera
n A-stroop f

64 amylase; diastase
f amylase f; diastase f
d Amylase f; Diastase f
e amilasa f; diastasa f
n amylase f; diastase f

65 angle of contact *(of the bagasse with a roller)*
f angle m de prise
d Greifwinkel m; Kontaktwinkel m
e ángulo m de contacto
n contacthoek m

66 angle of repose *(of dry sugar, bagasse)*
f talus m d'éboulement
d Schüttwinkel m; natürlicher Böschungs-
  winkel m
e ángulo m de resbalamiento
n natuurlijke hellingshoek m

67 angular velocity
f vitesse f angulaire
d Winkelgeschwindigkeit f
e velocidad f angular
n hoeksnelheid f

68 anion exchange resin; anion resin
f résine f échangeuse d'anions
d Anionenaustauscherharz n; Anionen-
  harz n
e resina f intercambiadora de aniones
n anionenuitwisselend hars n

69 annual meadowgrass; annual bluegrass *(US)*; **bluegrass** *(US)*
  (sb; weed)
f pâturin m annuel
d einjähriges Rispengras n; jähriges
  Rispengras n; jähriges Angergras n;
  Lommerrispengras n
e pelosa f; espiquilla f; cebadilla f;
  pelillo m; hierba f de punta; poa f anual
n straatgras n
l Poa annua

70 annual mercury; French mercury *(US)*; **garden mercury; herb mercury**
  (sb; weed)
f mercuriale f annuelle; vignette f
d einjähriges Bingelkraut n; Gartenbingel-
  kraut n; Hundskohl m; Merkurial-
  kraut n; Speckmelde f
e mercurial m
n eenjarig bingelkruid n
l Mercurialis annuus

* **annual nettle** → **2187**

* **annual sowthistle** *(US)* → **583**

* **annual yellow clover** → **2185**

71 annular space *(of vacuum pan)*
f espace m annulaire; intervalle m
  annulaire
d Ringraum m
e espacio m anular
n ringruimte f

72 anti-foaming agent; antifoam (agent); skimming medium; skimming agent; foam killer
f agent m antimousse; antimousse m
d Schaumbekämpfungsmittel n; Schaum-
  dämpfungsmittel n; Entschäumer m;
  Antischaummittel n
e medio m antispumante
n schuimwerend middel n

73 antifriction bearing sugar cane mill
f moulin m à cannes à roulements
d Wälzlager-Zuckerrohrmühle f

e molino *m* de caña de azúcar con
rodamientos de rodillos; molino *m*
cañero con rodamientos de rodillos
n rietsuikermolen *m* met lagers

* apc → 78

74 apical cavity
(sb; disease)
f cavité *f* dans la partie supérieure de la
racine
d Hohlköpfigkeit *f*
e caverna *f* apical
n kopuitholling *f*

75 apparent density
f densité *f* apparente
d scheinbare Dichte *f*
e densidad *f* aparente
n schijnbare dichtheid *f*

76 apparent dry matter
f matière *f* sèche apparente
d scheinbare Trockensubstanz *f*
e materia *f* seca aparente
n schijnbare droge stof *f*

77 apparent losses
f pertes *fpl* apparentes
d scheinbare Verluste *mpl*
e pérdidas *fpl* aparentes
n schijnbare verliezen *npl*

**78 apparent purity; apc; Pol percent spindle
Brix; app. pur.**
(The percentage proportion of sugar
determined by direct polarization on
solids, the solids being determined by
Brix hydrometer or by refractometer)
f pureté *f* apparente
d scheinbare Reinheit *f*
e pureza *f* aparente; pureza *f* en pol
porcentaje Brix
n schijnbare reinheid *f*

79 apparent solids
f solides *mpl* apparents
d scheinbare Feststoffe *mpl*
e sólidos *mpl* aparentes
n schijnbare vaste stoffen *fpl*

80 apparent temperature drop
f chute *f* apparente de température
d scheinbarer Temperaturabfall *m*; schein-
bares Temperaturgefälle *n*
e caída *f* aparente de temperatura
n schijnbare temperatuurdaling *f*

* app. pur. → 78

81 A-products
f produits *mpl* de premier jet
d A-Produkte *npl*
e productos *mpl* A
n A-produkten *npl*

* apron carrier → 2170

* apron conveyor → 2170

82 arc screen
f tamis *m* courbe
d Bogensieb *n*
e colador *m* parabólico; colador *m* de arco
n boogzeef *f*

83 area of evaporation; evaporating surface
(of evaporator vessel)
f surface *f* d'évaporation
d Heizfläche *f*
e superficie *f* de evaporación
n verdampingsoppervlak *n*

* areometer → 1285

84 armored scale
(sc; pest)
f diaspides *mpl*; diaspines *mpl*
d Deckelschildläuse *fpl*; schildtragende
Schildläuse *fpl*; gedeckelte Schildläuse
fpl; Austernschildläuse *fpl*
e diaspinos *mpl*
l Diaspididae

85 army cutworm (US)
(sb; pest)
l Chorizagrotis auxiliaris

86 armyworm (US); **American armyworm**
(sb; pest)
f chenille *f* légionnaire
d Heerwurm *m*
e cucumilla *f* ejército; gusano *m* militar;
gusano *m* soldado
l Pseudaletia unipuncta; Mythimma
unipuncta

87 arrow rot
(sc; fungous disease of inflorescence)
f pourriture *f* de la flèche
d Fusarium-Stengelfäule *f*
e pudrición *f* de la espiga
n toprot *n*

88 artificial circulation
(Ventilation)
f circulation *f* artificielle
d künstliche Zirkulation *f*; künstlicher
Umlauf *m*

e circulación *f* artificial
n kunstmatige circulatie *f*

89 asbestos
(Filter aid)
f amiante *f*
d Asbest *m*
e amianto *m*; asbesto *m*
n asbest *n*

90 ash
(The solid residue left after incineration
in the presence of oxygen)
f cendres *fpl*
d Asche *f*
e cenizas *fpl*
n as *f*

91 ash content
f teneur *f* en cendres
d Aschegehalt *m*
e contenido *m* de cenizas
n asgehalte *n*

92 ash gray blister beetle *(US)*
(sb; pest)
l Epicauta fabricii

**93 Association of the Professional
Organizations of the Sugar Trade for the
Countries of the EEC**
f Association *f* des Organisations
Professionnelles du Commerce des
Sucres pour les pays de la CEE;
ASSUC *f*
d Vereinigung *f* der Berufsorganisationen
des Zuckerhandels für die Länder der
EWG
e ASSUC *f*
n ASSUC *f*

94 A sugar; first sugar
f sucre *m* A; sucre *m* de premier jet
d A-Zucker *m*
e azúcar *m* A; azúcar *m* de primera;
azúcar *m* primera
n A-suiker *m*

95 A-sugar
(EEC)
f sucre *m* A
d A-Zucker *m*
e azúcar *m* A
n A-suiker *m*

96 asymmetric housing *(of cane mill)*
f chapelle *f* asymétrique
d asymmetrischer Ständer *m*
e virgen *f* asimétrica

* **A syrup** → 63

* **atmospheric tank** → 2076

97 atomizer
f atomiseur *m*; pulvérisateur *m*
d Düse *f*; Zerstäuber *m*
e aspersor *m* pulverizador
n verstuiver *m*

98 auto-filter
f auto-filtre *m*
d Autofilter *n*
e autofiltro *m*
n autofilter *n*

* **automatic feeder trap** → 100

99 automatic juice sampler
f échantillonneur *m* automatique de jus
d automatischer Saftprobenehmer *m*
e muestreador *m* automático de jugo;
tomador *m* automático de muestras de
jugo; muestreador *m* automático de
guarapo; tomador *m* automático de
muestras de guarapo
n automatische sapmonstertrekker *m*

**100 automatic monte-jus; michaelis;
automatic feeder trap**
f michaëlis *m*; purgeur *m* alimentateur
automatique; monte-jus *m* automatique
d automatischer Saftheber *m*;
automatische Saftpumpe *f*
e montajugos *m* automático; trampa *f* de
alimentación automática; Michaelis *m*
n automatische sappomp *f*; automatische
montejus *m*

**101 automatic pH-regulator; automatic pH-
controller**
f régulateur *m* automatique de pH
d automatischer pH-Regler *m*
e regulador *m* automático de pH
n automatische pH-regelaar *m*

**102 auxiliary cane carrier; cross cane
carrier; cross cane conveyor**
(Cane handling equipment)
f conducteur *m* de cannes auxiliaire;
conducteur *m* de cannes latéral
d zusätzlicher Rohrtransporteur *m*; Quer-
rohrtransporteur *m*; seitlicher Rohrtransporteur *m*
e conductor *m* auxiliar de cañas; conductor
m transversal de cañas
n dwarsliggende riettransporteur *m*

* **available sugar** → 1900

B

* **bachelor's-button** → 648

103 back boiling; recirculation of syrups; reintroduction of syrups
(The addition of lower-purity syrup to a pan boiling to adjust the purity to the desired figure)
f rentrée f d'égouts
d Rücknahme f von Abläufen
e entrada f de mieles
n stroopcirculatie f

* **back-end of the factory** → 2369

104 backing cloth; backing screen; backing liner; backing gauze; supporting screen
(in centrifugal basket)
f toile f de soutien
d Unterlagesieb n; Unterlagssieb n
e malla f de sostén; tela f de sostén; tamiz m reforzado; tela f de apoyo
n ondergaas n

105 back pressure
f contre-pression f
d Gegendruck m
e contrapresión f
n tegendruk m

106 back pressure steam turbine
f turbine f à vapeur à contre-pression
d Gegendruck-Dampfturbine f
e turbina f de vapor de contrapresión
n tegendruk-stoomturbine f

* **back roll(er)** *(of roller mill)* → 767

107 back-up chamber with flume gate valve
f compartiment m de retenue avec vanne dans le caniveau
d Rückstaukammer f mit Schwemmrinnen-schieber
e cámara f de embalse con compuerta en el canal de flotación

108 bacterial action
f action f bactérienne
d Bakterientätigkeit f
e acción f bacteriana
n bacteriewerking f

109 bacterial blight; black streak *(US)*
(sb; disease)
f Pseudomonas f
d Pseudomonas-Blattfleckenkrankheit f
e pseudomonas f

n Pseudomonas-bladvlekkenziekte f; bacteriële bladvlekkenziekte f

110 bacterial canker *(US)*
(sb; disease)
d bakterieller Wurzelkrebs m
n bacteriekanker m

* **bacterial canker** → 2723

111 bacterial disease
f maladie f bactérienne; bactériose f
d Bakterienkrankheit f; Bakteriose f
e enfermedad f bacteriana; bacteriosis f
n bacteriënziekte f; door bacteriën veroorzaakte ziekte f

* **bacterial gall** → 679

112 bacterial leaf gall
(sb; disease)
f tumeur f bactérienne
d bakterieller Blattumor m
e tumor m bacteriano
n knobbel m

* **bacterial leaf rot** → 114

113 bacterial mottle
(sc; bacterial disease)
f bigarrure f bactérienne
d bakterielle Scheckung f
e moteado m bacteriano
n bacteriële vlekkerigheid f

114 bacterial rot of leaves; bacterial leaf rot
(sb; disease)
f bactériose f aux feuilles; bactériose f des feuilles
d bakterielle Blattfäule f
e podredumbre f bacteriana de las hojas
n bacterieel bladrot n

115 bacterial soft rot
(sb; disease)
f pourriture f d'origine bactérienne; moisissure f grise de la racine
d bakterielle Schleimfäule f
e podredumbre f mucilaginosa
n bacterierot n

116 baffle
f chicane f
d Prallplatte f; Ablenkplatte f; Schikane f; Ablenkblech n; Prallblech n
e tabique m
n keerplaat f; stootplaat f; chicane f

117 bag v
f ensacher; mettre en sacs
d einsacken; absacken; sacken
e ensacar; encostalar
n in zakken plaatsen; in zakken doen;
   opzakken

* **bagacillo** → 713

* **bagacillo elevator** → 714

118 bagacillo fan
f ventilateur m à fine bagasse
d Feinbagassegebläse n; Feinbagasse-
   ventilator m
e ventilador m de bagacillo
n fijne-bagasseventilator m; fijne-
   ampasventilator m

* **bagacillo screen** → 715

* **bagacillo separator** → 716

119 bagasse; megass(e) *(British
   Commonwealth)*; **final bagasse; last-mill
   bagasse**
   (The residue of cane after crushing in
   one mill or a train of mills)
f bagasse f; mégasse f; bagasse f finale
d Bagasse f; Endbagasse f
e bagazo m; bagazo m final; bagazo m del
   último molino
n bagasse m; ampas m; eindbagasse m;
   eindampas m

* **bagasse blanket** → 130

120 bagasse board
   (By-product)
f panneau m de bagasse
d Bagassebrett n; Bagasseplatte f
e tablero m de bagazo
n bagasseplaat f; ampasplaat f

121 bagasse conveyor; bagasse carrier
f conducteur m de bagasse
d Bagassetransporteur m
e conductor m de bagazo
n bagassetransporteur m; ampascarrier m

122 bagasse dewatering
f déshydratation f de la bagasse
d Bagasseentwässerung f
e desaguado m del bagazo
n ontwatering f van de bagasse;
   ontwatering f van de ampas

* **bagasse diffusion** → 808

123 bagasse disintegrator
f désintégrateur m de bagasse
d Bagassedesintegrator m
e desintegrador m para bagazo
n bagassedesintegrator m; ampas-
   desintegrator m

124 bagasse elevator
f élévateur m de bagasse
d Bagasseelevator m
e elevador m de bagazo
n bagasse-elevator m; ampaselevator m

125 bagasse feeder
f alimentateur m de bagasse
d Bagassezuführer m
e alimentador m de bagazo
n bagassetoevoerinrichting f;
   ampastoevoerinrichting f

126 bagasse fibre
f fibre f de bagasse
d Bagassefaser f; Bagassefiber f
e fibra f de bagazo
n bagassevezel f; ampasvezel f

127 bagasse fibre board
f panneau m de fibre de bagasse
d Bagassefaserplatte f
e placa f de fibra de bagazo; tabla f de
   fibra de bagazo
n bagassevezelplaat f; ampasvezelplaat f

128 bagasse fuel
f combustible m de bagasse
d Bagassebrennstoff m
e combustible m de bagazo
n bagassebrandstof f; ampasbrandstof f

129 bagasse furnace
f four m à bagasse; foyer m à bagasse
d Bagasseofen m
e horno m de bagazo
n bagasseoven m; ampasoven m

130 bagasse layer; bagasse blanket
f couche f de bagasse
d Bagasseschicht f
e colchón m de bagazo; capa f de bagazo
n bagasselaag f; ampaslaag f

131 bagasse mill
f moulin m de repression
d Bagassemühle f
e molino m de bagazo
n bagassemolen m; ampasmolen m

* **bagasse opening** *(between top roller and
   back roller)* → 765

132 bagasse press
f presse f à bagasse
d Bagassepresse f
e prensa f de bagazo
n bagassepers f; ampaspers f

* **bagasse roll(er)** → 767

133 bagasse screen
f tamis m à bagasse
d Bagassesieb n
e tamiz m del bagazo
n bagassezeef f; ampaszeef f

134 bag filter
(A filter made of a series of fabric bags
hung vertically to spigots in a tube sheet
contained in a filter tank)
f filtre m à poches
d Beutelfilter n; Sackfilter n
e filtro m de bolsas
n zakfilter n

135 bagged sugar; packed sugar
f sucre m ensaché
d abgepackter Zucker m; gepackter
Zucker m
e azúcar m ensacado; azúcar m embolsado
n verpakte suiker m; suiker m in zakken

**136 bagging and filling weigher; bagging and
filling weighing machine**
f bascule f d'ensachage et de remplissage
d Absack- und Abfüllwaage f
e báscula f para ensacar y envasar
n zakkenvul- en afweegmachine f

137 bagging machine
f ensacheuse f
d Absackmaschine f
e máquina f de llenado de sacos; máquina
f llenadora de sacos; ensacadora f
n zakkenvulmachine f

138 bagging scale
f balance f ensacheuse
d Absackwaage f
e báscula f para ensacar
n weegschaal f voor het opzakken

139 bagging station
f poste m d'ensachage
d Absackstation f
e estación f de ensacado
n opzakstation n

140 bag of jute
f sac m de jute
d Jutesack m

e saco m de yute
n jutezak m

141 bag squasher; bag squeezer
f casseur m de sacs
d Sackzerquetscher m

* **Bahama grass** → 268

142 bale of bagasse
f balle f de bagasse
d Bagasseballen m
e bala f de bagazo
n baal f bagasse; baal f ampas

143 baling press
(Bagasse press)
f presse f à balles
d Ballenpresse f; Ballenpackpresse f
e prensa f embaladora
n balenpers f

144 ball mill
f broyeur m pulvérisateur à boulets
d Kugelmühle f
e molino m de bolas
n kogelmolen m

* **ball nettle** *(US)* → 1260

145 balls of massecuite
f grumeaux mpl de masse cuite
d Füllmasseknollen mpl; Füllmasseknoten
mpl
e grumos mpl de masa cocida
n masse cuiteklonters mpl;
vulmassaklonters mpl

146 band conveyor; belt conveyor
f transporteur m à courroie
d Bandförderer m; Gurtförderer m; Band-
transporteur m
e transportador m de correa; correa f
transportadora
n transportband m

147 banded flea beetle *(US)*
(sb; pest)
d Erdfloh m
l Systena taeniata

148 banded sclerotial disease
(sc; fungous disease)
d Blattscheidendürre f; Blattscheidenfäule f
e banda f esclerótica de la hoja
n rot n van de bladschede

149 band filter press
f filtre *m* à bande sous pression
d Bandfilterpresse *f*
e prensa *f* para cinta de filtro
n bandfilterpers *f*

* **barn grass** → 1722

* **barnyard grass** → 1722

* **barnyard millet** → 1722

**150 barometric column; barometric pipe;
barometric tail pipe; barometric leg
pipe; barometric tube; barometric leg
tube; barometric tail tube**
f colonne *f* barométrique; tube *m* de
condenseur barométrique
d Fallwasserrohr *n*; Fallrohr *n*
e columna *f* barométrica
n valwaterpijp *f*

151 barometric condenser; long condenser
f condenseur *m* barométrique
d barometrischer Kondensator *m*
e condensador *m* barométrico;
condensador *m* largo
n barometrische condensor *m*

* **barometric countercurrent condenser**
→ 665

* **barometric leg pipe** → 150

* **barometric leg tube** → 150

* **barometric pipe** → 150

* **barometric tail pipe** → 150

* **barometric tail tube** → 150

* **barometric tube** → 150

152 bar screen; grizzly
f grille *f* à barreaux
d Stabrost *m*; Stangenrost *m*
e parrilla *f* de barras
n staafzeef *f*

153 baryta
f baryte *f*
d Baryt *m*
e barita *f*
n bariet *n*

154 basal stem rot
(sc; fungous disease)
f pourriture *f* de la base de la tige

d Stengel- und Scheidenfäule *f*
e pudrición *f* de la base del tallo
n wortelstokziekte *f*

155 basic lead acetate
f acétate *m* basique de plomb
d basisches Bleiacetat *n*
e acetato *m* básico de plomo
n basisch loodacetaat *n*

156 basic price
(EEC)
f prix *m* de base
d Grundpreis *m*
e precio *m* de base
n basisprijs *m*

157 basic quantity
(EEC)
f quantité *f* de base
d Grundmenge *f*
e cantidad *f* de base
n basishoeveelheid *f*

158 basic quota
(EEC)
f quota *m* de base
d Grundquote *f*
e cuota *f* de base
n basisquotum *n*

159 basic regulation
(EEC)
f règlement *m* de base
d Grundverordnung *f*
e reglamento *m* de base
n basisverordening *f*

160 basic seed
(agr)
f semence *f* élite; graine *f* élite
d Elite-Saatgut *n*
n elite-zaad *n*

* **basket calandria** → 1090

* **basket calandria pan** → 1091

* **bastard alkanet** → 649

161 bastard sugar
f sucre *m* bâtard; bâtard *m*
d Bastardzucker *m*
n basterdsuiker *m*; bastaardsuiker *m*

162 batch carbona(ta)tion
f carbonatation *f* discontinue;
carbonatation *f* intermittente
d diskontinuierliche Saturation *f*;

diskontinuierliche Carbonatation *f*
e carbonatación *f* intermitente
n discontinue carbonatatie *f*

163 batch carbonator
f chaudière *f* à carbonater discontinue
d diskontinuierlicher Carbonatations-
behälter *m*
e carbonatador *m* discontinuo
n discontinue carbonatatieketel *m*

**164 batch centrifugal; discontinuous
centrifugal**
f centrifugeuse *f* discontinue; essoreuse *f*
discontinue; turbine *f* à sucre; centrifuge
f discontinue
d diskontinuierliche Zentrifuge *f*;
diskontinuierlich arbeitende Zentrifuge *f*;
periodische Zentrifuge *f*; periodisch
betriebene Zentrifuge *f*; chargenweise
arbeitende Zentrifuge *f*; periodisch
arbeitende Zentrifuge *f*
e centrífuga *f* discontinua; centrífuga *f* de
lote
n discontinue centrifuge *v*; discontinu
werkende centrifuge *f*

165 batch diffusion
f diffusion *f* discontinue; diffusion *f* en
vases
d diskontinuierliche Diffusion *f*
e difusión *f* discontinua
n discontinue diffusie *f*

* batch liming → 1333

166 batch scale
f balance *f* discontinue
d Chargenwaage *f*
e báscula *f* discontinua
n discontinue weegschaal *f*

**167 batch settler; intermittent settler;
intermittent subsider**
f décanteur *m* discontinu
d diskontinuierlich arbeitender
Dekanteur *m*; periodisch arbeitender
Absetzer *m*
e decantador *m* discontinuo; decantador *m*
intermitente; clasificador *m* intermitente
n discontinu decanteertoestel *n*;
discontinue decanteur *m*

* batch sulphitation → 836

* bath maceration → 1523

168 battery of centrifugals
f batterie *f* d'essoreuses; batterie *f* de

centrifuges
d Zentrifugenbatterie *f*
e batería *f* de centrífugas
n batterij *f* van centrifuges

**169 battery waste water; diffusion waste
water**
f eaux *fpl* résiduaires de diffusion; petites
eaux *fpl*; eaux *fpl* de diffusion
d Diffusionsablaufwasser *n*
e aguas *fpl* de difusión
n drukwater *n* van de diffusie

170 bayonet type cap; bayonet cap *(of cane
mill)*
f couvercle *m* à baïonnette
d Bajonettdeckel *m*
e cabezote *m* tipo bayoneta; tapa *f* de
bayoneta
n bajonetdeksel *n*

* bean aphid → 275

* bean aphis → 275

* bearbind → 997

* bearbine → 997

* beater *(of shredder)* **→ 1215**

171 beater disc *(of shredder)*
f disque *m* porte-marteaux
d Schlägerscheibe *f*
e disco *m* portamartillos
n hamerschijf *f*

* bee nettle → 1241

172 beet
f betterave *f*
d Rübe *f*
e remolacha *f*
n biet *f*; beet *f*
l Beta

* beet acceptance → 225

**173 beet acceptance on juice density; beet
reception on juice density**
f réception *f* des betteraves à la densité
d Rübenannahme *f* nach der Saftdichte
e recepción *f* de remoḻ...ias basada en la
densidad del jugo; recepción *f* de
remolachas basada en la densidad del
guarapo
n bietenreceptie *f* naar de sapdichtheid;
bietenontvangst *f* naar de sapdichtheid

**174 beet acceptance on sugar content basis;
beet reception on sugar content basis**
f réception *f* des betteraves à la richesse
saccharine; réception *f* saccharimétrique
des betteraves
d Rübenannahme *f* nach dem Zuckergehalt
e recepción *f* de remolachas basada en el
contenido de azúcar
n bietenreceptie *f* naar het suikergehalte;
bietenontvangst *f* naar het suikergehalte

**175 beet acceptance on weight basis; beet
reception on weight basis; beet
acceptance on tonnage basis; beet
reception on tonnage basis**
f réception *f* des betteraves au poids
d Rübenannahme *f* nach dem Gewicht
e recepción *f* de remolachas basada en el
peso
n bietenreceptie *f* naar het gewicht;
bietenontvangst *f* naar het gewicht

* **beet and mangold fly** → 1545

**176 beet armyworm *(US)*; sugar beet
armyworm *(US)***
(sb; pest)
f noctuelle *f* de la betterave
d Heerwurm *m*
e gardama *f* de la remolacha
l Spodoptera exigua

* **beet armyworm** → 1453

177 beet balance; beet weighing machine
f bascule *f* à betteraves
d Rübenwaage *f*
e báscula *f* de remolachas
n bietenweegschaal *f*; bietenbascule *f*

* **beet beetle** → 1872

178 beet bin
(A-frame structure built over beet flumes
for the short-term storage of beets at the
factory site)
f silo *m* à betteraves
d Rübensilo *m*
e silo *m* para la remolacha
n bietensilo *m*

179 beet brei analyzer; beet mush analyzer
f analyseur *m* de râpure
d Rübenbreianalysator *m*
e analizador *m* para pulpa de remolacha

* **beet bug** → 1416

180 beet campaign
f campagne *f* betteravière
d Zuckerrübenkampagne *f*; Rüben-
kampagne *f*
e campaña *f* remolachera; campaña *f* de
remolachas
n suikerbietcampagne *f*; bietenjaar *n*;
bietencampagne *f*

**181 beet carrion beetle; black carrion beetle
*(US)***
(sb; pest)
f silphe *m* de la betterave; silphe *m*
opaque (de la betterave)
d buckelstreifiger Aaskäfer *m*; Rübenaaskäfer *m*; buckelstreifiger Rübenaaskäfer *m*; grauhaariger Aaskäfer *m*;
buckliger Aaskäfer *m*; mattschwarzer
Aaskäfer *m*; brauner Rübenaaskäfer *m*
e silfo *m* opaco de la remolacha
n bietenaaskever *m*; doffe
bietenaaskever *m*; doffe bietenkever *m*;
matte bietenkever *m*
l Blitophaga opaca; Aclypea opaca

182 beet cell
(anat)
f cellule *f* de la betterave
d Rübenzelle *f*
e célula *f* de la remolacha
n bietcel *f*

183 beet cleaner
(agr)
f nettoyeur *m* de betteraves; décrotteur *m*
de betteraves
d Rübenreiniger *m*
e limpiadora *f* de remolachas
n bietenreiniger *m*

* **beet collecting centre** → 224

184 beet conveyor
f transporteur *m* de betteraves
d Rübentransporteur *m*
e transportador *m* de remolachas
n bietentransporteur *m*

185 beet crop; beet harvest
(agr)
f récolte *f* des betteraves
d Rübenernte *f*
e cosecha *f* de remolacha
n bietenoogst *m*

* **beet crop, sugar ~** → 2328

* **beet crowns** → 251

* **beet cultivation, sugar ~** → 2331

186 beet curly top virus
f virus *m* de la frisolée de la betterave
n "Beet Curly Top Virus" *n*

* **beet cutter** → 235

* **beet cyst nematode** *(US)* → 188

* **beet damping-off** → 280

187 beet drill; beet seeder *(US)*
(agr)
f semoir *m* à betteraves
d Rübendrillmaschine *f*
e sembradora *f* de remolachas
n bietenzaaimachine *f*

188 beet eelworm; beet cyst nematode *(US)*;
sugar beet eelworm; beet nematode;
sugar beet nematode *(US)*
(sb; pest)
f nématode *m* de la betterave; anguillule *f*
de la betterave
d Rübennematode *f*; Rübenälchen *n*
e nemátodo *m* de la remolacha; heterodera
f de la remolacha; anguílula *f* de la
remolacha
n bietencystenaaltje *n*
l Heterodera schachtii

189 beet elevator
f élévateur *m* à betteraves
d Rübenelevator *m*; Rübenaufzug *m*
e elevador *m* de remolachas
n bietenelevator *m*; bietenophaler *m*

190 beet end
(The section of the factory encompassing
the process from the beet flumes through
the juice purification. May also include
the evaporators, especially in older
factories not originally designed for the
extensive use of evaporator vapo(u)rs in
heating the vacuum pans)
f avant *m* de l'usine; avant-usine *f*
d Vorderbetrieb *m*
n voorfabriek *f*

* **beet factory** → 240

191 beet feeder; rotating beet feeder
f régulateur *m* d'alimentation en
betteraves; roue *f* régulatrice de
l'alimentation en betteraves; distributeur
m à alvéoles
d Zellenrad *n*
e alimentador *m* de remolachas; regulador

m de la alimentación con remolachas
n bietentoevoerrad *n*

192 beet flume
(A concrete-lined ditch or metal trough
designed for the hydraulic transport of
beets)
f caniveau *m* (à betteraves)
d Rübenschwemme *f*; Rübenschwemm-
rinne *f*; Rübenschwemmkanal *m*
e canal *m* de flotación (de remolachas);
saetín *m* para remolachas
n zwemgoot *f*; spoelgoot *f*

193 beet fragments
f morceaux *mpl* de betteraves; fragments
mpl de betteraves
d Rübenbruchstücke *npl*
e pedazos *mpl* de remolacha
n bietstukjes *npl*

194 beet gapper
(agr)
f éclaircisseuse *f* de betteraves
d Rübenverhacker *m*
e entresacador-aclarador *m* de remolachas
n bietendunner *m*

195 beet grower
f betteravier *m*; planteur *m* de betteraves;
cultivateur *m* de betteraves
d Rübenanbauer *m*; Rübenbauer *m*
e remolachero *m*; cultivador *m* de
remolachas; agricultor *m* de remolacha
n bietenplanter *m*; bietenteler *m*

* **beet growing, sugar ~** → 2331

196 beet grown for seed; seed beet
f betterave *f* porte-graines
d Samenrübe *f*
e remolacha *f* portagranos; portagranos *m*
n zaadbiet *f*

197 beet hanging
f accrochage *m* des betteraves
d Rübenverklemmung *f*

* **beet harvest** → 185

* **beet hoe** → 2162

198 beet hopper
f trémie *f* à betteraves
d Rübenbunker *m*; Rübentrichter *m*
e tolva *f* de remolachas
n bietenbunker *m*

199 beet juice
 f jus m de betterave
 d Rübensaft m
 e jugo m de remolacha; guarapo m de
   remolacha
 n bietsap n

200 beet knife
 (A rectangular piece of steel rolled or
 milled into a serrated shape for slicing
 beets into cossettes)
 f couteau m de coupe-racines
 d Schnitzelmesser n
 e cuchilla f de cortadora; cuchilla f de
   cortarraíces
 n bietenmes n

201 beet leaf beetle *(US)*
 (sb; pest)
 l Erynephala puncticollis

* **beet leaf bug** → **1416**

202 beet leaf chopper
 f désintégrateur m de feuilles de
   betteraves
 d Rübenblattzerkleinerer m
 e desmenuzador m de la hoja de
   remolacha
 n bietenbladdesintegrator m

203 beet leaf curl virus
 f virus m de la frisolée
 d Kräuselvirus n
 n "Beet Leaf Curl Virus "n

204 beet leaf drier
 f sécheur m de feuilles de betteraves
 d Rübenblatt-Trockner m
 e secador m de hojas de remolacha
 n bietenbladdroger m

205 beet leafhopper *(US)*
 (sb; pest)
 f cicadelle f de la betterave
 l Eutettix tenellus; Circulifer tenellus

* **beet leaf miner** *(US)* → **1545**

206 beet leaf preparation plant
 f installation f de traitement des feuilles
   de betterave
 d Rübenblattaufbereitungsanlage f
 e planta f de preparación de la hoja de
   remolacha
 n bietenbladbereidingsinstallatie f

* **beet leaf spot** → **461**

207 beet leaf washer
 f laveur m de feuilles de betterave
 d Rübenblattwäscher m; Rübenblatt-
   waschmaschine f
 e lavador m de hoja de remolacha
 n bietenbladwasmolen m

208 beet lifter; beet puller
 (agr)
 f souleveuse f de betteraves; arracheuse f
   de betteraves
 d Rübenroder m; Rübenrodepflug m;
   Rübenheber m
 e arrancadora f de remolacha; cavadora f
   de remolachas; elevadora f de
   remolachas; arado m remolachero;
   desarraigador m de remolacha
 n bietenrooier m; bietenrooimachine f

* **beet lifter and collector** → **1462**

* **beet lifter wheel** → **259**

* **beet lifting wheel** → **259**

209 beet marc
 (The water insoluble constituent of the
 sugar beet root)
 f marc m de betteraves
 d Rübenmark m
 e marco m de remolachas
 n bietenmerg n

210 beet mild yellowing
 (sb; disease)
 f jaunisse f modérée de la betterave
 d Vergilbungskrankheit f der Rübe; milde
   Vergilbung f
 e amarillez f moderada de la remolacha
 n vergelingsziekte f *(veroorzaakt door het
   zwakke vergelingsvirus)*

211 beet mild yellowing virus; BMYV
 f virus m de la jaunisse modérée;
   VJMB m; BMYV m
 d Vergilbungsvirus n; BMYV n; Rüben-
   vergilbungsvirus n; RVGV n
 e BMYV m
 n zwak vergelingsvirus n; "Beet Mild
   Yellowing Virus" n; "BMYV" n

* **beet molasses, sugar** ~ → **2332**

212 beet mosaic
 (sb; disease)
 f mosaïque f de la betterave
 d Rübenmosaik n; Mosaikkrankheit f der
   Rübe
 e mosaico m de la remolacha

   n mozaïek *f* van de biet; mozaïekziekte *f*
van de biet

213 beet mosaic virus
   f virus *m* de la mosaïque
   d Mosaikvirus *n*; Rübenmosaikvirus *n*
   e virus *m* del mosaico de la remolacha
   n bietenmozaïekvirus *n*

214 beet moth
   (sb; pest)
   f teigne *f* de la betterave
   d Rübenmotte *f*; Runkelrübenmotte *f*;
Rübenminiermotte *f*
   e tiña *f* de la remolacha; polilla *f*
minadora de la remolacha
   n bietemot *f*
   l Phthorimaea ocellatella

   * **beet mush analyzer** → 179

215 beet necrotic yellow vein virus; BNYVV
   f virus *m* de la rhizomanie; "beet necrotic
yellow vein virus" *m*; "BNYVV" *m*
   d "Beet Necrotic Yellow Vein Virus" *n*;
"BNYVV" *n*; Aderngelbfleckigkeits-
virus *n*
   e "beet necrotic yellow vein virus" *m*;
"BNYVV" *m*
   n "Beet Necrotic Yellow Vein Virus" *n*;
"BNYVV" *n*

   * **beet nematode** → 188

216 beet petiole-borer *(US)*
   (sb; pest)
   l Cosmobaris americana

217 beet pick-up loader
   (agr)
   f ramasseuse-chargeuse *f* de betteraves;
ramasseur-chargeur *m* de betteraves
   d Rübensammellader *m*
   e recogedora-cargadora *f* de remolachas
   n opraper-lader *m* van bieten

218 beet pile
   (A stack of beets usually destined for
long-term storage)
   f tas *m* de betteraves
   d Rübenhaufen *m*; Rübenmiete *f*; Rüben-
stapel *m*
   e pila *f* de remolachas
   n bietenhoop *m*

219 beet piling plant
   f installation *f* de stockage des betteraves
   d Rübenstapelanlage *f*

   e dispositivo *m* estibador de remolachas
   n bietenopslaginstallatie *f*

220 beet plough
   (agr)
   f souleveuse *f* de betteraves
   d Rodepflug *m*
   n bietenlichter *m*

   * **beet pricker** → 230

   * **beet puller** → 208

   * **beet pulp** → 939

221 beet pulp processing installation
   f installation *f* de traitement des pulpes
   d Pülpe- und Schnitzelaufbereitungs-
anlage *f*
   e planta *f* para la preparación de la pulpa
   n pulpverwerkingsinstallatie *f*

222 beet pump
   (A specially designed pump for elevating
beets and flume water)
   f pompe *f* à betteraves
   d Rübenpumpe *f*
   e bomba *f* de remolachas
   n bietenpomp *f*

223 beet rasp; beet saw
   f râpe *f* à betteraves; râpe *f* de réception
   d Rübenbreisäge *f*; Rübensäge *f*; Rüben-
fräse *f*
   e raspa *f* para remolachas
   n bietenrasp *f*

**224 beet receiving station; beet collecting
centre**
   (The station at which the growers' beets
are weighed, unloaded, usually dry
cleaned, and sampled)
   f centre *m* de réception des betteraves;
station *f* de réception des betteraves;
poste *m* de réception de betteraves
   d Rübenannahmestation *f*; Rübenabnahme-
station *f*; Rübenannahmezentrum *n*
   e estación *f* de recepción de remolachas
   n bietenontvangstplaats *f*;
bietenontvangstpost *m*

225 beet reception; beet acceptance
   f réception *f* des betteraves; prise *f* en
charge des betteraves
   d Rübenabnahme *f*; Rübenannahme *f*;
Rübenübernahme *f*
   e recepción *f* de remolachas
   n bietenontvangst *f*

* **beet reception on juice density** → 173

* **beet reception on sugar content basis** → 174

* **beet reception on tonnage basis** → 175

* **beet reception on weight basis** → 175

226 beet root; garden beet; red beet
  f betterave *f* rouge; betterave *f* potagère
  d rote Rübe *f*; Speiserübe *f*
  e remolacha *f* de mesa
  n rode biet *f*
  l Beta vulgaris ssp. esculenta

* **beetroot** → 1914

227 beet root tumor
  (sb; disease)
  f tumeur *f* à urophlyctis
  d Rübenkrebs *m*
  e lepra *f*
  n Urophlyctis-woekerziekte *f*

228 beet root weevil; sugar beet weevil
  (sb; pest)
  f charançon *m*; cléone *m* (de la betterave)
  d Rübenderbrüßler *m*; Rübenrüßler *m*
  e cleonus *m* de la remolacha
  l Bothynoderes punctiventris

229 beet rust
  (sb; disease)
  f rouille *f* de la betterave
  d Rübenrost *m*
  e roya *f* de la remolacha
  n bietroest *m*

230 beet sampler; beet pricker
  f échantillonneur *m* de betteraves
  d Rübenprobenehmer *m*; Rüpro *m*; Rüben-
     probenahmeanlage *f*; Rübenstecher *m*
  e sacamuestras *m* de remolachas;
     instalación *f* de toma de pruebas de
     remolacha
  n bietenmonsternemer *m*

231 beet sampling washer
  f laveur *m* d'échantillons de betterave
  d Rübenprobenwäsche *f*; Rübenproben-
     waschmaschine *f*
  e lavador *m* de muestras de remolacha
  n bietmonsterwasser *m*

* **beet saw** → 223

232 beet screw; beet scroll
  f hélice *f* à betteraves

  d Rübenschnecke *f*
  e husillo *m* de remolachas; tornillo *m* sin
     fin de remolachas
  n schroeftransporteur *m* voor bieten

* **beet seeder** *(US)* → 187

233 beet silo
  f silo *m* à betteraves
  d Rübensilo *m*
  e silo *m* de remolachas
  n bietensilo *m*

234 beet slice mincer
  f hache-cossettes *m*
  d Rübenschnitzelmühle *f*
  n snijdselhakmachine *f*

**235 beet slicer; beet slicing machine; beet
     cutter**
  (Apparatus which cuts the beets into
  slender strips)
  f coupe-racines *m*
  d Rübenschneidmaschine *f*;
     Rübenschnitzelmaschine *f*
  e cortadora *f* de remolachas; máquina *f*
     cortarraíces; máquina *f* cortadora de
     remolachas
  n bietensnijmolen *m*

* **beet slices** → 660

236 beet slicing
  f découpage *m* des betteraves
  d Zerschneiden *n* der Rüben; Rüben-
     schneiden *n*
  n snijden *n* van bieten

* **beet slicing machine** → 235

**237 beet slicing machine with horizontal
     knife disc; horizontal beet slicer; disc
     beet slicer; horizontal beet slicing
     machine**
  f coupe-racines *m* à plateau horizontal
  d Horizontalschneidmaschine *f*; Schneid-
     maschine *f* mit horizontaler Schneid-
     scheibe
  e cortadora *f* horizontal de remolachas;
     cortarraíces *m* con disco horizontal
  n snijmolen *m* met horizontale mesplaat;
     bietensnijmachine *f* met horizontale
     mesplaat

238 beet storage
  f mise *f* en stock des betteraves
  d Rübenlagerung *f*
  e almacenaje *m* de las remolachas
  n opstapeling *f* van de bieten

239 beet sugar
 f sucre *m* de betterave
 d Rübenzucker *m*
 e azúcar *m* de remolacha
 n bietsuiker *m*; beetsuiker *m*

 * **beet, sugar ~ → 2326**

240 beet sugar factory; sugar beet factory; beet factory
 f sucrerie *f* de betterave
 d Rübenzuckerfabrik *f*
 e azucarera *f* de remolacha; fábrica *f* de azúcar de remolacha
 n bietsuikerfabriek *f*; beetwortelsuikerfabriek *f*

241 beet sugar industry
 f industrie *f* sucrière de betteraves; industrie *f* sucrière betteravière
 d Rübenzuckerindustrie *f*
 e industria *f* azucarera de remolacha; industria *f* del azúcar de remolacha
 n bietsuikerindustrie *f*

242 beet sugar producing country
 f pays *m* producteur de sucre de betterave
 d Rübenzuckererzeugerland *n*
 e país *m* productor de azúcar de remolacha
 n bietsuikerproducerend land *n*

 * **beet tail catcher → 2458**

243 beet tail chopper; beet tail slicer
 f désintégrateur *m* de radicelles; broyeur *m* de radicelles
 d Rübenschwänzezerkleinerer *m*; Rübenschwänzezerreißer *m*
 e trituradora *f* de nabos de remolacha; triturador *m* de nabos de remolacha

244 beet tails
 (The lower, slender part of the beet root (anat))
 f queues *fpl* de betteraves; radicelles *fpl*
 d Rübenschwänze *mpl*
 e nabos *mpl* de remolacha; raicillas *fpl* de remolachas; colas *fpl* de remolachas
 n bietenstaartjes *npl*

 * **beet tail separator → 2458**

 * **beet tail slicer → 243**

245 beet tail washer
 f laveur *m* de radicelles
 d Rübenschwänzewaschmaschine *f*; Rübenschwänzewäsche *f*
 e lavador *m* de nabos de remolacha
 n wasmolen *m* voor bietenstaartjes

246 beet thinner; root thinner; down-the-row thinner
 (agr)
 f prédémarieuse *f*; éclaircisseuse *f*; distanceuse *f*
 d Rübenausdünner *m*; Auslichter *m*; Vereinzelungsmaschine *f*
 e raleadora *f* para remolachas; aclareadora *f* para remolachas
 n bietenrijendunmachine *f*

247 beet tipping plant
 f installation *f* de culbutement des betteraves
 d Rübenkippanlage *f*
 e dispositivo *m* basculador de remolachas
 n bietenkipinrichting *f*; bietenkipinstallatie *f*

248 beet top harvester; topper loader
 (agr)
 f décolleteuse-chargeuse *f*
 d Rübenköpfsammler *m*; Köpflader *m*; Wagenköpfer *m*
 e descoronadora-cargadora *f*
 n bietenverzamelkopper *m*

249 beet top harvester
 (agr)
 f arracheuse-décolleteuse *f* de betteraves
 d Rübenköpfroder *m*
 e arrancadora-descoronadora *f* de remolachas
 n bietenverzamelrooier *m*

250 beet topper
 (agr)
 f décolleteuse *f* de betteraves
 d Rübenköpfer *m*
 e descoronadora *f* de remolachas
 n bietenkopper *m*

 * **beet top rake → 2521**

251 beet tops; beet crowns
 (The beet leaves and petioles, which may or may not be accompanied by crowns or pieces of crown (anat))
 f collets *mpl* de betteraves
 d Rübenköpfe *mpl*
 e capacetas *fpl* de remolachas; cabezas *fpl* de remolachas
 n bietenkoppen *mpl*

252 beet top windrower
 (agr)

f groupeur *m* de verts
d Rübenblattsammler *m*
e rastrillo *m* amontonador de coronas

* **beet tortoise beetle** → 517

253 beet transport
f transport *m* de betteraves
d Rübentransport *m*
e transporte *m* de remolachas
n bietenvervoer *n*; bietentransport *n*

254 beet transporter
f transporteur *m* à betteraves
d Rübentransporteur *m*
e transportador *m* de remolachas
n bietentransporteur *m*

255 beet washer
(An apparatus for wet cleaning of the
beets)
f lavoir *m* à betteraves; laveur *m* à
betteraves
d Rübenwaschmaschine *f*; Rübenwäsche *f*
e lavadora *f* de remolachas; lavador *m* de
remolachas
n bietenwasmolen *m*

256 beet washing house
f lavoir *m* à betteraves; atelier *m* de
lavage des betteraves; station *f* de lavage
des betteraves; laverie *f*
d Rübenwäsche *f*; Rübenwaschhaus *n*
e casa *f* de lavado de remolacha
n bietenwasstation *n*

257 beet webworm *(US)*; **diamond spot pearl;
meadow moth**
(sb; pest)
f parée *f*; pyrale *f* de la betterave; tisseuse
f de la betterave
d Rübenzünsler *m*
e palomilla *f* de la remolacha; palometa *f*
grande
l Loxostege sticticalis; Phlyctaenodes
sticticalis

258 beet weevil
(sb; pest)
f charançon *m*
d Spitzsteißrüßler *m*; Klettenrüßler *m*;
spitzsteißiger Rübenrüßler *m*
l Tanymecus palliatus

* **beet weighing machine** → 177

**259 beet wheel; beet lifter wheel; beet lifting
wheel**
(A bucket wheel that elevates the beets

and, in some cases, water from the beet
flume)
f roue *f* à betteraves; roue *f* élévatrice
d Hubrad *n*; Rübenhubrad *n*; Rübenrad *n*
e rueda *f* elevadora de remolachas
n bietenopvoerrad *n*; bietenrad *n*

260 beet yard
f cour *f* à betteraves
d Rübenhof *m*
e patio *m* de remolacha
n bietenplein *n*

261 beet yellow net
(sb; disease)
f jaunisse *f* des nervures; "Beet Yellow
Net"
d Gelbnetzkrankheit *f*
e "Beet Yellow Net"
n "Beet Yellow Net"

262 beet yellow net virus
d Gelbnetzvirus *n*
n "Beet Yellow Net Virus"

263 beet yellows
(sb; disease)
f jaunisse *f* grave de la betterave
d Vergilbungskrankheit *f* der Rübe;
normale Vergilbung *f* der Rübe; viröse
Gelbsucht *f*
e amarillez *f* grave de la remolacha
n vergelingsziekte *f* *(veroorzaakt door het
sterke vergelingsvirus)*

264 beet yellows virus; BYV
f virus *m* de la jaunisse grave de la
betterave; BYV *m*
d Vergilbungsvirus *n*; BYV *n*;
Rübengelbvirus *n*; RGV *n*
e virus *m* de la amarillez de la remolacha;
BYV *m*
n sterk vergelingsvirus *n*; "BYV" *n*; biete-
vergelingsvirus *n*

* **beet yellow wilt disease** → 2735

265 beet yellow wilt virus
f virus *m* de la "Beet Yellow Wilt Disease"
d Gelbwelkevirus *n*
e virus *m* de la marchitez de la remolacha
n Yellow Wilt-virus *n*

* **beggar's lice** → 647

* **belt conveyor** → 146

266 belt-driven centrifugal
f centrifuge *f* à courroie; essoreuse *f* à

courroie
d Zentrifuge f mit Riemenantrieb
e centrífuga f accionada por correa
n centrifuge f met riemaandrijving

* **belvedere summer cypress** → 1411

267 **bentonite**
(Reagent)
f bentonite f
d Bentonit m
e bentonita f
n bentoniet n

268 **Bermuda grass; Bahama grass; devil grass; doob; sutch grass; star grass; dog's-tooth grass**
(sb; weed)
f cynodon m dactyle; chiendent m pied-de-poule; gros chiendent m; grand chiendent m; herbe f des Bermudes
d Hundszahngras n; Bermudagras n; Hundszahn m; Eckzahn m; Hundshirse f; Fingerhundszahn m
e hierba f Bermuda; grama f común; hierba f fina; pasto m Bermuda; zacate m; gramilla f; gramillón m; chepica f; hierba f de la virgen
n hondsgras n; handjesgras n
l Cynodon dactylon

269 **betacyanin formation**
(sb; disease)
f formation f de bétacyanine
d Betacyaninbildung f
e formación f de betacianina
n betacyaninevorming f

270 **betaine; trimethylglycocol**
f bétaïne f; triméthylglycocolle m
d Betain n; Trimethylglykokoll n
e betaína f; trimetil-glicocola f
n betaïne nf; trimethylglycocol n

271 **betaine nitrogen**
f azote m bétaïne; azote m bétaïque; N m bétaïne
d Betainstickstoff m
e nitrógeno m de betaína
n betaïnestikstof f

* **bibionid fly** → 1547

272 **biconvex downtake** *(of diametral circulation pan)*
f puits m biconvexe
d bikonvexes Rohr n
e tubo m biconvexo
n biconvexe buis f; biconvexe pijp f

* **big-sting nettle** → 2283

* **billbugs** *(US)* → 2661

273 **biochemical oxygen demand; BOD**
(Amount of oxygen required to biologically oxidize the organic matter in a sample over a period of time, usually five days)
f demande f biochimique en oxygène; D.B.O. f
d biochemischer Sauerstoffbedarf m; BSB m
e demanda f bioquímica del oxígeno; d.b.o. f
n biochemisch zuurstofverbruik n; B.Z.V. n

* **bird rape** → 2698

* **bitter clover** *(US)* → 2185

* **bitter dock** → 334

274 **bitumen-lined paper** *(for sugar storage)*
f papier m goudronné
d Teerpapier n
e papel m bituminado
n teerpapier n

275 **black bean aphid; bean aphid; bean aphis**
(sb; pest)
f puceron m noir des fèves; puceron m noir
d schwarze Bohnenlaus f; schwarze Bohnenblattlaus f; schwarze Rübenblattlaus f; schwarze Rübenlaus f
e pulgón m de la remolacha; pulgón m negro
n zwarte bonenluis f
l Aphis fabae; Doralis fabae

* **black beetles** *(US)* → 1975

276 **black bindweed; wild buckwheat; climbing buckwheat; dull-seed cornbind**
(sb; weed)
f renouée f liseron; vrillée f sauvage; faux-liseron
d Windenknöterich m
e corregüela f anual; polígono m trepador; corregüela f negra; enredadera f negra; pimentilla f trepadora; correhuela f anual; correhuela f negra
n zwaluwtong f; wilde boekweit f
l Polygonum convolvulus; Fallopia convolvulus (L.) Löve

277 black blister beetle *(US)*
(sb; pest)
l Epicauta pensylvanica

* **black carrion beetle** *(US)* → 181

**278 black cutworm; greasy cutworm; dark
sworth grass moth**
(sb; pest)
f noctuelle *f* ypsilon
d Ypsiloneule *f*
e noctuido *m* ypsilon; gusano *m* cortador
grasiendo; cucumilla *f* negra
l Agrotis ypsilon

**279 black grass; slender foxtail; mouse
foxtail; hunger grass**
(sb; weed)
f vulpin *m* des champs; queue *f* de renard;
vulpin *m*; queue *f* de rat; fenasse *f*; folle
farine *f*
d Ackerfuchsschwanz *m*; Ackerfuchs-
schwanzgras *n*; Mäusefuchsschwanz *m*
e alopecuro *m* de los campos; cola *f* de
zorra
n duist *f*
l Alopecurus agrestis; Alopecurus
myosuroides

280 black leg; root rot; beet damping-off
(sb; disease)
f pied *m* noir; maladie *f* du pied noir;
fonte *f* des semis
d Wurzelbrand *m*; Schwarzbeinigkeit *f*
e pie *m* negro
n wortelbrand *m*; bietenwortelbrand *m*;
bietenbrand *m*

**281 black medic(k); yellow trefoil; yellow
clover; hop clover**
(sb; weed)
f minette *f*; minette *f* dorée; luzerne *f*
lupuline; lupuline *f*; trèfle *m* jaune;
trèfle *m* houblonné; luzerne *f* houblon
d Hopfenklee *m*; Gelbklee *m*; Hopfenluzerne *f*; Lämmerklee *m*; Steinklee *m*;
Wolfsklee *m*
e mielga *f* azafranada; lupulina *f*; alfalfa *f*
lupulina; medicago *m* negro; mielga *f*
negra
n hopklaver *f*; hoprupsklaver *f*;
hopperupsklaver *f*
l Medicago lupulina

**282 black mustard; brown mustard; red
mustard**
(sb; weed)
f moutarde *f* noire
d schwarzer Senf *m*; Senfkohl *m*; Schwarz-

kohl *m*
e mostaza *f* negra
n zwarte mosterd *m*; bruine mosterd *m*
l Brassica nigra

283 black nightshade; common nightshade
(sb; weed)
f morelle *f* noire; tue-chien *m*;
amourette *f*
d schwarzer Nachtschatten *m*; Hühnertod *m*; Saukraut *n*
e tomtatillos *mpl*; hierba *f* mora;
morella *f*; solano *m* negro
n zwarte nachtschade *f*
l Solanum nigrum

284 black rot
(sc; fungous disease)
f cœur *m* noir; pourriture *f* noire
d Schwarzfäule *f*
e corazón *m* negro; guacatillo *m*
n zwartrot *n*

* **blackstrap** → 1053

285 blackstrap molasses
(Cane molasses containing 23.5% to
26.4% water and 48.5% to 53.4% total
sugars)
f "blackstrap molasses"
d "Blackstrap Molasses"
e melazas *fpl*
n "blackstrap molasses"

* **black streak** *(US)* → 109

286 black stripe
(sc; fungous disease)
f maladie *f* des stries noires
d Schwarzstreifigkeit *f*
e raya *f* negra
n zwarte-strepenziekte *f*

287 black woodvessel disease
(sb; disease)
f maladie *f* des vaisseaux noirs
d Pythium-Gelbsucht *f*
e enfermedad *f* de las tráqueas negras
n Pythium-geelzucht *f*

**288 bladder campion; catchfly; bladder
silene**
(sb; weed)
f silène *m* enflé; cornillet *m*
d aufgeblasenes Leimkraut *n*; Tauben-
kropf *m*
e colleja *f*
n blaassilene *m*

l Silene inflata Sm.; Silene vulgaris (Moench) Gar.

289 bladed rotor knife
(agr)
f rotor *m* à couteaux
d Messertrommel *f*
e cilindro *m* de cuchillos
n messentrommel *f*

290 bled vapo(u)r
f vapeur *f* de prélèvement
d Brüden *m*
e vapor *m* de los evaporadores
n stoom *m* van de verdamplichamen

291 blue *v*
f ajouter du bleu; bleuir; bleuter; azurer
d bläuen; blauen
e azular; azulear
n blauwen

* **bluebottle** → 648

292 blued sugar
f sucre *m* bleui; sucre *m* azuré
d gebläuter Zucker *m*
e azúcar *m* azulado
n geblauwde suiker *m*

* **bluegrass** *(US)* → 69

* **blue mallow** → 577

293 B massecuite; second massecuite
f masse *f* cuite B; masse *f* cuite de deuxième jet; masse *f* cuite deuxième jet; m c *f* II
d B-Füllmasse *f*; B-Kochmasse *f*
e masa *f* cocida B; masa *f* cocida de segunda; masa *f* cocida segunda
n B-masse cuite *f*; B-vulmassa *f*; masse cuite *f* B

294 B molasses; second molasses
f égout *m* B
d B-Ablauf *m*
e melaza *f* B; melaza *f* de segunda; miel *f* B
n B-stroop *f*

* **BMYV** → 211

* **BNYVV** → 215

* **BOD** → 273

295 boil *v* **by means of string-proof; boil** *v* **in**

f cuire au filet; cuire par preuve au filet; cuire à la plume
d auf Faden verkochen; zur Fadenprobe kochen; zähe kochen
e cocer por el cordón de jarabe; cocer al hilo
n op draad koken; inkoken

* **boiler** *(US)* → 1723

296 boiler feed-water
f eau *f* d'alimentation des chaudières
d Kesselspeisewasser *n*
e agua *f* de alimentación de calderas
n ketelvoedingswater *n*

297 boiler house
f chaufferie *f*
d Kesselhaus *n*
e casa *f* de caldera; sala *f* de calderas; instalación *f* de calderas
n ketelhuis *n*

* **boil** *v* **in** → 295

298 boiling
f bouillissage *m*; cuite *f*; cuisson *f*
d Kochen *n*; Verkochen *n*; Verkochung *f*
e cocción *f*; ebullición *f*
n koken *n*

* **boiling control apparatus** → 709

* **boiling house** → 1724

299 boiling house balance
f bilan *m* de la concentration
d Bilanz *f* der Kochstation
e balance *m* de la instalación de cocción
n balans *f* van het kookstation

300 boiling house control
f contrôle *m* de la concentration
d Kontrolle *f* der Kochstation
e control *m* de la instalación de cocción
n controle *f* van het kookstation

301 boiling house recovery; retention
(The ratio of the sucrose obtained in the sugar manufactured to that entering in the mixed juice)
f récupération *f* de la concentration
e recuperación *f* en la instalación de cocción; recuperación *f* en la instalación de calderas; retención *f*

302 boiling point
f point *m* d'ébullition
d Siedepunkt *m*; Kochpunkt *m*

e punto *m* de ebullición; punto *m* de cocción
n kookpunt *n*

303 boiling point rise; bpr; boiling point elevation; bpe
f élévation *f* du point d'ébullition; EPE *f*
d Siedepunktserhöhung *f*
e aumento *m* del punto de ebullición; a.p.e. *m*; elevación *f* del punto de ebullición; elevación *f* del punto de cocción
n kookpuntverhoging *f*

304 boiling scheme
(The overall plan of crystallization-separation stages in the sugar end, chiefly designated by the number of stages or "boilings" used)
f schéma *m* de cristallisation
d Kristallisationsschema *n*; Kochschema *n*
e esquema *m* de cristalización
n kristallisatieschema *n*; kookschema *n*

305 boiling stage
f étage *m* de cuisson
d Kochstufe *f*
e etapa *f* de cocimiento
n kooktrap *f*

* **boiling station** → 1724

306 boiling temperature
f température *f* d'ébullition
d Siedetemperatur *f*
e temperatura *f* de ebullición
n kooktemperatuur *f*

* **boiling to string-proof** → 2302

307 boiling under vacuum
f ébullition *f* sous vide
d Kochen *n* bei Unterdruck
e ebullición *f* al vacío
n koken *n* bij onderdruk

* **boil *v* off the massecuite** → 2488

308 boil *v* to grain
f grainer
d auf Korn kochen
e cocer sobre grano
n op grein koken

309 boil *v* up
f recuire
d aufkochen
e recocer
n opkoken

* **Bokhara clover** → 2679

310 bolt *v*; run *v* to seed; go *v* to seed
(agr)
f monter à graines
d schossen
e granar; semillar
n schieten; doorschieten

311 bolter
(agr)
f betterave *f* montée à graines
d Schosser *m*
n schieter *m*

312 bolting
(agr)
f montée *f* à graines
d Schossen *n*
n schieten *n*

313 bolting resistant
(agr)
f résistant à la montaison; résistant à la montée à graines
d schosserresistent; schoßresistent
n schietersresistent

314 bone char(coal); bone coal; bone black
(Charcoal prepared from selected bones of cattle)
f noir *m* animal; noir *m* d'os; charbon *m* d'os
d Tierkohle *f*; Knochenkohle *f*
e negro *m* animal; negro *m* de hueso; carbón *m* animal
n beenzwart *n*; beenderkool *f*

315 boron deficiency; heart rot
(sb; disease)
f carence *f* en bore; pourriture *f* de cœur
d Bormangel *m*; Herz- und Trockenfäule *f*
e deficiencia *f* de boro; carencia *f* de boro; podredumbre *f* de la corona
n boriumgebrek *n*; hartrot *n*

316 botrytis
(sb; disease)
f botrytis *f*
d Botrytis-Samenschwärze *f*
e botrytis *f*; botritis *f*
n botrytis *f*

317 bottom door of diffuser
f porte *f* inférieure du diffuseur
d unterer Deckel *m* des Diffuseurs
e capa *f* inferior de difusor
n onderdeksel *n* van diffusieketel

318 bottom door of vacuum pan
f vanne f de vidange d'appareil à cuire
d Ablaßventil n eines Kochapparates
e válvula f de descarga de tacho al vacío
n benedendeksel n van een kookpan

319 bottom roll(er); lower roll(er) *(of crusher, cane mill)*
f cylindre m inférieur
d unterer Roller m; untere Walze f; Unterroller m
e cilindro m inferior; rodillo m inferior; maza f inferior
n benedencilinder m; ondercilinder m; benedenrol f; onderrol f

* **bourbon scale** *(US)* → 534

* **bpe** → 303

* **bpr** → 303

320 B-products
f produits mpl de deuxième jet
d B-Produkte npl
e productos mpl B
n B-produkten npl

* **brassy(-toothed) flea beetle** → 1544

321 breakage of grain
(Crystallization)
f rupture f des grains
d Kristallbruch m
e rompimiento m de granos
n kristalbreuk f

* **breakdown of sucrose** → 2316

322 break v the vacuum
f casser le vide
d belüften
e romper el vacío
n het vacuüm verbreken

323 brei
(Fine beet particles; the product of a beet rasp, beet saw, or similar device)
f râpure f
d Brei m
e raspadura f
n raspsel m

324 bridge crane; gantry crane
(A cane crane)
f grue f à portique; portique m; pont-roulant m; grue f portique
d Brückenkran m; Portalkran m

e grúa f de pórtico; grúa-pórtico f
n brugkraan f; laadbrug f; portaalkraan f

* **bring v in the skipping** → 2488

325 briquette of bagasse
f briquette f de bagasse
d Bagassebrikett n
e ladrillo m de bagazo
n bagassebriket f; ampasbriket f

326 briquetting press
(A bagasse press)
f presse f à briqueter; briqueteuse f
d Brikettpresse f; Brikettierpresse f
e prensa f de ladrillos
n briketpers f

* **bristlegrass** *(US)* → 1195

327 bristle oat; lopside oat
(sb; weed)
f avoine f maigre; avoine f strigeuse; avoine f nerveuse
d Sandhafer m
n schrale haver m; zandhaver m; evene m
l Avena strigosa

* **bristle-stem hemp nettle** → 575

* **bristle thistle** → 2496

328 Brix
(The percent dry substance by hydrometry, using an instrument (Brix hydrometer) or table calibrated in terms of percent sucrose by weight in water solution)
f Brix m
d Brix m; Trockensubstanzgehalt m; TS-Gehalt m
e Brix m
n Brix m

329 Brix balance
f bilan m du Brix
d Brixbilanz f
e balance m del Brix
n Brixbalans f

* **Brix degree** → 761

330 Brix extraction
f extraction f de Brix
d Brixextraktion f
e extracción f de Brix
n Brixextractie f

331 Brix graph
f échelle f de Brix
d Brixskale f
e escala f del Brix; gráfica f del Brix
n Brixschaal f

332 Brix saccharometer
f saccharomètre m Brix
d Brixsaccharometer n
e sacarómetro m Brix
n Brixsaccharometer m

333 Brix spindle
f aréomètre m Brix
d Brixspindel f
e areómetro m Brix
n Brixweger m

334 broad-leaved dock; common dock; bitter dock; yellow dock
(sb; weed)
f rumex m à feuilles obtuses; patience f sauvage; oseille f à feuilles obtuses
d stumpfblättriger Ampfer m
e acedera f obtusifolia; aguja f de pastor; romaza f de hoja grande
n ridderzuring f
l Rumex obtusifolius

* **broad-leaved plantain → 1193**

335 brom(o)cresol
(pH control)
f bromecrésol m
d Bromkresol n
e bromocresol m
n broomcresol n

336 brom(o)phenol blue
(pH control)
f bromephénol m bleu
d Bromphenolblau n
e bromofenol m azul; azul m de bromofenol
n broomfenolblauw n

* **brown mustard → 282**

337 brown rot
(sc; fungous disease)
f pourriture f brune
d Braunfäule f
e pudrición f café
n bruinrot n

338 brown spot
(sc; fungous disease)
f maladie f des taches brunes
d Braunfleckigkeit f; braune Blattfleckenkrankheit f
e mancha f café
n bruine-vlekkenziekte f

339 brown stink bug *(US)*
(sb; pest)
l Euschistus servus

340 brown stripe; brown stripe disease
(sc; fungous disease)
f maladie f des stries brunes
d Braunstreifigkeit f
e raya f café
n bruine-strepenziekte f

341 brown sugar
(A marketable product consisting of fine sugar crystals lightly coated with a yellow or brown-colored syrup that contains a relatively high percentage of invert sugar, which keeps the product slightly moist)
f "brown sugar"; sucre m roux
d brauner Zucker m
e azúcar m castaño
n bruine suiker m

342 brushing *(the beets)*
f brossage m
d Bürsten n
n borstelen n

343 B sugar; second sugar
f sucre m B; sucre m de deuxième jet; sucre m deuxième jet
d B-Zucker m; Rohzucker m; RZ m
e azúcar m B; azúcar m de segunda; azúcar m segunda
n B-suiker m

344 B-sugar
(EEC)
f sucre m B
d B-Zucker m
e azúcar m B
n B-suiker m

345 bubble test
f preuve f au soufflé; essai m de soufflage
d Blasprobe f; Pustprobe f
e prueba f de soplado
n blaasproef f

346 bucket elevator
f élévateur m à godets; noria f
d Becherelevator m; Becheraufzug m
e elevador m de cangilones
n emmerladder f; Jacobsladder f

* **buckhorn plantain** *(US)* → 1979

347 **buffalo bur** *(US)*; **buffalo bur nightshade**
  (sb; weed)
  d Büffelklette *f*; Schnabelnachtschatten *m*
  l Solanum rostratum

348 **buffer solution**
  f solution *f* tampon
  d Pufferlösung *f*
  e solución *f* tampón
  n bufferoplossing *f*

349 **bugloss; small bugloss; wild bugloss**
  (sb; weed)
  f petite buglosse *f*; grisette *f*; lycopside *f*
    des champs
  d Wolfsauge *n*; Ackerkrummhals *m*
  e licopsis *f* arvensis
  n kromhals *m*; akkerkromhals *m*
  l Lycopsis arvensis

350 **building up the grain**
  f accroissement *m* du grain
  d Kornwachstum *n*
  e crecimiento *m* del grano
  n groei *m* van het grein

351 **build** *v* **up the massecuite; run** *v* **up the**
    strike
  f monter la cuite
  d die Kochmasse eindicken
  e crecer la masa cocida
  n de masse cuite indikken; de vulmassa
    indikken

352 **bulb eelworm; bulb nematode; stem-**
    and-bulb eelworm; stem nematode; stem
    eelworm
  (sb; pest)
  f anguillule *f* de la tige; nématode *m* de la
    tige; nématode *m* de la tige et du bulbe;
    anguillule *f* des tiges
  d Stockälchen *n*; Stengelälchen *n*;
    Rübenkopfälchen *n*
  e anguílula *f* del tallo; anguílula *f* del
    centeno
  n stengelaaltje *n*
  l Ditylenchus dipsaci

353 **bulbous persicaria**
  (sb; weed)
  f renouée *f* noueuse
  d Ampferknöterich *m*
  e polígono *m* para perdíz
  n knopige duizendknoop *m*
  l Polygonum lapathifolium ssp.
    lapathifolium

354 **bulk density** *(of the cane on the carrier)*
  f densité *f* apparente
  e densidad *f* aparente
  n stortgewicht *n*

* **bulk, in ~** → 1298

355 **bulk raw sugar**
  f sucre *m* brut en vrac
  d loser Rohzucker *m*
  e azúcar *m* crudo a granel
  n ruwe suiker *m* in bulk

356 **bulk storage**
  f stockage *m* en vrac
  d Lagerung *f* in loser Form
  e almacenamiento *m* a granel; almacenaje
    m a granel
  n opslag *m* in bulk

357 **bulk sugar**
  f sucre *m* en vrac
  d loser Zucker *m*
  e azúcar *m* a granel
  n suiker *m* in bulk; bulksuiker *m*

358 **bulk sugar silo**
  f silo *m* à sucre en vrac
  d Silo *m* für losen Zucker
  e silo *m* para azúcar a granel
  n silo *m* voor bulksuiker

* **bull nettle** *(US)* → 1260

* **bull thistle** *(US)* → 2215

359 **bur franseria**
  (sb; weed)
  l Franseria discolor

360 **burned cane**
  f canne *f* brûlée
  d gebranntes Rohr *n*
  e caña *f* quemada
  n gebrand riet *n*

* **burned juice** *(US)* → 1705

* **burning nettle** *(US)* → 2187

* **burnt lime** → 1876

* **burweed marsh elder** *(US)* → 1551

* **butter-and-eggs** → 2734

361 **buttercup; butterflower; goldcup;**
    kingcup; crowfoot
  (sb; weed)

   f renoncule *f*
   d Hahnenfuß *m*; Butterblume *f*;
     Ranunkel *f*
   e ranúnculo *m*
   n boterbloem *f*; hanevoet *m*; hanepoot *m*
   l Ranunculus sp.

   * **butter print** → **2612**

   * **Büttner drier** → **2564**

362 by-product
   f sous-produit *m*
   d Nebenprodukt *n*
   e subproducto *m*
   n bijprodukt *n*

   * **BYV** → **264**

C

363 cabbage moth; cabbage dot
(sb; pest)
f noctuelle *f* du chou; brassicaire *f*;
omicron *m* nébuleux
d Kohleule *f*; gemeine Hochstaudenflur-
Blättereule *f*
e noctuido *m* de la col; noctua *f* de la col
n kooluil *m*
l Mammestra brassicae; Barathra brassicae

364 cabbage thrips
(sb; pest)
d Kohlrüben-Blasenfuß *m*; Frühjahrs-
Ackerblasenfuß *m*
l Thrips anguisticeps

* **cake → 1016**

365 caking of sugar
f agglomération *f* du sucre
d Zusammenbacken *n* des Zuckers
e aterronamiento *m* del azúcar
n samenbakken *n* van de suiker

366 calandria
(A heating element used in certain types
of evaporator bodies and vacuum pans,
consisting of a drum- or lens-shaped
body traversed by vertical tubes with the
steam or heating vapo(u)r in the space
surrounding the tubes)
f faisceau *m* (tubulaire); calandre *f*;
chambre *f* de vapeur à faisceau
tubulaire; chambre *f* de chauffe
d Röhrenheizkörper *m*; Heizkammer *f*
e calandria *f*; cuerpo *m* tubular; cámara *f*
de calentamiento
n stoomtrommel *f*

367 calandria evaporator
f évaporateur *m* à calandre; évaporateur
m à faisceau
d Vertikalrohrverdampfer *m* mit Innenheiz-
kammer
e evaporador *m* de calandria
n verdamper *m* met stoomtrommel

368 calandria pan
f appareil *m* à cuire à faisceau
d Kochapparat *m* mit Heizkammer
e tacho *m* al vacío de calandria
n kookpan *f* met stoomtrommel

369 calcining zone; combustion zone *(of lime
kiln)*
f zone *f* de décomposition

d Brennzone *f*
e zona *f* de calcinación
n ontledingszone *f*; ontledingsruimte *f*

370 calcium deficiency
(sb; disease)
f carence *f* en calcium
d Kalkmangel *m*
e deficiencia *f* de calcio; carencia *f* de
calcio
n calciumgebrek *n*; kalkgebrek *n*

* **calcium saccharate process → 2279**

* **calfs snout → 1456**

371 California bur clover *(US)*
(sb; weed)
f luzerne *f* hérissée
d rauher Schneckenklee *m*; gezähnelter
Schneckenklee *m*; steifhaariger
Schneckenklee *m*
l Medicago hispida

* **California dandelion → 441**

372 calorific value *(of bagasse)*
f pouvoir *m* calorifique
d Heizwert *m*
e valor *m* calorífico; V.C. *m*
n verbrandingswaarde *f*; calorische
waarde *f*

373 calorimeter
f calorimètre *m*
d Kalorimeter *n*
e calorímetro *m*
n warmtemeter *m*; calorimeter *m*

* **calorisator → 806**

374 campaign; season
(The period in which the beets/canes are
processed)
f campagne *f*
d Kampagne *f*
e zafra *f*; campaña *f*
n campagne *f*

* **Canada thistle** *(US)* **→ 676**

* **canary grass → 2650**

375 candle filter
f filtre *m* à bougies
d Kerzenfilter *n*
e filtro *m* de bujías
n kaarsfilter *n*

376 candy *(noun)*; **candy sugar**
f sucre *m* candi; candi *m*
d Kandiszucker *m*; Kandis *m*
e azúcar *m* cande; azúcar *m* candil
n kandij *f*; kandijsuiker *m*

377 candy *v*
f cristalliser; se candir
d auskristallisieren
e cristalizar
n uitkristalliseren

* **candy sugar** → 376

378 cane bundle
f botte *f* de cannes; paquet *m* de cannes; faisceau *m* de cannes
d Rohrbündel *n*
e paquete *m* de cañas; bulto *m* de cañas
n rietbundel *m*

379 cane carrier; cane conveyor
f conducteur *m* de cannes; chaîne *f* à cannes
d Rohrtransporteur *m*; Rohrförderband *n*
e conductor *m* de cañas
n riettransporteur *m*; rietcarrier *m*

380 cane chopper
(agr)
f tronçonneuse *f* de cannes
d Häckselmaschine *f* für Zuckerrohr
e divisora *f* de tallos; troceadora *f* de cañas
n hakselmachine *f* voor suikerriet

* **cane conveyor** → 379

381 cane crane
f grue *f* à cannes
d Zuckerrohrkran *m*
e grúa *f* cañera
n suikerrietkraan *f*

382 cane crusher
f défibreur *m*
d Crusher *m*; Vorbrecher *m*
e desfibradora *f* de caña; desfibradora *f* para caña
n crusher *m*; kneuzer *m*

383 cane cutter
f coupe-cannes *m*
d Rohrschnitter *m*
e cortacañas *f*
n canecutter *m*

384 cane diffuser
f diffuseur *m* de cannes
d Rohrdiffuseur *m*; Rohrextraktionsapparat *m*
e difusor *m* de cañas
n rietdiffuseur *m*

* **cane diffusion** → 809

385 cane disintegrator
f défibreuse *f* de cannes
d Rohrentfaserer *m*
e desintegrador *m* de cañas
n rietrafelaar *m*

386 cane elevator
f élévateur *m* de cannes
d Rohrelevator *m*
e elevador *m* de cañas
n rietelevator *m*

387 cane grab
f grappin *m* à cannes
d Zuckerrohrgreifer *m*
e cuchara *f* de cañas
n suikerrietgrijper *m*

388 cane grower
f cultivateur *m* de canne (à sucre)
d Zuckerrohranbauer *m*; Zuckerrohr-farmer *m*
e agricultor *m* de caña (de azúcar); cañero *m*
n suikerrietteler *m*

389 cane handling equipment
f engins *mpl* de manutention des cannes
d Rohrhandhabungsgeräte *npl*
e máquinas *fpl* para el manejo de las cañas

390 cane juice
f jus *m* de canne
d Rohrsaft *m*
e jugo *m* de caña; guarapo *m* de caña
n rietsap *n*

391 cane knives
f coupe-cannes *m*
d Rohrmesser *npl*; Rohrschneider *mpl*
e cuchillas *fpl* cañeras; cuchillas *fpl* para caña; cuchillas *fpl* cortacañas
n rietsnijmessen *npl*; rietmessen *npl*

392 cane knives with horizontal blades; knife set with horizontal blades
f coupe-cannes *m* à lames horizontales
d Rohrmesser *npl* mit horizontal angeordneten Schneiden
e cuchillas *fpl* cañeras con hojas horizontales

n rietmessen *npl* met horizontaal
aangebrachte sneden

393 cane layer
 f couche *f* de cannes
 d Rohrschicht *f*
 e colchón *m* de cañas
 n rietlaag *f*

394 cane leveller
 f égaliseur *m* de couche de cannes
 d Zuckerrohrschichtbegrenzer *m*
 e nivelador *m* de caña
 n suikerrietlaagbegrenzer *m*

395 cane loader
 (agr)
 f chargeuse *f* de cannes
 d Zuckerrohrlader *m*
 e cargadora *f* de cañas
 n suikerrietlader *m*

396 cane mill
 f moulin *m* à cannes
 d Zuckerrohrmühle *f*
 e molino *m* cañero; molino *m* de caña de
 azúcar
 n suikerrietmolen *m*

 * **cane molasses, sugar ~ → 2345**

 * **cane opening** *(between front roller and
 top roller)* → 978

397 cane preparation
 f préparation *f* de la canne
 d Rohraufbereitung *f*
 e preparación *f* de la caña
 n rietvoorbewerking *f*; rietvoorbereiding *f*

398 cane preparator
 f appareil *m* de préparation de la canne;
 instrument *m* de préparation de la
 canne; engin *m* de préparation de la
 canne
 d Rohraufbereitungsapparat *m*
 e aparato *m* empleado en la preparación
 de la caña; instrumento *m* de
 preparación de la caña
 n rietvoorbewerkingswerktuig *n*;
 rietvoorbereidend werktuig *n*;
 rietvoorbewerker *m*

399 cane rake
 f rateau *m* à cannes
 d Rohrrechen *m*
 e rastrillo *m* de cañas
 n riethark *f*

400 cane reception
 f réception *f* des cannes
 d Rohrannahme *f*
 e recepción *f* de cañas
 n rietontvangst *f*; rietreceptie *f*

401 cane refinery
 f raffinerie *f* de canne; raffinerie *f* de
 sucre de canne
 d Rohrzuckerraffinerie *f*
 e refinería *f* de caña
 n rietsuikerraffinaderij *f*;
 rietsuikerraffineerderij *f*

 * **cane roll(er) → 980**

402 cane sampling
 f échantillonnage *m* de la canne
 d Rohrprobenahme *f*
 e toma *f* de muestras de caña
 n rietmonsterneming *f*

403 cane shredder
 f shredder *m*
 d Rohrentfaserer *m*; Shredder *m*;
 Zerfaserer *m*
 e desmenuzadora *f*; desfibradora *f*
 n shredder *m*; rafelmachine *f*; rafelaar *m*

404 cane sugar
 f sucre *m* de canne
 d Rohrzucker *m*
 e azúcar *m* de caña
 n rietsuiker *m*

 * **cane, sugar ~ → 2339**

405 cane sugar factory; cane sugar mill; mill
 f sucrerie *f* de canne
 d Rohrzuckerfabrik *f*
 e azucarera *f* de caña; fábrica *f* de azúcar
 de caña; ingenio *m* azucarero
 n rietsuikerfabriek *f*

406 cane sugar industry
 f industrie *f* du sucre de canne
 d Rohrzuckerindustrie *f*
 e industria *f* del azúcar de caña; industria
 f azucarera de caña
 n rietsuikerindustrie *f*;
 rietsuikernijverheid *f*

 * **cane sugar mill → 405**

407 cane sugar producing country
 f pays *m* producteur de sucre de canne
 d Rohrzuckererzeugerland *n*
 e país *m* productor de azúcar de caña
 n rietsuikerproducerend land *n*

408 cane topper
(agr)
f écimeuse *f* de cannes
d Entgipfelmaschine *f* für Zuckerrohr
e desmochadora *f*; eliminadora *f* de
 cogollos
n topper *m*

409 cane variety
f variété *f* de canne
d Rohrsorte *f*; Rohrvarietät *f*
e variedad *f* de caña
n rietsoort *f*; rietvariëteit *f*

410 cane yard
f plate-forme *f*
d Rohrabladeplatz *m*; Rohrplatz *m*;
 Rohrlagerplatz *m*; Rohrannahmeplatz *m*
e patio *m*
n rietopslagplaats *f*; rietontvangstplaats *f*

411 cap *(of cane mill)*
f chapeau *m*
d Deckel *m*
e cabezote *m*; tapa *f*; cabezal *m*
n deksel *n*

* **capeweed → 441**

412 capsid bug
(sb; pest)
f punaise *f*
d Kartoffelwanze *f*
e chinche *f* de la hoja
l Calocoris norvegicus

* **capsid bug** *(US)* → 574

413 caramel
f caramel *m*
d Karamel *m*
e caramelo *m*
n caramel *f*; karamel *f*

414 caramel
f teinture *f* de caramel; couleurs *fpl* de
 sucre
d Zuckerfarbe *f*; Zuckercouleur *f*; Kulör *f*
e caramelo *m*; azúcar *m* quemado
n caramel *f*

415 caramelization
f caramélisation *f*
d Karamelisierung *f*; Karamelisation *f*;
 Karamelbildung *f*
e caramelización *f*
n caramelisatie *f*; caramelvorming *f*

416 caramelize *v*
(To form products from sugar by burning
 sugar or similar carbohydrates)
f se caraméliser
d karamelisieren
e caramelizar
n carameliseren

417 carbonatation; carbonation *(US)*
(The defecation process in which the
 removal by filtration of precipitated
 calcium carbonate formed by the
 addition of lime and carbon dioxide to an
 impure sugar liquor results in an overall
 reduction in the level of impurities in
 that liquor)
f carbonatation *f*
d Karbonatation *f*; Carbonatation *f*;
 Saturation *f*
e carbonatación *f*
n carbonatatie *f*

418 carbona(ta)tion factory
f sucrerie *f* à carbonatation
d Carbonatationsfabrik *f*
e fábrica *f* con carbonatación
n carbonatatiefabriek *f*; carbonaterende
 fabriek *f*

419 carbona(ta)tion juice; carbonated juice
f jus *m* carbonaté; jus *m* de
 carbonatation; jus *m* trouble
d Carbonatationssaft *m*; Schlammsaft *m*
e jugo *m* carbonatado; jugo *m* de
 carbonatación
n carbonatatiesap *n*; gecarbonateerd sap *n*;
 vuilsap *n*

420 carbona(ta)tion juice distributor
f distributeur *m* de jus carbonaté
d Schlammsaftverteiler *m*; Carbonatations-
 saftverteiler *m*
e distribuidor *m* de jugo carbonatado
n carbonatatiesapverdeler *m*; vuilsapverdeler *m*

421 carbona(ta)tion juice pump
f pompe *f* à jus trouble
d Schlammsaftpumpe *f*; Carbonatationssaft-
 pumpe *f*
e bomba *f* de jugo turbio
n vuilsappomp *f*; carbonatatiesappomp *f*

422 carbona(ta)tion juice return;
 carbona(ta)tion juice recycling
f recyclage *m* du jus trouble; recyclage *m*
 du jus carbonaté
d Schlammsaftrücknahme *f*;
 Carbonatationssaftrücknahme *f*

e recirculación f del jugo carbonatado;
recirculación f del jugo de carbonatación;
retorno m del jugo carbonatado; retorno
m del jugo de carbonatación
n terugbrenging f van het gecarbonateerde
sap; terugbrenging f van het carbonatatiesap; terugbrenging f van het vuilsap

* **carbona(ta)tion tank** → 426

423 carbonate *(noun)*
f carbonate m
d Carbonat n; Karbonat n
e carbonato m
n carbonaat n

424 carbonate v
(To introduce a gas rich in CO_2 into the
juice)
f carbonater
d carbonatieren; karbonatieren
e carbonatar
n carbonateren

425 carbonate v **batchwise**
f carbonater en discontinu
d diskontinuierlich carbonatieren
e carbonatar discontinuamente
n intermitterend carbonateren; discontinu
carbonateren

* **carbonated juice** → 419

* **carbonation** *(US)* → 417

426 carbonator; carbona(ta)tion tank;
carbonating tank
f chaudière f à carbonater; chaudière f de
carbonatation; appareil m de
carbonatation; caisse f à carbonatation;
bac m à carbonater
d Carbonatationsapparat m;
Carbonatationsgefäß n; Carbonateur m
e carbonatador m; tanque m de
carbonatar; tanque m de carbonatación
n carbonatatieketel m; carbonateur m;
carbonatatiebak m

427 carbon dioxide
f gaz m carbonique; dioxyde m de
carbone; acide m carbonique
d Kohlendioxid n; Kohlensäure f
e gas m de ácido carbónico; dióxido m de
carbono; ácido m carbónico
n koolzuur n

428 carbon dioxide washer
f laveur m à gaz carbonique
d Kohlensäurewäscher m

e lavador m de ácido carbónico
n koolzuurwasser m

429 Carolina grasshopper *(US)*; **Carolina**
locust *(US)*
(sb; pest)
l Dissosteira carolina

430 carried-over quantity
(EEC)
f quantité f à reporter
d Übertragungsmenge f
e cantidad f de report
n over te dragen hoeveelheid f

431 carrier beet harvester
(agr)
f décolleteuse-arracheuse-nettoyeuse-
chargeuse-transporteuse f
d Rübenvollerntemaschine f; Rüben-
kombine f
e cosechadora f integral de remolacha
n bunkerrooier m

432 carrier with scraper chain
(A cane carrier)
f conducteur m à chaînes à raclettes
d Kratzerkettentransporter m
e conductor m con cadenas de rascadores
n schraapband m

433 carting off beet harvester
(agr)
f décolleteuse-arracheuse-débardeuse f
d Bunkerköpfroder m; Rübenvollernter m
e cosechadora f de remolacha
n kopper-bunkerrooier m

434 cartridge filter
f filtre m à cartouches
d Patronenfilter n
e filtro m de cartuchos
n patroonfilter n

435 cassonade
f cassonade f
d Farin m
e azúcar m bruto; azúcar m moreno;
azúcar m sin refinar
n cassonade f

436 castor sugar; caster sugar *(US)*
(A fine-grained granulated sugar, so
called because it is used in table castors
with perforated tops)
f sucre m en poudre
d Puderzucker m
e azúcar m en polvo
n poedersuiker m

437 **catalyst**
(A substance which accelerates a reaction
without itself being changed)
f catalyseur *m*
d Katalysator *m*
e catalizador *m*
n katalysator *m*

438 **catch-all; save-all; entrainment**
separator; entrainment arrestor;
entrainment catcher
(A device for separating entrained matter
from the vapo(u)r leaving a vacuum pan
or evaporator body)
f désucreur *m*; dessucreur *m*;
séparateur *m*; ralentisseur *m*; vase *m* de
sûreté
d Saftfänger *m*; Saftabscheider *m*
e separador *m*; separador *m* de arrastre(s);
cogedor *m* de jugo
n sapvanger *m*

* **catchfly** → 288

* **catchweed** → 509

* **catchweed bedstraw** → 509

439 **cation exchanger**
f échangeur *m* de cations
d Kationenaustauscher *m*
e cambiador *m* de cationes
n kationenuitwisselaar *m*

440 **cation exchange resin; cation resin**
f résine *f* échangeuse de cations
d Kationenaustauscherharz *n*; Kationen-
harz *n*
e resina *f* intercambiadora de cationes
n kationenuitwisselend hars *n*

441 **cat's ear; California dandelion;**
capeweed; gosmore
(sb; weed)
f porcelle *f*; porcelle *f* enracinée
d gemeines Ferkelkraut *n*
e hierba *f* de halcón
n gewoon biggekruid *n*
l Hypochaeris radicata

* **cat-tail millet** → 2730

442 **caustic soda**
(Scale removal)
f soude *f* caustique
d kaustische Soda *f*; Ätznatron *n*
e sosa *f* cáustica
n bijtende soda *m*; natriumhydroxyde *n*;
natronloog *f*

* **CCT** → 572

443 **cell-less drum filter**
f filtre *m* à tambour rotatif sans cellules
d zellenloses Trommelfilter *n*
e filtro-tambor *m* sin células
n trommelfilter *n* zonder cellen

444 **cell-type drum filter**
f filtre *m* rotatif à cellules
d Zellentrommelfilter *n*
e filtro-tambor *m* con células
n cellentrommelfilter *n*

445 **cell-type vacuum drum filter**
f filtre *m* rotatif sous vide de type
multicellulaire
d Vakuum-Zellentrommelfilter *n*
e filtro *m* rotatorio al vacío de tipo
multicelular
n vacuüm-cellentrommelfilter *n*

446 **cellulose**
f cellulose *f*
d Zellulose *f*; Cellulose *f*
e celulosa *f*
n cellulose *f*; celstof *f*

* **centipedes** → 1658

* **central downtake** → 448

* **central tube** → 448

447 **central vacuum**
f vide *m* central
d Zentralvakuum *n*
e vacío *m* central
n centraal vacuüm *n*

* **centre of crystallization** → 697

448 **centre well; central downtake; central**
tube; central well
(Communicating tube of relatively large
diameter traversing the axis of a
calandria which permits a convective
circulation of the liquid being heated; i.e.,
upward through the tubes and downward
through the centre well)
f puits *m* central
d Mittelrohr *n*; Zentralrohr *n*
e tubo *m* central; vía *f* de descenso
n middenbuis *f*; centrale buis *f*

449 **centrifugal; centrifugal machine;**
machine; fugal; centrifugal drier;
centrifuge
(The machine used to separate sugar

crystals retained on a rotating basket
containing a mesh, from the syrup)
f essoreuse f; essoreuse f centrifuge;
  turbine f; centrifuge fm; turbine f
  centrifuge essoreuse; centrifugeuse f
d Zentrifuge f
e centrífuga f; máquina f centrífuga;
  centrifugadora f; centrífugo m
n centrifuge f; centrifugaalmachine f

* **centrifugalability** → 1141

450 **centrifugal basket**
  f tambour m de centrifuge; panier m de
    centrifuge
  d Zentrifugentrommel m; Zentrifugen-
    korb m; Schleuderkorb m
  e tambor m de centrífuga; canasto f de
    centrífuga
  n centrifugetrommel f

* **centrifugal basket with level bottom**
  → 1082

451 **centrifugal basket with steep conical
  bottom**
  f tambour m de centrifuge avec fond
    conique en pente forte; panier m de
    centrifuge avec fond conique en pente
    forte
  d Steilbodenzentrifugentrommel f; Steil-
    konuszentrifugentrommel f
  e tambor m de centrífuga con el fondo en
    forma de cono muy inclinado; canasto m
    de centrífuga con el fondo en forma de
    cono muy inclinado
  n centrifugetrommel f met steile
    kegelvormige bodem

452 **centrifugal compressor**
  f compresseur m centrifuge
  d Kreiselverdichter m
  e compresor m centrífugo
  n centrifugaalcompressor m

* **centrifugal drier** → 449

453 **centrifugal force**
  f force f centrifuge
  d Fliehkraft f; Zentrifugalkraft f
  e fuerza f centrífuga
  n centrifugale kracht f;
    middelpuntvliedende kracht f

454 **centrifugalling; fugalling; centrifuging;
  centrifuge separation; purging**
  f essorage m; turbinage m
  d Zentrifugieren n; Schleudern n;
    Abschleudern n; Zentrifugation f

e centrifugado m; purgado m;
  centrifugación f
n centrifugeren n

* **centrifugal machine** → 449

455 **centrifugal machine for washing sugar**
  f centrifuge f de clairçage
  d Deckschleuder f
  e centrífuga f de lavado
  n dekcentrifuge f

456 **centrifugal operator; sugar drier;
  machine man; fugal operator**
  f turbineur m; ouvrier m turbineur
  d Bedienungsmann m der Zentrifuge
  e operador m de centrifuga
  n bedieningsman m van centrifuge;
    centrifugist m

* **centrifugal plough** → 832

457 **centrifugal pump**
  f pompe f centrifuge
  d Kreiselpumpe f; Zentrifugalpumpe f
  e bomba f centrífuga
  n centrifugaalpomp f

458 **centrifugal separator; centrifugal-type
  separator**
  f séparateur m centrifuge
  d Trennschleuder f; Zentrifugal-
    separator m
  e separador m centrífugo
  n centrifugale afscheider m;
    centrifugaalseparator m

459 **centrifugal station**
  f station f de turbinage
  d Zentrifugenstation f
  e departamento m de centrífugas
  n centrifugestation n

* **centrifugal sugar** → 710

* **centrifugal-type separator** → 458

* **centrifuge** → 449

460 **centrifuge v; spin v; machine v; purge v**
  f centrifuger; essorer; turbiner
  d zentrifugieren; schleudern; abschleudern
  e centrifugar; purgar
  n centrifugeren

* **centrifuge separation** → 454

* **centrifuge wash** → 2638

* **centrifuging** → 454

461 cercospora leaf spot; beet leaf spot
(sb; disease)
f cercosporiose f
d Cercospora-Blattfleckenkrankheit f
e cercosporiosis f; cercospora f de la
remolacha
n Cercospora-bladvlekkenziekte f

462 certificate of origin
(EEC)
f certificat m d'origine
d Ursprungszeugnis n
e certificado m de origen
n certificaat n van oorsprong

* **chadlock** → 2696

**463 chain type elevator; open web elevator;
elevating web**
(agr)
f tablier m à chaînes
d Siebkette f
e cadena f embocadora
n zeefketting f

464 chamber filter press
f filtre-presse m à chambres
d Kammerfilterpresse f
e filtro-prensa m de cámaras
n kamerfilterpers f

465 change in colo(u)r
f changement m de couleur
d Farbumschlag m; Farbveränderung f
e cambio m de color
n kleuromslag m

466 changing the knives
f changement m des couteaux
d Auswechseln n der Messer
e cambiado m de cuchillas
n verwisseling f van de messen

467 char cistern; char filter *(US)*
(The vessel containing the bone charcoal
through which the liquor percolates in
the decolo(u)rizing stage)
f filtre m à noir animal
d Knochenkohlefilter n
e filtro m de carbón animal
n beenzwartfilter n; beenderkoolfilter n

468 char distributor
f distributeur m de noir animal
d Tierkohleverteiler m; Knochenkohle-
verteiler m
e distribuidor m de carbón animal

n beenzwartverdeelinrichting f;
beenderkoolverdeelinrichting f

469 char elevator
f élévateur m de noir animal
d Knochenkohle-Elevator m
e elevador m de carbón animal
n beenzwartelevator m;
beenderkoolelevator m

* **char filter** *(US)* → 467

470 charging *(of centrifugal)*
f chargement m
d Füllung f
e carga f
n vulling f

471 charging speed *(of centrifugal)*
f vitesse f de chargement
d Fülldrehzahl f
e velocidad f al cargar
n toerental n bij het vullen

472 charhouse
f atelier m de noir animal
d Knochenkohlehaus n; Knochenkohle-
station f
e estación f de carbón animal
n beenzwartstation n;
beenderkoolstation n

473 char kiln
f four m à noir animal
d Knochenkohle-Ofen m
e horno m de carbón animal
n beenzwartoven m; beenderkooloven m

* **charlock** → 2696

* **cheese** *(slang) (of cane mill)* → 1282

474 cheeseweed *(US)*; **little mallow** *(US)*
(sb; weed)
l Malva parviflora

475 chemical losses
f pertes fpl chimiques
d chemische Verluste mpl
e pérdidas fpl químicas
n chemische verliezen npl

476 chemical oxygen demand; COD
(Its determination provides a measure of
the oxygen equivalent of that portion of
the organic matter in a sample which is
susceptible to oxidation by a strong
chemical oxidant)
f demande f chimique en oxygène;

D.C.O. *f*
d chemischer Sauerstoffbedarf *m*; CSB *m*
e demanda *f* química del oxígeno
n chemisch zuurstofverbruik *n*; C.Z.V. *n*

477 chevron *(on crusher roller)*
f chevron *m*
d Rille *f*; Keep *f*; Kerbe *f*
e chevrón *m*
n keep *f*

478 chevrons in ace of diamonds *(on crusher roller)*
f chevrons *mpl* en as de carreau
d rautenförmige Rillen *fpl*; rautenförmige Keepen *fpl*; rautenförmige Kerben *fpl*
e chevrones *mpl* en "diamante"
n ruitvormige kepen *fpl*

479 chevrons point upwards *(on crusher roller)*
f chevrons *mpl* pointe en haut
d Rillen *fpl* mit der Spitze nach oben; Keepen *fpl* mit der Spitze nach oben; Kerben *fpl* mit der Spitze nach oben
e chevrones *mpl* con la punta hacia arriba
n kepen *fpl* met de punt naar boven

* **chickweed** → **568**

* **China jute** → **2612**

* **Chinese pursley** → **2078**

* **chingma abutilon** → **2612**

* **chips** → **660**

480 chlorotic streak; fourth disease *(Java)*; **pseudo scald** *(Australia)*
(sc; virus disease)
f stries *fpl* chlorotiques; maladie *f* des stries chlorotiques
d chlorotische Streifenkrankheit *f*
e raya *f* clorótica
n chlorotische strepen *fpl*

* **chokeless pump** → **2577**

481 chop *v*
f désintégrer; tronçonner
d zerkleinern
e desmenuzar
n hakken; in stukjes snijden

482 chopper-harvester
(agr)
f récolteuse-tronçonneuse *f* de cannes
d Erntemaschine *f* mit Häcksler

e cosechadora *f* divisora de tallos; cosechadora *f* troceadora
n oogstmachine *f* met hakselaar

483 chopping loading harvester; cut load harvester; chopper-harvester-loader
(agr)
f récolteuse-tronçonneuse-chargeuse *f* de cannes
d Erntemaschine *f* mit Häcksler und Lader
e cosechadora *f* de caña; combinada *f*
n oogstmachine *f* met hakselaar en lader

484 chopping roller
(agr)
f rouleau *m* tronçonneur
d Häckselrolle *f*
e rodillo *m* cortador de caña
n hakselrol *f*

* **chufa** → **1682**

* **chufa flat sedge** → **1682**

485 chute
f goulotte *f*
d Schurre *f*; Rutsche *f*
e canal *m* inclinado
n goot *f*

* **CIBE** → **1334**

486 cicada; harvest fly; locust
(sc; pest)
f cigale *f* chanteuse
d Singzirpe *f*; Singzikade *f*
e cicádido *m*; cigarra *f*
n zangkrekel *m*; zingcicade *f*
l Cicada

487 cif-price
(eco)
f prix *m* caf; prix *m* cif
d cif-Preis *m*
e precio *m* cif
n cif-prijs *m*

* **circular ribbon** → **1977**

488 circulation evaporator
f évaporateur *m* à circulation
d Umlaufverdampfer *m*
e evaporador *m* de circulación (natural)
n verdamper *m* met natuurlijke circulatie

489 circulation juice
f jus *m* de circulation
d Zirkulationssaft *m*; Umlaufsaft *m*
e jugo *m* de circulación; guarapo *m* de

circulación
n circulatiesap n

490 circulation juice heater
f réchauffeur m pour jus de circulation
d Zirkulationssaftwärmer m; Umlaufsaft-
wärmer m
e calentador m de jugo de circulación;
calentador m de guarapo de circulación
n circulatiesapverwarmer m

491 circulation juice pump
f pompe f à jus de circulation
d Zirkulationssaftpumpe f; Umlaufsaft-
pumpe f
e bomba f del jugo de circulación; bomba f
del guarapo de circulación
n circulatiesappomp f

492 circulation juice tank
f bac m à jus de circulation
d Umlaufsaftbehälter m; Zirkulationssaft-
behälter m
e tanque m de jugo de circulación;
depósito m de jugo de circulación;
tanque m de guarapo de circulación;
depósito m de guarapo de circulación
n circulatiesapbak m

493 circulation pump
f pompe f de circulation
d Zirkulationspumpe f
e bomba f de circulación
n circulatiepomp f

* **clammy chickweed** → 1643

* **clamp** v → 910

494 clamp aphid
(sb; pest)
f puceron m des silos
d Mietenlaus f
l Hyperomizus staphyleae;
Rhopalosiphonimus staphyleae

495 clamp rot; storage rot
(sb; disease)
f pourriture f de silo; pourriture f en silo
d Mietenfäule f; Mietenkopffäule f; Kopf-
und Mietenfäule f
e podredumbre f de silo
n kuilrot n

496 clarification
(The process of removing undissolved
materials from a liquid. Specifically,
removal of solids either by settling or
filtration)

f clarification f
d Klärung f
e clarificación f
n klaring f

497 clarification aid; clarifying agent
f adjuvant m de clarification; agent m de
défécation
d Klärhilfsmittel n
e aditivo m de clarificación
n klaarhulpmiddel n

* **clarification apparatus** → 500

* **clarification pan** → 500

498 clarification station; clarifier station
f poste m de décantation; station f de
clarification
d Klärstation f
e departamento m de decantación;
departamento m de clarificadores
n klaarstation n

499 clarified juice; clear juice; clarifier juice
(The finished product of the clarification
process)
f jus m clair; jus m clarifié; jus m léger
d geklärter Saft m; Klarsaft m; Klärsaft m
e guarapo m clarificado; guarapo m claro;
jugo m decantado; guarapo m de
defecadoras; jugo m clarificado; jugo m
claro; jugo m de defecadoras
n schoonsap n; klaarsap n

**500 clarifier; clarification pan; clarification
apparatus**
(Apparatus for the elimination by
sedimentation of suspended solids from a
turbid liquid)
f décanteur m; clarificateur m
d Dekanteur m; Klärgefäß n
e clarificadora f; decantador m;
clarificador m
n klaarpan f; clarificateur m;
decanteertoestel n; decanteur m

* **clarifier juice** → 499

* **clarifier station** → 498

501 clarify v
f clarifier
d klären; abklären
e clarificar
n klaren

* **clarifying agent** → 497

* **clasping pepperweed** → 2729

502 classifier
f classificateur *m*
d Klassierer *m*
e aparato *m* clasificador; clasificadora *f*;
clasificador *m*
n klasseerapparaat *n*

503 cleaner-loader
(agr)
f décrotteuse *f*
d Reinigungslader *m*
e cargadora-limpieza *f*
n reiniger-lader *m*

**504 cleaning drum; cleaning cage;
cylindrical cleaning drum**
(agr)
f tambour *m* décrotteur; tambour *m*
nettoyeur; tambour *m* de nettoyage
d Reinigungstrommel *f*
e tambor *m* de limpieza; tambor *m*
cilíndrico de limpieza
n reinigingstrommel *f*

505 clear filtrate
f filtrat *m* clair
d klares Filtrat *n*
e filtrado *m* claro
n klaar filtraat *n*

* **clear juice** → 499

506 clear juice box *(of clarifier)*
f boîte *f* à jus clair
d Klarsaftkasten *m*
e caja *f* de jugo claro; caja *f* de guarapo
claro
n klaarsapkist *f*

507 clear juice tank
f bac *m* à jus clarifié
d Klarsaftbehälter *m*
e depósito *m* de jugo claro; depósito *m* de
jugo decantado; depósito *m* de guarapo
claro
n klaarsapbak *m*

508 clear-winged grasshopper *(US)*
(sb; pest)
l Camnula pellucida

**509 cleavers; hariff; clivers; catchweed; goose
grass; catchweed bedstraw**
(sb; weed)
f gratteron *m*; gaillet *m* gratteron
d Klebkraut *n*; Klettenlabkraut *n*; Kleblab-
kraut *n*

e amor *m* de hortelano; hierba *f* del amor
n kleefkruid *n*
l Galium aparine

510 click beetle
(sb; pest)
f taupin *m* rayé; taupin *m* des moissons
d Saatschnellkäfer *m*
e elátero *m* del trigo
l Agriotes lineatus

511 click-beetle
(sb; pest)
f taupin *m* *(adulte)*; maréchal *m*; tape-
marteau *m*
d Saatschnellkäfer *m*
e elatérido *m*
l Agriotes sp.

512 click beetles
(sc; pest)
f forgerons *mpl*; taupins *mpl*; élatéridés
mpl; notopèdes *mpl*
d Schnellkäfer *mpl*; Springkäfer *mpl*;
Feldschnellkäfer *mpl*; Springfeldkäfer
mpl
e elatéridos *mpl*; escarabajos *mpl* de los
gusanos de alambre; escarabajos *mpl* de
resorte
n kniptorren *fpl*; springkevers *mpl*;
springtorren *fpl*
l Elateridae

* **climbing buckwheat** → 276

* **climbing film effect** *(in evaporator)*
→ 1008

**513 climbing film evaporator; rising film
evaporator**
f évaporateur *m* à grimpage; évaporateur
m à flot montant
d Steigstromverdampfer *m*; Kletterfilmverdampfer *m*
e evaporador *m* de película ascendente;
evaporador *m* de flujo ascendente;
evaporador *m* de película elevada
n klimverdampapparaat *n*; verdamptoestel
n volgens Kestner

514 climb *v* up
f grimper
d steigen; klettern
e subir
n stijgen

515 clipping machine
f machine *f* à couper le sucre
d Knippmaschine *f*

e máquina f para cortar azúcar
n knipmachine f

* clivers → 509

516 clod crusher; clod breaker
  f émottoir m
  d Schollenbrecher m; Zerkrümler m
  e rulo m desterronador
  n kluitenbreker m

517 clouded tortoise beetle; beet tortoise
  beetle; clouded shield beetle
  (sb; pest)
  f casside f nébuleuse; casside f de la
    betterave; casside f de l'arroche
  d nebliger Schildkäfer m; Rübenschild-
    käfer m; Nebelschildkäfer m
  e casida f nebulosa; tortuguilla f de la
    remolacha
  n gevlekte schildpadtor f;
    bietenschildpadtor f; schilpadtorretje n
    van de bieten
  l Cassida nebulosa

518 cloudy filtrate
  (First juice obtained from continuous
  rotary vacuum filter)
  f filtrat m trouble
  d trübes Filtrat n
  e filtrado m turbio
  n troebel filtraat n

519 cloudy juice
  f jus m trouble
  d trüber Saft m; Trübe f
  e jugo m turbio; guarapo m turbio
  n troebel sap n

520 cloudy sword-grass; sword-grass moth
  (US)
  (sb; pest)
  f noctuelle f antique; antique f; brunâtre f
  d Scharteneule f; graues Moderholz n;
    graue Moderholzeule f; Feldkräutergras-
    flur-Moderholzeule f
  l Calocampa exoleta

521 clover cutworm (US)
  (sb; pest)
  f noctuelle f; ver m gris
  d Klee-Eule f; Meldenflur-Blättereule f
  l Scotogramma trifolii

522 clover leafhopper (US)
  (sb; pest)
  l Aceratagallia sanguinolenta

523 cluster; seed ball (US)
  (agr)
  f glomérule m
  d Knäuel nm
  e glomérulo m
  n kluwen n

524 clustered sugar
  f grumeau m de sucre
  d Zuckerknoten m; Zuckerklumpen m
  e grumo m de azúcar
  n suikerklonter m

525 C magma
  f magma m C
  d C-Magma n
  e magma m C
  n C-magma n

526 C massecuite; third massecuite
  f masse f cuite C; masse f cuite de
    troisième jet; masse f cuite troisième jet;
    m c f III
  d C-Füllmasse f; C-Kochmasse f
  e masa f cocida C; masa f cocida de
    tercera; masa f cocida tercera
  n C-masse cuite f; C-vulmassa f; masse
    cuite f C

527 C molasses; third molasses
  f égout m C
  d C-Ablauf m
  e melaza f C; melaza f de tercera; miel f C
  n C-stroop f

528 coagulant
  f coagulant m
  d Koagulationsmittel n; Koagulans n;
    Koagulator m
  e coagulante m
  n coaguleermiddel n

529 coagulate v
  f coaguler
  d koagulieren
  e coagular
  n coaguleren

530 coagulation
  f coagulation f
  d Koagulieren n; Koagulation f
  e coagulación f
  n coagulatie f

* coarse → 2019

531 coarse sugar; semolina sugar
  f sucre m semoule

d Grießzucker *m*
n griessuiker *m*; grove suiker *m*

* **coat** v *(US)* → **1740**

* **coated seed** *(US)* → **1741**

* **Cobb's disease of sugar cane** → **1211**

532 **cockchafer; common cockchafer;
     European cockchafer** *(US)***; European
     common cockchafer** *(US)*
     (sb; pest)
f hanneton *m* (commun); hanneton *m*
   vulgaire
d Maikäfer *m*; gemeiner Maikäfer *m*;
   Feldmaikäfer *m*; Maikolbenkäfer *m*
e abejorro *m* (común); cochorro *m*;
   melolonta *f*
n meikever *m*
l Melolontha melolontha; Melolontha
   vulgaris

533 **cocksfoot; orchard grass** *(US)***; cockspur;
     rough cock's-foot**
     (sb; weed)
f dactyle *m* aggloméré; dactyle *m*
   pelotonné; patte *f* de lièvre; herbe *f* des
   vergers
d Knaulgras *n*; Wiesenknäuelgras *n*;
   gemeines Knaulgras *n*; gemeines Knäuelgras *n*; rauhes Knäuelgras *n*; Hundsgras *n*
e dactilo *m* apelotonado; dactilo *m* ramoso;
   dactilo *m* aglomerado; grama *f* en
   jopillos; pasto *m* ovillo
n kropaar *f*
l Dactylis glomerata

* **cock's-foot** → **1722**

   , 1722

* **cockspur grass** → **1722**

* **coco grass** → **1686**

534 **coconut scale; bourbon scale** *(US)*
     (sc; pest)
f cochenille *f* du cocotier
d Kokospalmenschildlaus *f*
e cochinilla *f* del cocotero; guagua *f* común
   del cocotero
l Aspidiotus destructor

535 **co-current barometric condenser**
f condenseur *m* barométrique à co-
   courants; condenseur *m* barométrique à
   courants parallèles

d barometrischer Gleichstromkondensator *m*
e condensador *m* barométrico de corriente
   paralela
n barometrische gelijkstroomcondensor *m*

536 **co-current condenser; parallel-current
     condenser**
f condenseur *m* à co-courants; condenseur
   m à courants parallèles
d Gleichstromkondensator *m*
e condensador *m* de corriente paralela
n gelijkstroomcondensor *m*

* **COD** → **476**

537 **coefficient of heat transmission;
     coefficient of heat transfer**
f coefficient *m* de transmission de chaleur
d Wärmedurchgangszahl *f*
e coeficiente *m* de transmisión de calor
n warmtedoorgangscoëfficiënt *m*

538 **coefficient of variation; CV**
     (Coefficient of variation of the screen
     opening through which sugar crystals
     will just pass)
f coefficient *m* de variation
d Variationskoeffizient *m*; CV-Wert *m*
e coeficiente *m* de variación; C.V. *m*
n variatiecoëfficiënt *m*; CV-waarde *f*

539 **coil (vacuum) pan**
f appareil *m* à cuire à serpentins
d Kochapparat *m* mit Heizschlangen;
   Verdampfungskristallisator *m* mit Heizschlangen
e tacho *m* al vacío de serpentines
n serpentijnkookpan *f*; spiraalkookpan *f*

540 **cold digestion**
f digestion *f* à froid
d kalte Digestion *f*
e digestión *f* en frío
n koude digestie *f*

541 **cold filtrate**
     (The liquid effluent from the filters that
     separate the cold saccharate)
f filtrat *m* froid
d kaltes Filtrat *n*
e filtrado *m* frío
n koud filtraat *n*

542 **cold imbibition**
f imbibition *f* à froid
d kalte Imbibition *f*
e imbibición *f* fría
n koude imbibitie *f*

543 cold liming
f chaulage *m* à froid
d kalte Kalkung *f*
e encalado *m* en frío; alcalinización *f* en
   frío; alcalización *f* en frío
n koude kalking *f*

544 cold raw juice
f jus *m* brut froid
d kalter Rohsaft *m*
e jugo *m* crudo frío; guarapo *m* crudo frío
n koud ruwsap *n*

545 cold saccharate
f saccharate *m* froid
d kaltes Saccharat *n*
e sacarato *m* frío
n koud saccharaat *n*

546 cold sulphitation
f sulfitation *f* à froid
d kalte Sulfitation *f*
e sulfitación *f* en frío
n koude sulfitatie *f*

547 collar rot
   (sc; fungous disease)
f pourriture *f* de collier
d Kragenfäule *f*
e pudrición *f* del tronco
n kraagrot *n*

548 collecting flume
f caniveau *m* collecteur
d Sammelschwemmkanal *m*; Sammel-
   schwemmrinne *f*
e canal *m* colector de flotación
n verzamelzwemgoot *f*;
   verzamelspoelgoot *f*

549 collecting pipe *(of evaporator vessel)*
f collecteur *m*
d Sammelrohr *n*
e colector *m*
n verzamelpijp *f*

550 collective packing machine
f fardeleuse *f*
d Sammelpackmaschine *f*
e empaquetadora-colectora *f*
n wikkelmachine *f*

551 collector drum
   (agr)
f tambour *m* ramasseur
d Sammeltrommel *f*
e tambor *m* agrupador; tolva *f* de parrilla
   de raíces
n verzameltrommel *f*

* **collembolan** → 2238

* **collembolous insect** → 2238

552 colloid
f colloïde *m*
d Kolloid *n*
e coloide *m*
n colloïde *fn*

553 colloidal substance
f substance *f* colloïdale
d kolloidale Substanz *f*
e sustancia *f* coloidal
n colloïdale stof *f*

* **colorant** *(US)* → 557

554 colorimeter
f colorimètre *m*
d Kolorimeter *n*
e colorímetro *m*
n colorimeter *m*

555 colo(u)r determination
f détermination *f* de la couleur
d Farbbestimmung *f*
e determinación *f* del color
n kleurbepaling *f*

556 colo(u)red sugar
f sucre *m* additionné de colorants
d gefärbter Zucker *m*
e azúcar *m* adicionado de colorantes
n suiker *m* met toegevoegde kleurstoffen

557 colo(u)ring matter; colorant *(US)*
f matière *f* colorante; colorant *m*
d Farbstoff *m*
e materia *f* colorante; colorante *m*
n kleurstof *f*; kleurmiddel *n*

* **colo(u)r removal** → 741

558 colo(u)r type
f type *m* de couleur
d Farbtyp *m*
e tipo *m* de color
n kleurtype *n*

559 coltsfoot; foalfoot; common coltsfoot
   (sb; weed)
f tussilage *m* pas-d'âne; pas-d'âne *m*;
   tussilage *m*; pied *m* de poulain; pas-
   d'âne *m* commun
d Huflattich *m*; gemeiner Huflattich *m*;
   Pferdefuß *m*; Quirinkraut *n*
e tusílago *m*; fárfara *f*; uña *f* de caballo
n klein hoefblad *n*; hoevebladen *npl*;

hoeven *mpl*; paardeklauw *m*
l Tussilago farfara

* comb → 2061

560 "combing" shredder
(Maxwell type)
f shredder-peigneur *m*
d Shredder *m* mit Kammleiste
e desfibradora *f* peinadora
n shredder *m* met kamlijst

561 comb-type bar *(of shredder)*
f peigne *m*
d Kammleiste *f*
e peine *m*
n kamlijst *f*

562 combustion chamber *(of boiler)*
f chambre *f* de combustion
d Verbrennungsraum *m*
e cámara *f* de combustión
n verbrandingsruimte *f*;
verbrandingskamer *f*

563 combustion gases
f gaz *mpl* de combustion
d Verbrennungsgase *npl*
e gases *mpl* de combustión
n verbrandingsgassen *npl*

* combustion zone *(of lime kiln)* → 369

564 commercial sugar
f sucre *m* commercial
d Handelszucker *m*; kommerzieller
Zucker *m*
e azúcar *m* comercial; azúcar *m* de envase
n handelssuiker *m*

565 common agricultural policy
(EEC)
f politique *f* agricole commune; PAC *f*
d gemeinsame Agrarpolitik *f*
e política *f* agrícola común
n gemeenschappelijke landbouwpolitiek *f*

**566 common alkanet; alkanet; common
bugloss; ox-tongue**
(sb; weed)
f buglosse *f*; langue *f* de boeuf; buglosse *f*
officinale
d gemeine Ochsenzunge *f*; echte Ochsenzunge *f*; gebräuchliche Ochsenzunge *f*;
gewöhnliche Ochsenzunge *f*
e buglosa *f*; lengua *f* de buey
n gewone ossetong *f*
l Anchusa officinalis

* common amaranth → 1914

* common beet → 1543

* common burdock *(US)* → 1454

567 common capsid
(sb; pest)
f capside *m*
d Wiesenwanze *f*
e chinche *f* Lygus pratensis
l Lygus pratensis

* common capsid bug *(US)* → 2470

568 common chickweed; chickweed
(sb; weed)
f mouron *m* blanc; mouron *m* des oiseaux;
morgeline *f*; moyenne stellaire *f*
d Vogelmiere *f*; Mäusedarm *m*; Vogel-
Sternmiere *f*; Gänsemiere *f*; Hühnerkraut *n*; Vogelkraut *n*; Sternblume *f*
e hierba *f* gallinera; hierba *f* de los
canarios; pamplina *f*; hierba *f* pajarera;
capiquí *m*
n vogelmuur *f*
l Stellaria media (L.) Vill.

* common cockchafer → 532

569 common cocklebur *(US)*
(sb; weed)
l Xanthium pensylvanicum

* common coltsfoot → 559

* common corn cockle → 646

570 common crane fly
(sb; pest)
f tipule *f* potagère
d Kohlschnake *f*
e tipula *f* de las huertas
n koollangpootmug *f*
l Tipula oleracea

**571 common crane fly; marsh crane fly;
allied crane fly; crane fly; leatherjacket**
(sb; pest)
f tipule *f* des prairies
d Wiesen-Erd-Schnake *f*; Sumpfschnake *f*;
Wiesenschnake *f*
e zancudo *m* Tipula paludosa
l Tipula paludosa *(imago)*

572 Common Customs Tariff; CCT
(EEC)
f Tarif *m* Douanier Commun; TDC *m*
d Gemeinsamer Zolltarif *m*; GZT *m*

e Arancel *m* de Aduanas Común; Tarifa *f*
Aduanera Común; AAC *m*
n Gemeenschappelijk Douanetarief *n*;
GDT *n*

* **common daisy** → **731**

* **common dandelion** → **733**

* **common dock** → **334**

* **common earwig** → **892**

* **common forget-me-not** → **1000**

* **common froghopper** *(adult)* → **1558**

573 **common fumitory; fumitory; earth smoke**
(sb; weed)
f fumeterre *f* officinale; fumeterre *f*
d gemeiner Erdrauch *m*; echter
Erdrauch *m*; Feldraute *f*; Katzen-
kerbel *m*; Taubenkerbel *m*
e sangre *f* de Cristo; capa *f* de reina;
palomilla *f*; fumaria *f*; gitanillas *fpl*
n gewone duivenkervel *f*; aardrook *m*
l Fumaria officinalis

574 **common green capsid; capsid bug** *(US)*;
common green capsid bug
(sb; pest)
f capside *m*; punaise *f*; punaise *f* verte
des pousses; capside *m* des cultures
fruitières
d Futterwanze *f*; grüne Futterwanze *f*
e chinche *f* del manzano
l Lygus pabulinus

* **common groundsel** → **1203**

575 **common hemp nettle; day nettle; hemp
nettle; downy hemp nettle; bristle-stem
hemp nettle**
(sb; weed)
f ortie *f* royale; ortie *f* épineuse; galéope
m tétrahit
d gemeiner Hohlzahn *m*; gewöhnlicher
Hohlzahn *m*; stechender Hohlzahn *m*
e cáñamo *m* bastardo; galeopsis *m* erizado;
ortiga *f* real
n gewone hennepnetel *f*
l Galeopsis tetrahit

576 **common horsetail; toadpipe; field
horsetail**
(sb; weed)
f prêle *f* des champs; queue *f* de cheval;
queue-de-rat *f*
d Ackerschachtelhalm *m*; Ackerkannen-

kraut *n*; Kannenkraut *n*; Pferde-
schwanz *m*; Scheuerkraut *n*; Zinn-
kraut *n*
e equiseto *m* menor; cola *f* de caballo
menor
n hermoes *f*; heermoes *f*; akkerpaarde-
staart *m*; heringmoes *f*
l Equisetum arvense

* **common lamb's-quarters** *(US)* → **970**

577 **common mallow** *(US)*; **dwarf mallow; low
mallow; blue mallow**
(sb; weed)
f mauve *f* naine; mauve *f* des chemins;
fromageon *m*
d Käsepappel *f*; Wegmalve *f*
e malva *f* enana
n klein kaasjeskruid *n*; kleine malve *f*
l Malva neglecta Wallr.

578 **common mallow; high mallow; wild
mallow**
(sb; weed)
f mauve *f* sauvage; mauve *f* sylvestre;
grande mauve *f*
d wilde Malve *f*; Hanfpappel *f*; Wald-
malve *f*
e malva *f* común
n groot kaasjeskruid *n*
l Malva sylvestris

579 **Common Market Sugar Organization**
(EEC)
f organisation *f* commune du marché du
sucre
d gemeinsame Marktorganisation *f* für
Zucker; gemeinsame Marktordnung *f* für
Zucker; EG-Zuckermarktordnung *f*
e organización *f* común del mercado del
azúcar
n gemeenschappelijke marktordening *f*
voor suiker

* **common melilot** → **2728**

580 **common milkweed** *(US)*
(sb; weed)
f asclépiade *f* à ouate; herbe *f* à l'ouate;
asclépiade *f* de Syrie
d Seidenpflanze *f*; echte Seidenpflanze *f*;
syrische Seidenpflanze *f*
n zijdeplant *f*
l Asclepias syriaca; Asclepias cornuta

* **common mouse-ear** → **1643**

* **common nettle** → **2283**

* **common nightshade** → 283

* **common orache** → 2236

* **common plantain** → 1193

* **common poppy** → 652

* **common purslane** → 1868

581 common ragweed *(US)*; **ragweed** *(US)*
(sb; weed)
f ambroisie f; ambrosia f; ambroisie f à
feuilles d'armoise
d hohe Ambrosie f; beifußblättrige
Ambrosie f; beifußblättriges Traubenkraut n
e ambrosía f
n alsemambrosia f
l Ambrosia artemisiifolia

* **common ragwort** → 1884

* **common reed** → 1930

* **common reed grass** → 1930

* **common rush** → 2199

* **common Russian thistle** → 2025

* **common silver Y** → 1156

* **common silver Y moth** → 1156

582 common sorrel; garden sorrel; sour dock
(sb; weed)
f grande oseille f sauvage; grande
oseille f; oseille f commune; oseille f
acide; surette f
d Sauerampfer m; großer Sauerampfer m;
großer Ampfer m; französischer
Spinat m; Wiesensauerampfer m
e acedera f común; agrilla f; vinagrera f
n veldzuring f
l Rumex acetosa

**583 common sowthistle; milk sowthistle;
annual sowthistle** *(US)*; **hare's lettuce;
smooth sowthistle**
(sb; weed)
f laiteron m maraîcher; laiteron m annuel;
laiteron m potager; lait m d'âne
d Kohlgänsedistel f; gemeine Gänsedistel f; Gartengänsedistel f; Moos-
distel f; Saudistel f
e lechuguilla f silvestre; cerraja f común;
cerraja f anual

n gewone melkdistel m
l Sonchus oleraceus

**584 common speedwell; veronica; drug
speedwell; health speedwell**
(sb; weed)
f véronique f officinale; véronique f mâle;
thé m d'Europe
d echter Ehrenpreis m; Wald-
Ehrenpreis m
e verónica f común; verónica f macho
n mannetjes-ereprijs m
l Veronica officinalis

**585 common St. John's wort; Saint-John's-
wort; Klamath weed; perforated St.
John's wort**
(sb; weed)
f millepertuis m perforé; herbe f de la
Saint-Jean; millepertuis m commun
d Tüpfel-Johanniskraut n; echtes Johannis-
kraut n; durchlöchertes Hartheu n;
getüpfeltes Hartheu n; Tüpfelhartheu n;
Hexenkraut n
e hipericón m; hierba f de San Juan;
corazoncillo m
n Sint-Janskruid n
l Hypericum perforatum

586 common stork's bill; common storksbill
(US); **alfilaria; alfileria; alfilerilla; pin
gras; hemlock stork's bill**
(sb; weed)
f bec m de héron; érodion m à feuilles de
cigüe; érodium m à feuilles de cigüe;
bec-de-grue m; aiguille f
d Reiherschnabel m; Schierlings-Reiher-
schnabel m; gemeiner Reiherschnabel m; schierlingsblättriger Reiherschnabel m
e pico m de cigüeña; aguja f
n gewone reigersbek m; kraanhals m
l Erodium cicutarium

* **common sunflower** → 2420

* **common toadflax** → 2734

* **common velvet grass** → 2737

* **common vole** → 613

* **common yellow oxalis** → 2585

587 Community sugar; EEC-sugar
(EEC)
f sucre m communautaire
d Gemeinschaftszucker m; EG-Zucker m

e azúcar *m* comunitario
n gemeenschapssuiker *m*; EG-suiker *m*

* **compass lettuce** → 1831

* **complete cane harvester** → 2691

588 **complete sugar-beet harvester**
(agr)
f machine *f* combinée
d Vollerntemaschine *f*; kombinierte
Maschine *f*; Rübenvollernter *m*
e máquina *f* completa; cosechadora *f* de
remolacha; cosechadora *f* integral de
remolacha
n gecombineerde bietenoogstmachine *f*

589 **compound clarification**
f clarification *f* composée
d kombinierte Reinigung *f*; kombinierte
Klärung *f*
e clarificación *f* compuesta
n gecombineerde klaring *f*

590 **compound imbibition; compound
saturation**
f imbibition *f* composée
d kombinierte Imbibition *f*; mehrfache
Imbibition *f*
e imbibición *f* compuesta
n gecombineerde imbibitie *f*; meervoudige
imbibitie *f*

591 **compressed air**
f air *m* comprimé
d Druckluft *f*; Preßluft *f*
e aire *m* comprimido
n perslucht *f*

592 **compressed bagasse**
f bagasse *f* comprimée
d zusammengepreßte Bagasse *f*;
komprimierte Bagasse *f*
e bagazo *m* comprimido
n samengeperste bagasse *m*;
samengeperste ampas *m*

593 **compression ratio**
f rapport *m* de compression
d Kompressionsverhältnis *n*
e relación *f* de compresión
n samenpersingsverhouding *f*

* **concentrate** *v* → 2482

594 **concentration**
f concentration *f*
d Eindicken *n*

e concentración *f*
n indikken *n*

595 **concentrator**
(An evaporator body for precisely
adjusting the concentration of thick juice
or standard liquor destined for long-term
storage, or of CSF or for preparing
graining charges)
f concentrateur *m*
d Eindicker *m*; Eindickapparat *m*;
Konzentrator *m*
e concentrador *m*
n indikker *m*

* **concentrator** → 35

596 **condensate**
(Water obtained by condensation of
steam or vapo(u)r in surface condensers,
or in the steam or vapo(u)r chests of
heating or evaporating vessels)
f eaux *fpl* condensées
d Kondensat *n*
e aguas *fpl* condensadas; condensados
mpl
n condensaat *n*

597 **condensate flash tank**
f détendeur *m* d'eaux condensées
d Kondensatentspanner *m*
e expandidor *m* de condensado
n condensaatontspanner *m*

598 **condensate manifold**
f collecteur *m* d'eaux condensées
d Kondensatsammelleitung *f*; Kondensat-
sammelrohr *n*
e colector *m* de aguas condensadas;
colector *m* de condensados
n condensaatverzamelpijp *f*

599 **condensate pump**
f pompe *f* à eaux condensées
d Kondensatpumpe *f*
e bomba *f* de aguas condensadas; bomba *f*
de condensados
n condensaatpomp *f*

600 **condensate suction pipe**
f tuyau *m* d'adduction des eaux
condensées
d Kondensatsaugleitung *f*; Kondensat-
ansaugleitung *f*
e tubo *m* de succión de condensados; tubo
m de succión de aguas condensadas
n condensaatzuigleiding *f*;
condensaatzuigpijp *f*

601 condenser
(Apparatus for the condensation of steam
or vapo(u)r, generally using water as the
cooling medium)
f condenseur *m*
d Kondensator *m*
e condensador *m*
n condensor *m*

602 condenser water
(A mixture of condensate and cooling
water produced by a direct-contact
condenser)
f eau *f* de condenseur; eau *f* de
condensation
d Kondensatorwasser *n*; Fallwasser *n*
e agua *f* de condensador
n condensatiewater *n*

603 conductimetric ash; conductometric ash
f cendres *fpl* conductimétriques
d konduktimetrische Asche *f*; kondukto-
metrische Asche *f*; Leitfähigkeitsasche *f*
e ceniza *f* conductométrica
n conductometrische as *f*

604 conductivity meter; conductometer
f conductivimètre *m*; conductomètre *m*
d Leitfähigkeitsmeßgerät *n*; Leitfähigkeits-
messer *m*
e conductibilímetro *m*; conductivímetro *m*
n geleidbaarheidsmeter *m*

* **conductometric ash → 603**

605 cone-press
(Silver diffusion)
f cône-presse *f*; presse *f* à cônes
d Kegelpresse *f*
e prensa *f* de conos
n kegelpers *f*

* **confectioner's glucose → 2248**

606 conglomerate
(A cluster of mutually intergrown
crystals)
f conglomérat *m*
d Konglomerat *n*; Kristalaggregat *n*;
Agglomerat *n*
e conglomerado *m*; aglomerado *m*
n conglomeraat *n*

607 conical cage-wheel
(agr)
f roue-cage *f* tronconique
d konische Siebtrommel *f*
e tambor *m* cónico de rejilla
n kegelvormige zeeftrommel *f*

608 connecting pipe; interconnecting pipe
(between evaporator vessels)
f tuyau *m* d'intercommunication; tuyau *m*
d'interconnexion
d Verbindungsrohr *n*
e tubo *m* de intercomunicación; tubo *m* de
comunicación; tubo *m* de interconexión
n verbindingspijp *f*

609 constant ratio mill; mill with fixed ratio
f moulin *m* à réglage autostable; moulin
m à réglage stable
d Mühle *f* mit fester Einstellung; Mühle *f*
mit unveränderlicher Einstellung
n molen *m* met vaste instelling; molen *m*
met onveranderlijke instelling

610 constant setting mill; self-setting mill
f moulin *m* à auto-réglage
d selbsteinstellende Mühle *f*
e molino *m* autoregulable
n molen *m* met automatische instelling

611 consumption sugar
f sucre *m* de consommation
d Verbrauchszucker *m*
e azúcar *m* de consumo
n consumptiesuiker *m*

612 consumption sugar value
(eco)
f valeur *f* en sucre de consommation
d Verbrauchszuckerwert *m*
e valor *m* de azúcar de consumo
n consumptiesuikerwaarde *f*

**613 continental field mouse; common vole;
field vole**
(sb; pest)
f campagnol *m*; petit campagnol *m*;
campagnol *m* des champs; campagnol *m*
commun
d Feldmaus *f*; gemeine Feldmaus *f*
e ratón *m* del campo; ratón *m* campesino;
topillo *m* campesino; ratilla *f* del campo
n veldmuis *f*
l Microtus arvalis

614 continuous carbona(ta)tion
f carbonatation *f* continue; carbonatation *f*
en continu
d kontinuierliche Carbonatation *f*
e carbonatación *f* continua
n continue carbonatatie *f*

615 continuous centrifugal
f centrifugeuse *f* continue; essoreuse *f*
continue; centrifuge *f* continue
d kontinuierliche Zentrifuge *f*; Konti-

Zentrifuge *f*; kontinuierlich arbeitende
Zentrifuge *f*
e centrífuga *f* continua
n continue centrifuge *f*

* **continuous clarifier** → **623**

* **continuous decanter** → **623**

616 continuous diffuser; continuous extractor
f diffuseur *m* continu; extracteur *m*
continu
d kontinuierlicher Extraktionsapparat *m*;
kontinuierlich arbeitender Extraktionsapparat *m*
e difusor *m* continuo
n continue diffuseur *m*

* **continuous diffusion** → **618**

617 continuous evaporator
f évaporateur *m* continu
d Durchlaufverdampfer *m*
e evaporador *m* continuo
n continue verdamper *m*

**618 continuous extraction; continuous
diffusion**
f extraction *f* continue; diffusion *f*
continue
d kontinuierliche Extraktion *f*;
kontinuierliche Diffusion *f*
e extracción *f* continua; difusión *f*
continua
n continue extractie *f*; continue diffusie *f*

* **continuous extractor** → **616**

619 continuous filter
f filtre *m* continu
d kontinuierlich arbeitendes Filter *n*
e filtro *m* continuo
n continu werkend filter *n*

620 continuous filtration
f filtration *f* continue
d kontinuierliche Filtration *f*
e filtración *f* continua
n continue filtratie *f*

621 continuous liming
f chaulage *m* continu
d kontinuierliche Kalkung *f*;
kontinuierliche Scheidung *f*
e alcalinización *f* continua; alcalización *f*
continua; encalado *m* continuo
n continue kalking *f*

**622 continuous rotating vacuum filter;
continuous rotary vacuum filter**
f filtre *m* rotatif continu sous vide
d kontinuierliches Dreh-Vakuum-Filter *n*;
kontinuierlich arbeitendes Dreh-Unter-
druck-Filter *n*
e filtro *m* rotativo continuo al vacío
n continu roterend vacuümfilter *n*

**623 continuous settler; continuous subsider;
continuous clarifier; continuous decanter**
f décanteur *m* continu; clarificateur *m*
d kontinuierlicher Dekanteur *m*;
kontinuierlicher Absetzer *m*;
kontinuierlich arbeitender Dekanteur *m*
e decantador *m* continuo; clarificador *m*
continuo; sedimentador *m* continuo
n continu decanteertoestel *n*; continue
decanteur *m*

624 continuous sulphitation
f sulfitation *f* continue
d kontinuierliche Sulfitation *f*
e sulfitación *f* continua
n continue sulfitatie *f*

625 continuous vacuum pan
f appareil *m* à cuire continu; cristallisateur
m continu
d kontinuierlicher Kochapparat *m*;
kontinuierlicher Eindampf-
kristallisator *m*; kontinuierlicher
Verdampfungskristallisator *m*
e cristalizador *m* continuo de
concentración; evapocristalizador *m*
continuo
n continu kookapparaat *n*; continue
kookpan *f*

* **contract for delivery** → **764**

626 contraction (*of sugar*)
f contraction *f*
d Kontraktion *f*
e contracción *f*
n samentrekking *f*; contractie *f*

* **contract of delivery** → **764**

627 contribution
(EEC)
f cotisation *f*
d Abgabe *f*
e cotización *f*
n heffing *f*

* **conveying belt** → **630**

628 conveying chute
  f goulotte f de transport
  d Förderrutsche f; Förderschurre f
  e canal m transportador
  n glijgoot f

629 conveyor
  f transporteur m
  d Förderer m; Transporteur m
  e transportador m
  n transporteur m

630 conveyor band; conveyor belt; conveying belt
  f courroie f transporteuse; bande f transporteuse; ruban m transporteur
  d Förderband n; Transportband n
  e correa f transportadora; correa f de transporte
  n transportband m

631 conveyor belt with transverse ribs
  f courroie f transporteuse à lattes transversales
  d Förderband n mit Querrippen
  e correa f de transporte con nervios transversales
  n transportband m met dwarsribben

632 conveyor bucket
  f godet m de transporteur
  d Förderbecher m
  e cangilón m de transporte; cubo m de transporte
  n transporteuremmer m

633 conveyor trough
  f gouttière f à secousses; goulotte f
  d Förderrinne f
  e conducto m transportador
  n transportgoot f

634 cooler
  f refroidisseur m
  d Kühler m
  e enfriador m; refrigerador m
  n koeler m

* **cooler crystallizer → 637**

635 cooling bath
  f bain m de refroidissement; bain m réfrigérant
  d Kühlungsbad n
  e baño m refrigerante
  n koelbad n

636 cooling crystallization
  f cristallisation f par refroidissement forcé
  d Kühlungskristallisation f
  e cristalización f por enfriamiento
  n kristallisatie f in koeltroggen; kristallisatie f met geforceerde koeling

637 cooling crystallizer; cooler crystallizer
  f malaxeur m refroidisseur; malaxeur m de refroidissement
  d Kühlkristallisator m; Kühlmaische f
  e cristalizador m por refrigeración; cristalizador-enfriador m
  n koeltrog m

638 cooling disc (of Werkspoor crystallizer)
  f disque m refroidisseur
  d Kühlscheibe f
  e disco m de enfriamiento
  n koelschijf f

639 cooling surface
  f surface f de refroidissement
  d Kühlfläche f
  e superficie f refrigerante
  n koelend oppervlak n

640 cooling tower
  (Vacuum equipment)
  f tour f de réfrigération; réfrigérant m à fascines
  d Kühlturm m
  e torre f de enfriamiento; torre f de refrigeración
  n koeltoren m

641 cooling water
  f eau f de refroidissement
  d Kühlwasser n
  e agua f de enfriamiento; agua f de refrigeración; agua f refrigerante
  n koelwater n

642 cooling water pump
  f pompe f à eau de refroidissement
  d Kühlwasserpumpe f
  e bomba f para agua refrigerante
  n koelwaterpomp f

643 cooling zone (of lime kiln)
  f zone f de refroidissement
  d Kühlzone f
  e zona f de enfriamiento
  n afkoelingszone f; afkoelingsruimte f

644 copper deficiency
  (sb; disease)
  f carence f en cuivre
  d Kupfermangel m
  e deficiencia f de cobre; carencia f de

cobre
n kopergebrek n

* **cornbirne** → 997

* **corn bluebottle** → 648

* **corn buttercup** → 647

* **corn campion** → 646

645 **corn chamomile; field chamomile; corn camomile; field camomile**
(sb; weed)
f fausse camomille f; anthémide f des champs; anthémis f des champs; oeil m de vache
d Ackerhundskamille f; Ackerkamille f; falsche Kamille f
e manzanilla f silvestre; manzanilla f de los campos
n valse kamille f; wilde kamille f; akkerkamille f
l Anthemis arvensis

* **corn chrysanthemum** → 650

646 **corn cockle; corn campion; crown-of-the-field; common corn cockle**
(sb; weed)
f couronne f des blés; nielle f des blés; nielle f des champs; lychnide f des moissons
d Kornrade f; Rade f; Ackerkrone f; Kornnelke f; Roggenrose f
e neguillón m; yetón m
n bolderik m; gemene bolderik m
l Agrostemma githago

647 **corn crowfoot; corn buttercup; beggar's lice; goldlocks**
(sb; weed)
f renoncule f des champs
d Ackerhahnenfuß m
e ranúnculo m arvense; ranúnculo m silvestre; ranúnculo m de los campos; abrepuños m
n akkerboterbloem f; akkerhanevoet m; kroonranonkel f
l Ranunculus arvensis

648 **cornflower; bluebottle; corn bluebottle; hurt-sickle; bachelor's-button**
(sb; weed)
f bleuet m; bluet m; centaurée f bleue; barbeau m; aubifoin m
d Kornblume f; Tremse f; Zyane f
e azulejo m; aciano m; clavel m de San Juan; liebrecilla f

n korenbloem f
l Centaurea cyanus

* **corn grass** → 2144

649 **corn gromwell; field gromwell; stoneseed; bastard alkanet**
(sb; weed)
f buglosse f des champs; grémil m des champs; herbe f aux perles; lithosperme m des champs
d Ackersteinsame f; Bauernschminke f; Brennkraut n; Steinhirse f
e onoquiles m; borraja f silvestre; yuyo m moro; mijo m del sol; abremanos m
n ruw parelzaad n
l Lithospermum arvense

650 **corn marigold; field marigold; yellow ox-eye daisy; corn chrysanthemum**
(sb; weed)
f chrysanthème m des moissons; janne f; marguerite f; marguerite f des blés
d Saatwucherblume f
e margarita f del trigo; margarita f de los sembrados; ojos mpl de los sembrados
n gele ganzebloem f; wilde goudsbloem f; gele goudsbloem f
l Chrysanthemum segetum

* **corn mayweed** → 2059

651 **corn mint; field mint; tule mint** *(US)*
(sb; weed)
f menthe f des champs; baume f des champs
d Ackerminze f; Feldminze f; Kornminze f
e menta f silvestre; menta f Japonesa; menta f de burro
n akkermunt f; veldmunt f
l Mentha arvensis

652 **corn poppy; field poppy; red poppy; common poppy**
(sb; weed)
f coquelicot m; pavot m sauvage; pavot m coquelicot
d Klatschmohn m; Klappermohn m; Klatschrose f; Kornrose f; Feldmohn m; Feuerblume f; Kornmohn m; wilder Mohn m
e amapola f; ababol m; rosella f; amapola f común; amapola f de los campos; ababa f; flor f de chivo
n gewone klaproos f
l Papaver rhoeas

* **corn snapdragon** → 1456

653 **corn sowthistle; perennial sowthistle;
field sowthistle**
(sb; weed)
f laiteron *m* des champs
d Ackergänsedistel *f*; Ackersaudistel *f*;
Milchdistel *f*
e cerraja *f*; lechuguilla *f*; hierba *f* del
sacre
n akkermelkdistel *m*; moesmelkdistel *m*
l Sonchus arvensis

654 **corn spurry; corn spurrey**
(sb; weed)
f spergule *f* des champs; spargoule *f*;
spargoule *f* des champs
d Feld-Spark *m*; gemeiner Spörgel *m*;
Ackerspark *m*; Ackerspörgel *m*; Acker-
knöterich *m*; Mariengras *n*
e esparcilla *f*; espérgula *f* de los campos
n gewone spurrie *f*
l Spergula arvensis

* **corn thistle** → 676

655 **corn toadflax**
(sb; weed)
f linaire *f* des champs
d Ackerleinkraut *n*
e linaria *f* amarilla de hoja estrecha
n blauwe leeuwebek *m*
l Linaria arvensis

656 **cossette conveyor; slices conveyor**
f transporteur *m* à cossettes
d Schnitzeltransporteur *m*; Schnitzel-
förderer *m*
e transportador *m* de cosetas
n snijdseltransporteur *m*

657 **cossette elevator; slices elevator**
f élévateur *m* à cossettes
d Schnitzelelevator *m*
e elevador *m* de cosetas
n snijdselelevator *m*; snijdselophaler *m*

658 **cossette mixer**
f malaxeur *m* de cossettes
d Schnitzelmaische *f*
e macerador *m* de cosetas
n snijdselmengmachine *f*

659 **cossette pump**
f pompe *f* à cossettes
d Schnitzelpumpe *f*
e bomba *f* de cosetas
n snijdselpomp *f*

660 **cossettes; chips; beet slices**
(Slender strips of beets cut by the slicers)

f cossettes *fpl*
d Schnitzel *npl*; Rübenschnitzel *npl*
e cosetas *fpl*
n snijdsels *npl*; bietsnijdsels *npl*

661 **cossette scalder**
f installation *f* d'échaudage des cossettes
d Schnitzelbrühanlage *f*
e instalación *f* para la escaldadura de
cosetas
n snijdselbroei-installatie *f*

* **cossettes with V-shaped cross sections**
→ 1981

662 **cotton aphid; melon aphid; melon and
cotton aphid**
(sc; pest)
f puceron *m* du melon (et du cotonnier)
d grüne Gurkenblattlaus *f*; grüne
Baumwollblattlaus *f*; grüne Melonenblattlaus *f*
e pulgón *m* del algodonero; pulgón *m* del
algodón; pulgón *m* de los melones;
pulgón *m* del melón; áfido *m* del algodón
l Aphis gossypii

663 **cotton root knot nematode** *(US)*
(sb; pest)
d Baumwoll-Gallenälchen *n*; Baumwoll-
Wurzelgallenälchen *n*
l Meloidogyne incognita-acrita

664 **couch grass; twitch; quack grass** *(US)*;
quick grass *(US)*; **quitch grass** *(US)*;
scutch grass *(US)*; **twitch grass;
witchgrass; white couch**
(sb; weed)
f chiendent *m* rampant; chiendent *m*;
chiendent *m* commun; chiendent *m*
ordinaire; froment *m* rampant
d gemeine Quecke *f*; kriechende Quecke *f*
e grama *f* del Norte; agropiro *m* común;
grama *f* oficinal
n kweek *f*
l Agropyron repens; Triticum repens

* **counterblade, knife-box** ~ → 1407

665 **countercurrent barometric condenser;
barometric countercurrent condenser**
f condenseur *m* barométrique à contre-
courants
d barometrischer Gegenstromkondensator *m*
e condensador *m* barométrico de
contracorriente
n barometrische tegenstroomcondensor *m*

666 countercurrent carbona(ta)tion
  f carbonatation *f* à contre-courant
  d Gegenstromcarbonatation *f*
  e carbonatación *f* en contracorriente
  n tegenstroomcarbonatatie *f*

667 countercurrent condenser
  f condenseur *m* à contre-courants
  d Gegenstromkondensator *m*
  e condensador *m* de contracorriente;
    condensador *m* en contracorriente
  n tegenstroomcondensor *m*

668 countercurrent cossette mixer
  f malaxeur *m* de cossettes à contre-
    courant
  d Gegenstromschnitzelmaische *f*
  e macerador *m* de cosetas en
    contracorriente
  n tegenstroom-snijdselmengmachine *f*

669 countercurrent extraction
  f extraction *f* à contre-courant
  d Gegenstromextraktion *f*
  e extracción *f* en contracorriente
  n tegenstroomextractie *f*

670 countercurrent flow; counterflow
  f circulation *f* méthodique
  d Gegenstrom *m*
  e circulación *f* en contracorriente
  n tegenstroom *m*

671 counter knife
  f contre-lame *f*; contre-couteau *m*
  d Gegenmesser *n*
  e contracuchillo *m*
  n tegenmes *n*

672 cow cockle; soapwort; cowherb
    (sb; weed)
  f saponaire *f* des vaches; vaccaire *f*;
    vaccaria *f*
  d Kuhkraut *n*; Saat-Kuhkraut *n*
  e hierba *f* de la vaca
  n koekruid *n*
  l Vaccaria pyramidata

  * **cow cress → 1004**

673 C-products
  f produits *mpl* de troisième jet
  d C-Produkte *npl*
  e productos *mpl* C
  n C-produkten *npl*

674 crane fly; tipula; leatherjacket *(larva)*
    (sb; pest)
  f tipule *f*

  d Schnake *f*
  e tipula *f*
  n langpootmug *f*
  l Tipulidae

  * **crane fly → 571**

  * **cranesbill → 857**

675 creeping buttercup; creeping crowfoot
    (sb; weed)
  f renoncule *f* rampante
  d kriechender Hahnenfuß *m*
  e ranúnculo *m* rastrero; botonera *f*; botón
    m de oro
  n kruipende boterbloem *f*
  l Ranunculus repens

  * **creeping fat hen → 2236**

676 creeping thistle; Canada thistle *(US)*;
    corn thistle
    (sb; weed)
  f chardon *m* des champs; cirse *m* des
    champs; chardon *m* vulgaire; sarrête *f*
    des champs
  d Ackerkratzdistel *f*; Ackerdistel *f*; Hafer-
    distel *f*
  e cardo *m*; cardo *m* condidor; cardo *m*
    cundidor; abrepuños *m*
  n akkerdistel *m*
  l Cirsium arvense

  * **creeping wart cress → 2447**

677 crinkle
    (sb; disease)
  f frisolée *f*
  d Kräuselkrankheit *f*
  e rizadura *f* virosa; rizado *m*
  n krulziekte *f*

  * **cross cane carrier → 102**

  * **cross cane conveyor → 102**

  * **cross-wise complete harvesters → 1728**

  * **crowdweed → 1004**

  * **crowfoot → 361**

678 crown cleaner
    (agr)
  f moulinet *m* rotatif
  d Putzschleuder *f*
  e limpiadora *f* centrífuga
  n bladveger *m*

679 crown gall; bacterial gall
(sb; disease)
f galle f; tumeur f bactérienne (du collet
et des racines)
d Wurzelkropf m; Bakterienkrebs m;
Kronengalle f
e agalla f del cuello; tumor m bacteriano
de la raíz; hernia f de la raíz; agalla f de
corona
n wortelknobbel m

* **crown-of-the-field** → 646

680 crown rot; storage rot *(US)*; **phoma root
rot** *(US)*
(sb; disease)
f pourriture f sèche de la racine
d Phoma-Trockenfäule f am Rübenkörper
e podredumbre f seca de la raíz
n droog wortelrot n

* **crude glucose** → 2247

681 crush v
f défibrer
d zerquetschen; brechen; zerkleinern
e desfibrar
n kneuzen

682 crusher *(of bales of bagasse)*
f concasseur m
d Brecher m
e trituradora f
n breker m

* **crusher, cane ~** → 382

683 crusher roll(er)
f cylindre m de défibreur
d Brecherroller m; Brecherwalze f;
Crusherroller m; Crusherwalze f
e cilindro m de desmenuzadora; rodillo m
de desmenuzadora; maza f de
desmenuzadora
n crushercilinder m; kneuzercilinder m;
crusherrol f; kneuzerrol f

* **crushing season** → 1596

684 crush v **the cane**
f déchiqueter la canne
d das Rohr zerquetschen; das Rohr
zerkleinern
e desmenuzar la caña
n het riet kneuzen

685 crystal
f cristal m
d Kristall m

e cristal m
n kristal n

686 crystal content
f teneur f en cristaux
d Kristallgehalt m
e contenido m de cristales
n kristalgehalte n

687 crystal growth
f croissance f des cristaux
d Kristallwachstum n; croissance f
cristalline
e crecimiento m de los cristales
n kristalgroei m

688 crystallizable sugar
f sucre m cristallisable
d kristallisierbarer Zucker m
e azúcar m cristalizable
n kristalliseerbare suiker m

689 crystallization
(The total weight of sucrose in the solid
product of the centrifugals, including the
dissolved sucrose in any syrup adhering
to the crystals, expressed as a percentage
of the total weight of sucrose, both
crystallized and dissolved, in the
massecuite spun)
f cristallisation f
d Kristallisation f
e cristalización f
n kristallisatie f

690 crystallization aid
f adjuvant m de cristallisation
d Kristallisationshilfsmittel n
e auxiliar m de cristalización
n kristallisatiehulpmiddel n

691 crystallization in motion
f cristallisation f en mouvement
d Kristallisation f in Bewegung
e cristalización f en movimiento
n kristallisatie f in beweging

**692 crystallization rate; velocity of
crystallization**
f vitesse f de cristallisation
d Kristallisationsgeschwindigkeit f; Kristall-
wachstumsgeschwindigkeit f
e velocidad f de cristalización
n kristallisatiesnelheid f

693 crystallization stage
f étage m de cristallisation
d Kristallisationsstufe f

e etapa f de cristalización
n kristallisatiestap m; kristallisatietrap m

694 crystallize v
f cristalliser
d kristallisieren
e cristalizar
n kristalliseren

**695 crystallized sugar; crystallized refined
sugar**
f sucre m cristallisé
d Kristallzucker m
e azúcar m cristalizado; azúcar m en
cristales
n kristalsuiker m

696 crystallizer
(Apparatus for continuing the
crystallization of sugar in a massecuite
after its discharge from a vacuum pan,
by providing retention time, stirring, and
cooling at a controlled rate)
f cristalliseur m; cristallisoir m;
cristallisateur m
d Kristallisationsapparat m; Kristallisier-
maische f
e cristalizador m
n kristallisoir m

* **crystallizer pan** → 1424

697 crystal nucleus; centre of crystallization
f amorce f de cristallisation; germe m
cristallin; germe m de cristal
d Kristallkeim m; Kristallisationskeim m;
Kristallisationskern m; Impfling m;
Kristallisationszentrum n
e núcleo m de cristal
n kristalkern f; kristallisatiekern f

698 crystal size
f taille f des cristaux; dimension f des
cristaux
d Kristallgröße f
e tamaño m de cristales
n kristalgrootte f

699 crystal yield
f rendement m en cristaux
d Kristallausbeute f
e rendimiento m en cristales
n kristalopbrengst f

**700 crystal yield of a massecuite; exhaustion
of a massecuite**
(Proportion of crystal recovered from the
massecuite, expressed as percentage of
its sucrose content)

f épuisement m d'une masse cuite
d Kristallausbeute f einer Füllmasse
e rendimiento m en cristales de una masa
cocida
n kristalopbrengst f van een masse cuite

701 C sugar; third sugar
f sucre m C; sucre m de troisième jet;
sucre m troisième jet
d C-Zucker m
e azúcar m C; azúcar m de tercera
n C-suiker m

702 C-sugar
(EEC)
f sucre m C
d C-Zucker m
e azúcar m C
n C-suiker m

* **Cuba grass** → 1366

703 cube cutting machine
f cassoir m à sucre
d Knippmaschine f; Zuckerknipp-
maschine f
e máquina f para cortar azúcar en cubos;
máquina f para cortar cuadradillos
n tablettenknipmachine f

704 cube press; cubing press; cuber *(US)*
f mouleuse f pour morceaux de sucre;
presse f à comprimés
d Würfelzuckerpresse f
e prensa f por azúcar en cubos y tabletas
n suikertablettenpers f

705 cube sugar
f sucre m en morceaux
d Würfelzucker m
e azúcar m en cuadradillos; azúcar m en
terrones; azúcar m en pancitos; azúcar
m en cubos
n suikerklontje n

706 cube-sugar plant
f installation f pour sucre en morceaux;
installation f pour la fabrication de sucre
en morceaux
d Würfelzuckeranlage f
e instalación f para el azúcar en cubos;
instalación f para cuadradillos; planta f
por azúcar en cubos y tabletas
n suikertabletteninstallatie f

* **cubing press** → 704

* **cuckoo-spit insect** → 1558

* **cuckoo spit (insects)** → 1134

* **cuckoospittles** → 1134

707 **cucumber mosaic; dwarf**
(sb; disease)
f mosaïque *f* du concombre
d Gurkenmosaik *n*
e mosaico *m* del pepino
n komkommermozaïek *f*

708 **cucumber mosaic virus**
(sb)
f virus *m* de la mosaïque du concombre
d Gurkenmosaikvirus *n*
e virus *m* del mosaico del pepino
n virus *n* van de komkommermozaïek

709 **cuitometer; boiling control apparatus; vacuum pan control instrument**
f cuitomètre *m*
d Kochkontrollapparat *m*
e cuitómetro *m*; instrumento *m* para control de tacho
n kookcontrole-apparaat *n*

* **culmicolous smut** → 1703

* **cure all** → 2058

710 **cured sugar; spun sugar; centrifugal sugar**
f sucre *m* centrifugé; sucre *m* turbiné
d abgeschleuderter Zucker *m*; zentrifugierter Zucker *m*
e azúcar *m* centrifugado; azúcar *m* centrífugo
n gecentrifugeerde suiker *m*

711 **curled dock; curly dock; yellow dock**
(sb; weed)
f rumex *m* crépu; patience *f* crépue; oseille *f* crépue
d krauser Ampfer *m*; krausblättriger Ampfer *m*
e acedera *f*; lengua *f* de vaca; hidrolápato *m* menor; romaza *f* crespa
n krulzuring *f*
l Rumex crispus

712 **curly top**
(sb; disease)
f "curly top"
d "Curly Top"
e "Curly Top"
n "Curly Top"

713 **cush cush; bagacillo; fine bagasse**
f fine bagasse *f*; folle bagasse *f*

d Feinbagasse *f*; Bagacillo *f*; "Cush Cush"
e bagacillo *m*; bagazo *m* fino
n fijne bagasse *m*; fijne ampas *m*

714 **cush cush elevator; bagacillo elevator; fine bagasse elevator**
f élévateur *m* de folle bagasse; élévateur *m* cush cush
d Cush-Cush-Elevator *m*; Feinbagasse-Elevator *m*
e elevador *m* de bagacillo
n elevator *m* van fijne bagasse; elevator *m* van fijne ampas

715 **cush cush screen; bagacillo screen; fine bagasse screen**
f épulpeur *m*; tamiseur *m*; séparateur *m*; tamis *m* à fine bagasse
d Cush-Cush-Sieb *n*; Feinbagassesieb *n*
e pachaquil *m*; tamiz *m* de bagacillo
n fijne-bagassezeef *f*; fijne-ampaszeef *f*

716 **cush cush separator; bagacillo separator; fine bagasse separator**
f séparateur *m* de folle bagasse
d Feinbagasseabscheider *m*
e separador *m* de bagacillo
n fijne-bagasseafscheider *m*; fijne-ampasafscheider *m*

717 **cutleaf nightshade** *(US)*
(sb; weed)
l Solanum triflorum

* **cut load harvester** → 483

718 **cut-over line; cut-over pipe** *(between vacuum pans)*
f tuyauterie *f* de coupage
e tubería *f* de pases

719 **cutter-topper-buncher**
(agr)
f coupeuse-écimeuse-botteleuse *f* de cannes
d Schneid-, Entgipfel- und Binde-maschine *f*
e cortadora-desmochadora-atadora *f* de caña
n snijder-topper-zelfbinder *m*

720 **cutter-topper-windrower**
(agr)
f coupeuse-écimeuse-andaineuse *f* de cannes
d Schneid-, Entgipfel- und Schwadlege-maschine *f*
e cortadora-desmochadora-hileradora *f* de

caña
n snijder-topper-zwadlegger *m*

721 cut *v* **the strike**
   f couper la cuite; transvaser la cuite
   e dar cortes

722 cutting blade
   (agr)
   f lame *f* coupante
   d Schneidmesser *n*
   e cuchilla *f* cortadora
   n snijmes *n*

723 cutting knives
   f coupe-cannes *m* tronçonneur
   d Schneidmesser *npl*
   e cuchillas *fpl* cortadoras
   n snijmessen *npl*

724 cutworm; white-line dart moth
   (sb; pest)
   f noctuelle *f* des moissons
   d Wintersaateule *f*; Saateule *f*
   e noctuido *m* común de las mieses;
     noctuido *m* de los sembrados
   l Agrotis segetum; Euxoa segetum

   * **cutworm** → 2703

   * **CV** → 538

725 cyanophyllum scale
   (sc; pest)
   l Hemiberlesia cyanophylli; Abgrallapsis
     cyanophylli

726 cyclone
   (A mechanical classifying device
   normally used to remove solid particles
   from a gas stream by means of
   centrifugal force)
   f cyclone *m*
   d Zyklon *m*
   e ciclón *m*
   n cycloon *m*

727 cyclone separator
   f séparateur *m* cyclone
   d Zyklonabscheider *m*
   e separador *m* ciclón
   n cycloonafscheider *m*

728 cyclon washer *(for beet tails)*
   f laveur *m* à cyclone
   d Zyklonwäsche *f*
   e lavadora *f* de ciclón
   n cycloonwasmolen *m*

729 cylindrical basket *(of centrifugal)*
   f panier *m* cylindrique
   d zylindrischer Korb *m*; zylindrischer
     Trommel *f*
   e canasto *f* cilíndrico
   n cilindrische trommel *f*

   * **cylindrical cleaning drum** → 504

730 cyst nematodes
   (sc; pest)
   f nématodes *mpl* des racines
   d Zystenälchen *npl*
   e heteroderas *fpl*
   n cystenaaltjes *npl*
   l Heterodera spp.

D

731 daisy; common daisy; English daisy
(US); garden daisy
(sb; weed)
f pâquerette *f*; pâquerette *f* vivace; petite
marguerite *f*
d Gänseblümchen *n*; Maßliebchen *n*; mehr-
jähriges Gänseblümchen *n*; aus-
dauerndes Gänseblümchen *n*; Samt-
röschen *n*
e margarita *f*; vellorita *f*; bellorita *f*;
chirivita *f*
n madeliefje *n*; maagdelief *m*;
meizoentje *n*
l Bellis perennis

732 damaged beet
f betterave *f* blessée
d beschädigte Rübe *f*
e remolacha *f* dañada
n beschadigde biet *f*

* **damping** → **909**

733 dandelion; milk gowan; common
dandelion
(sb; weed)
f pissenlit *m*; dandelion *m*; dent-de-lion *f*;
pissenlit *m* commun
d gemeiner Löwenzahn *m*; gemeine Kuhblume *f*; Löwenzahn *m*; Kuhblume *f*;
Kettenblume *f*; Pfaffenkraut *n*; echter
Löwenzahn *m*
e diente *m* de león; amargón *m*;
taraxacón *m*
n paardebloem *f*
l Taraxacum officinale Web.

734 darkling beetle (US); darkling ground
beetle (US)
(sb; pest)
l Blapstinus fuligenosus

735 dark-sided cutworm
(sb; pest)
l Euxoa messoria

* **dark sworth grass moth** → **278**

* **dart moth** → **2565**

* **day nettle** → **575**

736 dead-burnt lime; dead lime
f chaux *f* morte; chaux *f* éteinte
d totgebrannter Kalk *m*

e cal *f* apagada
n gebluste kalk *m*

737 deaerate *v (the turbid liquid)*
f désaérér
d entlüften
e desairear
n ontluchten

738 dealkalinization
f désalcalinisation *f*
d Entalkalisierung *f*
e desalcalinización *f*
n verzadigen *n*

* **decalcification** → **762**

739 decalcify *v*; **delime** *v*
f décalcifier; déchauler
d entkalken
e descalcificar; desencalar
n ontkalken

740 decantation
f décantation *f*
d Dekantierung *f*; Absetzen *n*;
Eindicken *n*
e decantación *f*
n decantering *f*; indikken *n*

* **decanter** → **2484**

* **decolo(u)r** *v* → **742**

741 decolo(u)ration; decolo(u)rization;
colo(u)r removal
f décoloration *f*
d Entfärbung *f*
e decoloración *f*; descoloración *f*;
descoloramiento *m*
n ontkleuring *f*

* **decolo(u)ring carbon** → **744**

742 decolo(u)rize *v*; **decolo(u)r** *v*
f décolorer
d entfärben
e decolorear; descolorar; decolorar
n ontkleuren

743 decolo(u)rizing agent; decolo(u)rizer
f décolorant *m*
d Entfärbungsmittel *n*; Entfärber *m*
e agente *m* para de(s)colorar;
descolorante *m*; decolorante *m*
n ontkleuringsmiddel *n*

744 decolo(u)rizing carbon; decolo(u)ring
carbon

f charbon m décolorant
d Entfärbungskohle f
e carbón m para de(s)colorar; carbón m
   descolorante
n ontkleuringskool f

745 decolo(u)rizing power
f pouvoir m décolorant
d Entfärbungsvermögen n
e poder m para de(s)colorar
n ontkleuringsvermogen n; ontkleurende
   kracht f

746 decolo(u)rizing resin
f résine f décolorante
d Entfärbungsharz n
e resina f descolorante
n ontkleuringshars n

747 decomposition losses
f pertes fpl par décomposition
d Zersetzungsverluste mpl
e pérdidas fpl por descomposición
n ontledingsverliezen npl

748 defecate v
(To purify juices by adding massive doses
of calcium oxide or hydroxide)
f déféquer
d kalken; scheiden
e defecar
n kalken

749 defecated juice; limed juice
f jus m chaulé; jus m défequé
d Kalkungssaft m; gekalkter Saft m;
   geschiedener Saft m; Scheidesaft m
e jugo m defecado; jugo m alcalinizado;
   jugo m encalado; guarapo m defecado;
   guarapo m encalado; guarapo m
   alcalinizado
n gekalkt sap n; defecatiesap n;
   gedefequeerd sap n

* **defecating pan → 754**

* **defecating tank → 754**

750 defecation
f défécation f
d Defäkation f
e defecación f
n defecatie f

751 defecation factory
f sucrerie f à défécation
d Defäkationsfabrik f
e fábrica f con defecación

n defecatiefabriek f; defequerende
   fabriek f

* **defecation pan → 754**

**752 defecation slime; defecation mud;
defecation scum**
f écume f de défécation
d Scheideschlamm m
e espuma f de defecación
n defecatieschuim n

* **defecation tank → 754**

* **defecation with dry lime → 877**

753 defecation with milk of lime
f chaulage m du jus avec du lait de chaux;
   défécation f par lait de chaux
d Naßscheidung f; Kalkmilchscheidung f;
   nasse Scheidung f
e encalado m con lechada de cal;
   defecación f con lechada de cal;
   defecación f húmeda; defecación f por
   lechada de cal reciente
n kalkzetting f met kalkmelk; defecatie f
   met kalkmelk

**754 defecator; defecation pan; defecation
tank; defecating pan; defecating tank**
f défécateur m
d Scheidepfanne f
e defecador m
n defecatiepan f

755 defeco-carbona(ta)tion
(Simultaneous liming and carbonatation)
f épuration f calco-carbonique; chaulage-
   carbonatation m synchrones;
   carbonatation f calco-carbonique;
   purification f calco-carbonique; calco-
   carbonatation f
d Kalkungscarbonatation f; Scheide-
   saturation f; Defäkosaturation f;
   simultane Kalkungskarbonatation f
e calco-carbonatación f
n kalkingscarbonatatie f

756 deficiency disease
f maladie f de carence; maladie f
   carentielle
d Mangelkrankheit f
e enfermedad f por carencia; enfermedad f
   carencial
n gebreksziekte f

**757 deflocculation agent; deflocculating
agent; deflocculant**
f agent m antifloculant; agent m de

dispersion
d Dispersionsmittel *n*; Dispergiermittel *n*
e dispersante *m*
n dispersiemiddel *n*

758 defoamed juice; skimmed juice
f jus *m* démoussé
d entschäumter Saft *m*
e jugo *m* desespumado; guarapo *m* desespumado
n afgeschuimd sap *n*; ontschuimd sap *n*

759 defrothing
f démoussage *m*
d Entschäumung *f*
e desespumación *f*
n ontschuiming *f*; afschuiming *f*

760 degree Baumé
f degré *m* Baumé
d Grad *m* Baumé
e grado *m* Baumé
n graad *m* Baumé

761 degree Brix; Brix degree
f degré *m* Brix
d Grad *m* Brix; Grad *m* Balling; Brixgrad *m*
e grado *m* Brix
n graad *m* Brix; Brixgraad *m*

* **dehardening** → 2197

* **delime** *v* → 739

762 deliming; decalcification
f déchaulage *m*; décalcification *f*
d Entkalkung *f*
e descalcificación *f*; desencalado *m*
n ontkalking *f*

763 deliming agent
f anticalcaire *m*
d Entkalkungsmittel *n*
e desencalante *m*
n ontkalkingsmiddel *n*

764 delivery contract; contract of delivery; contract for delivery
(EEC)
f contrat *m* de livraison
d Liefervertrag *m*
e contrato *m* de suministro
n leveringsovereenkomst *f*

765 delivery opening; discharge opening; bagasse opening *(between top roller and back roller)*
f ouverture *f* arrière; ouverture *f* de sortie
d Austragsöffnung *f*; Austragsspaltweite *f*; Öffnung *f* der Mühle an der Austragsseite
e abertura *f* de salida; abertura *f* trasera
n afvoeropening *f*

766 delivery plate *(at last mill)*
f table *f* de sortie
d Austragstisch *m*
e plataforma *f* de salida
n afvoerplaat *f*

767 delivery roll(er); discharge roll(er); bagasse roll(er); back roll(er) *(of roller mill)*
f cylindre *m* de sortie
d Austragsroller *m*; Austragswalze *f*
e cilindro *m* de salida; rodillo *m* bagacero; bagacero *m*
n bagassecilinder *m*; ampascilinder *m*; afvoercilinder *m*; achtercilinder *m*

768 delivery setting *(of the rollers of a mill)*
f réglage *m* de sortie
d Austragseinstellung *f*
e ajuste *m* de salida
n instelling *f* van de afvoer

769 delivery side *(of cane mill; crusher; shredder)*
f côté *m* sortie
d Austragsseite *f*
e lado *m* de salida
n afvoerzijde *f*

770 demerara
(Type of sugar)
f demerara *m*
d Demerara *m*
e demerara *m*
n demerara *m*

771 demineralization
f déminéralisation *f*
d Entsalzung *f*
e desmineralización *f*
n ontzouting *f*

772 demineralized juice
f jus *m* déminéralisé
d entsalzter Saft *m*
e jugo *m* desmineralizado; guarapo *m* desmineralizado
n ontzout sap *n*

773 denaturation
(Alteration of the beet-cell protoplasm, usually through the effect of temperature, with the result that the cell-

wall permeability and the coagulation of
the cell-protein matter are increased)
f dénaturation f
d Denaturierung f
e desnaturalización f
n denaturatie f

774 denaturation
(EEC)
f dénaturation f
d Denaturierung f
e desnaturalización f
n denaturering f

775 denaturation premium
(EEC)
f prime f de dénaturation
d Denaturierungsprämie f
e prima f de desnaturalización
n denatureringspremie f

776 denature v *(the beet-cell protoplasm)*
f dénaturer
d denaturieren
e desnaturalizar
n denatureren

777 densimeter
f densimètre m
d Dichtemesser m; Densimeter n
e densímetro m
n dichtheidsmeter m; densimeter m

778 deposit
(EEC)
f caution f
d Kaution f
e depósito m
n cautie f

779 depulp v
f épulper
d entpülpen
e despulpar
n de pulp verwijderen

780 depulping
f épulpage m
d Entpülpung f
e despulpado m
n pulpverwijdering f

781 descaling; scale removal
f détartrage m
d Belagentfernung f
e desincrustación f
n verwijdering f van de afzetsels;
  verwijdering f van de incrustaties

* **desiccation** → 878

782 desugar v; **desugarize** v
f dessucrer; désucrer
d entzuckern
e extraer el azúcar
n ontsuikeren

* **desugaration of molasses** → 785

783 desugaring; desugarizing; desugarization
f dessucrage m; désucrage m
d Entzuckerung f
e desazucarado m; desazucaración f
n ontsuikering f

* **desugaring of molasses** → 785

* **desugarize** v → 782

784 desugarized pulp
f pulpes fpl désucrées; pulpes fpl
  dessucrées
d entzuckerte Schnitzel npl
e pulpa f desazucarada
n ontsuikerde pulp f

* **desugarizing** → 783

**785 desugarizing of molasses; desugaring of
molasses; desugaration of molasses**
f dessucrage m de la mélasse; désucrage
  m de la mélasse
d Melasseentzuckerung f
e desazucarado m de la melaza;
  desazucaración f de la melaza
n ontsuikering f van melasse

786 desuperheater
f désurchauffeur m
d Heißdampfkühler m
e desobrecalentador m; enfriador m de
  vapor recalentado
n stoomkoeler m

787 desweeten v
f désucrer; dessucrer; laver
d entzuckern; waschen; auswaschen
e desazucarar; lavar
n ontsuikeren; wassen; uitwassen

* **detention time** → 1960

788 deteriorated beet
f betterave f altérée
d alterierte Rübe f
e remolacha f alterada; remolacha f
  deteriorada
n beschadigde biet f

789 deteriorated cane
f canne f altérée
d alteriertes Zuckerrohr n
e caña f deteriorada; caña f alterada
n beschadigd suikerriet n

* **determined losses** → 5

* **devil grass** → 268

790 dew point
f point m de rosée
d Taupunkt m
e punto m de rocío
n dauwpunt n

791 dextran
f dextrane m
d Dextran n
e dextrana f
n dextran n

* **dextrorotary sugar** → 793

792 dextrorotation
f rotation f dextrogyre
d Rechtsdrehung f
e rotación f dextrogira
n rechtsdraaiing f

793 dextrorotatory sugar; dextrorotary sugar
f sucre m dextrogyre
d rechtsdrehender Zucker m
e azúcar m dextrógiro; azúcar m dextra-
rotatorio
n rechtsdraaiende suiker m

* **dextrose** → 1170

* **D.I.** → 821

794 diametral circulation calandria *(of
calandria pan)*
f calandre f à circulation diamétrale
e calandria f de circulación diametral

795 diametral circulation pan
(A calandria pan)
f appareil m à cuire à circulation
diamétrale
e tacho m al vacío con circulación
diametral

* **diamond spot pearl** → 257

796 diaphragm pump
f pompe f à membrane
d Membranpumpe f

e bomba f de diafragma
n membraanpomp f

* **diastase** → 64

**797 diatomaceous earth; infusorial earth;
diatomite; kieselguhr**
(A filter aid)
f terre f à diatomées; diatomite f; terre f
d'infusoires; kieselguhr m
d Diatomeenerde f; Infusorienerde f;
Kieselgur f; Diatomit m
e tierra f de diatomeas; tierra f de
infusorios; kieselgur m; diatomita f
n diatomeeënaarde f; diatomiet n;
infusoriënaarde f; kiezelgoer n

798 differential amount
(EEC)
f montant m différentiel
d Differenzbetrag m; Präferenzbetrag m
e montante m diferencial
n differentieel bedrag n

799 differential grasshopper *(US)*;
differential locust *(US)*
(sb; pest)
l Melanoplus differentialis

800 differential levy
(EEC)
f cotisation f différentielle
d Differenzabgabe f; Präferenzabgabe f
e cotización f diferencial
n differentiële bijdrage f

* **differential locust** *(US)* → 799

801 diffuser; extractor
(Apparatus for obtaining sugar-bearing
juice from the cossettes by a combination
of leaching and cell-membrane dialysis)
f diffuseur m; extracteur m
d Diffuseur m; Extrakteur m
e difusor m
n diffuseur m; diffusieketel m

* **diffusion** → 952

* **diffusion after mills** → 808

* **diffusion apparatus** → 954

**802 diffusion apparatus with bottom
discharge**
f appareil m de diffusion avec vidange par
le bas
d Diffuseur m mit unterer Entleerung
e aparato m de difusión con descarga

inferior
n diffusieapparaat n met onderlossing

803 diffusion battery
f batterie f de diffusion; batterie f de
  diffuseurs; batterie f d'extracteurs
d Diffusionsbatterie f
e bateriá f de difusión; batería f de
  difusores
n diffusiebatterij f

804 diffusion coefficient
f coefficient m de diffusion
d Diffusionskoeffizient m
e coeficiente m de difusión
n diffusiecoëfficiënt m

805 diffusion juice
f jus m de diffusion
d Diffusionssaft m
e jugo m de difusión; guarapo m de
  difusión
n diffusiesap n

806 diffusion-juice heater; calorisator
f calorisateur m; réchauffeur m du jus de
  diffusion
d Rohsaftvorwärmer m; Kalorisator m
e precalentador m de guarapo de difusión;
  calorizador m
n calorisator m

807 diffusion losses; extraction losses
  (Sucrose leaving the process in the
  pressed pulp, when the press water is
  being returned to the diffuser; otherwise,
  sucrose leaving the process in the wet
  pulp. Usually expressed as percent on
  beets or percent on sucrose entered)
f pertes fpl à la diffusion; pertes fpl à
  l'extraction
d Diffusionsverluste mpl; Extraktions-
  verluste mpl
e pérdidas fpl en difusión; pérdidas fpl en
  extracción
n diffusieverliezen npl; extractieverliezen
  npl

**808 diffusion of bagasse; bagasse diffusion;
  diffusion after mills**
f diffusion f de la bagasse; diffusion f
  après moulins
d Bagassediffusion f
e difusión f del bagazo
n bagassediffusie f; ampasdiffusie f

**809 diffusion of cane; cane diffusion; pure
  diffusion**
f diffusion f de la canne; diffusion f pure

d Zuckerrohrdiffusion f
e difusión f de la caña
n rietdiffusie f

* diffusion plant → 957

810 diffusion press water
f eaux fpl de pression de diffusion
d Diffusionsdruckwasser n
e agua f de prensado de difusión
n diffusiedrukwater n

811 diffusion process; extraction process
f procédé m de diffusion; procédé m
  d'extraction
d Diffusionsverfahren n; Extraktions-
  verfahren n
e procedimiento m de difusión;
  procedimiento m de extracción
n diffusieprocédé n; extractieprocédé n;
  diffusieproces n; extractieproces n

812 diffusion screen; diffusion strainer
f tôle f perforée du diffuseur
d Diffuseursieb n
e criba f de difusor
n zeefplaat f van diffusieketel

**813 diffusion temperature; extraction
  temperature**
f température f de diffusion; température
  f d'extraction
d Diffusionstemperatur f; Extraktions-
  temperatur f
e temperatura f de difusión; temperatura f
  de extracción
n diffusietemperatuur f; extractie-
  temperatuur f

814 diffusion time; extraction time
f durée f de la diffusion; durée f de
  l'extraction
d Diffusionszeit f; Extraktionszeit f
e duración f de la difusión; duración f de
  la extracción
n diffusieduur m; extractieduur m

* diffusion tower → 958

* diffusion waste water → 169

815 diffusion water; extraction water
f eau f de diffusion; eau f d'extraction
d Diffusionswasser n; Extraktionswasser n
e agua f de difusión; agua f de extracción
n diffusiewater n; extractiewater n

**816 diffusion water return; diffusion water
  recycling; reutilization of diffusion waste**

**water; extraction water return;
extraction water recycling**
f recyclage m des eaux de diffusion;
reprise f des eaux de diffusion; recyclage
m des eaux d'extraction; reprise f des
eaux d'extraction
d Diffusionswasserrücknahme f;
Extraktionswasserrücknahme f
e retorno m de aguas de difusión; retorno
m de aguas de extracción
n terugname f van diffusiewater;
terugbrenging f van diffusiewater;
terugname f van extractiewater;
terugbrenging f van extractiewater

817 diffusion zone; extraction zone
f zone f de diffusion; zone f d'extraction
d Diffusionszone f; Extraktionszone f
e zona f de difusión; zona f de extracción
n diffusiezone f; extractiezone f

* **dig** v *(US; beet)* → **1460**

818 digester
f digesteur m
d Digester m
e digestor m
n digestor m

* **digger-loader** *(US)* → **1461**

* **digger wheel** *(US)* → **2005**

* **digging disc** → **1464**

* **digging share** → **1465**

819 diluted juice
f jus m dilué
d verdünnter Saft m
e jugo m diluido; guarapo m diluido
n verdund sap n

* **dilute juice** → **1609**

820 dilution
(Quantity of imbibition water which
enters the mixed juice, per cent of cane)
f dilution f
d Verdünnung f
e dilución f; agua f de dilución
n verdunning f

821 dilution indicator; D.I.; safety factor
(Australia) (old)
(Moisture per cent non-pol)
f indice m de dilution; I.D. m
d Verdünnungsfaktor m

e indicador m de dilución
n verdunningsfactor m

822 diploid seed
(agr)
f semence f diploïde; graine f diploïde
d diploides Saatgut n
e semilla f diploide
n diploïde zaad n

823 direct pulp drying
f séchage m des pulpes par gaz de foyer
indépendant
d Schnitzelfeuertrocknung f
e secado m de pulpa con fuego directo
n pulpdroging f met verse stoom

824 dirt and trash separator
f décrotteur m
d Erdabscheider m
e separador m de tierra
n aardeafscheider m

825 dirt percentage; dirt tare
(Beet reception)
f pour-cents mpl de tare; déchet m pour-
cents; tare f de terre; pourcentage m de
terre; tare-terre f
d Schmutzprozente npl; Erdanhang m;
Schmutzanteil m; Schmutzanhang m
e tara f tanto por ciento; porcentaje m de
suciedad; tara f de tierra
n grondtarra f; tarra f aan aarde

826 dirt separator
f classificateur m de déchets
d Schmutzklassierer m
e clasificador m de suciedades
n vuilklasseerapparaat n

* **dirt tare** → **825**

* **dirty water** → **2119**

827 disaccharide
f disaccharide m
d Disaccharid n
e disacarido m
n disaccharide n

* **disc beet slicer** → **237**

828 discharge mixer
f malaxeur m d'attente
d Ausfüllmaische f
e cristalizador m de descarga
n wachtkristallisoir m

* **discharge opening** → **765**

* **discharge plough** → 832

829 discharger; unloader
f déchargeur m
d Ablader m
e descargador m
n losser m; aflader m

830 discharger; unloader *(in centrifugal)*
f déchargeur m
d Ausräumer m
e descargador m; descargadora f
n ruimer m

* **discharge roll(er)** → 767

831 discharging *(of centrifugal)*
f vidange f
d Entleerung f
e descarga f
n ledigen n

**832 discharging plough; discharge plough;
centrifugal plough**
f charrue f de déchargement; soc m
d Räumpflug m; Pflugschar f; Räum-
schar f
e arado m de descarga
n suikerploeg m

833 discharging speed
f vitesse f de déchargement
d Ausräumdrehzahl f; Räumdrehzahl f
e velocidad f de descarga
n toerental n bij het lossen

**834 disc of the slicer; disc of the slicing
machine**
f plateau m de coupe-racines
d Schneidscheibe f einer Schnitzel-
maschine
e plato m de cortarraíces; disco m
cortador
n mesplaat f van snijmolen

* **discontinuous centrifugal** → 164

835 discontinuous filtration
f filtration f discontinue
d diskontinuierliche Filtration f
e filtración f discontinua
n discontinue filtratie f; intermitterende
filtratie f

**836 discontinuous sulphitation; batch
sulphitation**
f sulfitation f discontinue
d diskontinuierliche Sulfitation n

e sulfitación f discontinua
n intermitterende sulfitatie f

837 discount *(for beets)*
(EEC)
f réfaction f
d Abschlag m
e refacción f
n rafactie f; refactie f

838 disintegrator
f broyeur m; désintégrateur m
d Schleudermühle f; Desintegrator m;
Schlagmühle f
e desintegrador m
n breker m

839 displacement pump
f pompe f à déplacement
d Verdrängerpumpe f
e bomba f de desplazamiento
n verdringerpomp f

* **dissolver** → 1572

**840 distributing mixer; distributor-mixer;
distribution mixer**
f malaxeur-distributeur m
d Verteilermaische f; Verteilertrog m
e mezclador-distribuidor m; distribuidor-
mezclador m
n verdeelgoot f; verdeeltrog m

841 distributor
f distributeur m
d Verteiler m
e distribuidor m
n verdeler m

* **distributor-mixer** → 840

842 division of a beet knife
f nombre m de divisions d'un couteau de
coupe-racines; pas m d'un couteau de
coupe-racines
d Teilung f eines Schnitzelmessers
e paso m de cuchilla cortadora; división f
de cuchilla cortadora
n steek m van een bietenmes

843 dodder; love vine; strangle weed
(sb; weed)
f cuscute f; barbe f de moine
d Seide f; Teufelszwirn m; Flechtgras n
e cuscuta f
n warkruid n
l Cuscuta sp.

* **dog camomile** → 2286

* **dog chamomile** → 2286

* **dog fennel** → 2286

* **dog nettle** → 2187

* **dog's-tooth grass** → 268

* **doob** → 268

844 **dosing pump**
  f pompe f doseuse
  d Dosierpumpe f
  e bomba f para dosificación; bomba f
     dosificadora; bomba f de dosificar
  n doseerpomp f

845 **double carbona(ta)tion**
  f double carbonatation f; carbonatation f
     double
  d doppelte Carbonatation f; zweimalige
     Carbonatation f; zweistufige
     Carbonatation f
  e doble carbonatación f
  n dubbele carbonatatie f

846 **double crusher**
  f double défibreur m
  d Doppelcrusher m
  e desmenuzadora f doble; doble
     desmenuzadora f
  n dubbele crusher m; dubbele kneuzer m

847 **double-drum rock catcher; double-drum
     stone catcher**
  f épierreur m à double tambour
  d Doppel-Trommel-Steinabscheider m;
     Doppel-Trommel-Steinfänger m
  e separador m de piedras de doble tambor
  n steenvanger m met dubbele trommel;
     stenenvanger m met dubbele trommel

848 **double effect; double effect evaporator**
  f double effet m
  d Zweifachverdampfer m; Zweikörper-
     verdampfer m
  e doble efecto m
  n verdampapparaat n met twee
     verdamplichamen; verdampapparaat n à
     double effet

849 **double effect evaporation**
  f évaporation f à double effet
  d zweistufige Verdampfung f; Zweifach-
     verdampfung f; Zweikörper-
     verdampfung f
  e evaporación f de doble efecto
  n tweevoudige verdamping f;
     tweetrapsverdamping f

* **double effect evaporator** → 848

850 **double imbibition**
  f imbibition f double
  d doppelte Imbibition f
  e imbibición f doble
  n dubbele imbibitie f

851 **double knives**
  f couteaux mpl doubles
  d Doppelmesser npl
  e cuchilla f doble
  n dubbele messen npl

852 **double purging**
  f double turbinage m
  d zweimaliges Zentrifugieren n;
     zweistufiges Abschleudern n
  e purgado m doble; doble purga f
  n centrifugeren n in twee trappen

853 **double refined sugar**
  f sucre m double raffiné; sucre m
     biraffiné
  d Doppelraffinade f
  e azúcar m doble refinado
  n dubbel geraffineerde suiker m

854 **double screw press**
  f presse f à double hélice
  d Doppelschneckenpresse f
  e prensa f helicoidal doble
  n pers f met dubbele schroef

855 **double simple imbibition**
  f imbibition f simple double
  d zweistufige einfache Imbibition f
  e imbibición f simple doble
  n dubbele eenvoudige imbibitie f;
     eenvoudige imbibitie f in twee trappen

856 **double sulphitation**
  f double sulfitation f
  d zweistufige Sulfitation f
  e sulfitación f doble
  n dubbele sulfitatie f

857 **dove's foot cranesbill; dove's foot
     geranium; cranesbill**
     (sb; weed)
  f géranium m mou; bec m de cigogne;
     géranium m à feuilles molles
  d weicher Storchschnabel m
  e geranio m de los caminos
  n zachte ooievaarsbek m
  l Geranium molle

* **down-the-row thinner** → 246

* **downy hemp nettle** → 575

858 **downy mildew**
(sb; disease)
f mildiou *m*
d falscher Mehltau *m*
e mildiú *m*
n valse meeldauw *m*

859 **downy mildew; downy mildew disease**
(sc; fungous disease)
f mildiou *m*
d falscher Mehltau *m*
e enfermedad *f* del añublo lanoso
n valse meeldauw *m*

* **dredging pump** → 2577

860 **dried beet**
f betterave *f* séchée
d Trockenrübe *f*
e remolacha *f* secada
n droge biet *f*

861 **dried pulp; dried beet pulp; dry pulp**
f pulpes *fpl* séchées; pulpes *fpl* sèches
d Trockenschnitzel *npl*
e pulpa *f* seca
n gedroogde pulp *f*; droge pulp *f*

862 **drier; dryer**
f sécheur *m*; séchoir *m*
d Trockner *m*
e secadero *m*
n droger *m*

* **drilling to a stand** → 2213

863 **drink**
f charge *f*; alimentation *f*; alimentation *f*
par gorgées
d Einzug *m*; Zuzug *m*
e alimentación *f*; aspiración *f*
n intrekken *n*

* **drink of syrup** → 1323

864 **dropping the strike; dropping the
skipping**
(Discharging the batch of finished
massecuite from the vacuum pan)
f lâchage *m* de la cuite; vidange *f* de la
masse cuite
d Ablassen *n* des Suds
e descarga *f* de la masa cocida
n ledigen *n* van de kookpan; aftrekken *n*
van het kooksel

865 **drop plate**
f plateau *m* à gouttes
d Tüpfelplatte *f*
e meseta *f* de gotas
n druppelplaat *f*

866 **drop *v* the strike**
(To discharge the contents of a vacuum
pan)
f lâcher la cuite
d den Sud ablassen
e vaciar la templa; caer la templa
n het kooksel aftrekken

* **drug speedwell** → 584

867 **drum**
f tambour *m*
d Trommel *f*
e tambor *m*
n trommel *f*

* **drum-cutter** *(US)* → 870

868 **drum elevator**
(agr)
f cage *f* d'écureuil
d Siebrad *n*
e tambor *m* de descarga de rejilla
n zeefrad *n*

869 **drum filter**
f filtre *m* à tambour
d Trommelfilter *n*
e filtro *m* rotativo
n trommelfilter *n*

870 **drum-slicer; drum-cutter** *(US)*
f coupe-racines *m* à tambour
d Trommelschneidmaschine *f*; Vertikal-
schneidmaschine *f*
e cortadora *f* vertical de remolachas
n trommelsnijmolen *m*

871 **dry *v***
f sécher
d trocknen
e secar
n drogen

872 **dry crushing**
f pression *f* sèche
d Trockenpressung *f*
e presión *f* seca
n droge persing *f*

873 **dry-crushing bagasse**
f bagasse *f* de pression sèche
d trockengepreßte Bagasse *f*

e bagazo *m* de presión seca
n drooggeperste bagasse *f*; drooggeperste ampas *m*

874 dry-crushing extraction; extraction by dry crushing
f extraction *f* de pression sèche
d Extraktion *f* mittels Trockenpressung
e extracción *f* de presión seca
n extractie *f* met droge persing

875 dry-crushing juice
f jus *m* de pression sèche
d trockengepreßter Saft *m*
e jugo *m* de presión seca
n drooggeperst sap *n*

876 dry-crushing mill
f moulin *m* de pression sèche
d mit Trockenpressung arbeitende Mühle *f*
e molino *m* de presión seca
n met droge persing werkende molen *m*

877 dry defecation; dry liming; defecation with dry lime
f chaulage *m* du jus avec de la chaux en roche concassée; chaulage *m* roche; chaulage *m* à la chaux en roche; défécation *f* à la chaux vive
d Trockenscheidung *f*; trockene Scheidung *f*
e encalado *m* con trozos de cal viva; defecación *f* seca; defecación *f* por cal viva en terrones
n kalkzetting *f* met stukken; defecatie *f* met kluitkalk

* **dryer** → **862**

878 drying; desiccation
f dessiccation *f*; séchage *m*
d Trocknung *f*; Austrocknung *f*
e secado *m*; desecación *f*
n droging *f*; uitdroging *f*

879 drying drum
f tambour *m* sécheur; tambour *m* de séchage
d Trockentrommel *f*
e tambor *m* secador; tambor *m* de secado
n droogtrommel *f*

880 drying kiln
f four *m* à sécher; four *m* sécheur; séchoir *m*; étuve *f*
d Trocknungsofen *m*; Trockenstube *f*; Darre *f*
e horno *m* para el secado
n droogoven *m*

881 drying oven
f étuve *f*
d Trockenschrank *m*; Trockenofen *m*
e estufa *f*
n droogstoof *f*

* **drying with chimney gas** → **1853**

* **dry liming** → **877**

* **dry pulp** → **861**

882 dry rot
(sb; disease)
f pourriture *f* sèche; pourriture *f* du coeur
d Herz- und Trockenfäule *f*
e podredumbre *f* seca
n droog rot *n*

883 dry rot
(sc; fungous disease)
f pourriture *f* sèche
d Trockenfäule *f*
e pudrición *f* seca
n droog rot *n*

884 dry substance
(The water-soluble, moisture-free matter present)
f matière *f* sèche; M.S. *f*
d Trockensubstanz *f*; TS *f*; Tr.S. *f*
e materia *f* seca
n droge stof *f*

* **dry substance** → **2526**

885 dry-substance content
(Percent water-soluble, moisture-free matter as determined by drying or other direct means of determining moisture or solids content)
f teneur *f* en matière sèche
d Trockensubstanzgehalt *m*; TS-Gehalt *m*
e contenido *m* de materia seca
n droge-stofgehalte *n*

886 dry top rot
(sc; fungous disease)
f pourriture *f* sèche de la pointe
d Gipfeltrockendürre *f*
e pudrición *f* seca del cogollo
n droog toprot *n*

887 dry unloading *(of beets)*
f déchargement *m* à sec
d Trockenentladung *f*
e descarga *f* en seco
n droge lossing *f*

* **dull-seed cornbind** → 276

888 dust box
(A device to remove sugar dust from air,
usually employing water sprays; a dust
collector)
f dépoussiéreur *m*
d Entstauber *m*
e separador *m* de polvo; despolvoreador *m*
n stofvanger *m*

889 dust removal plant
f installation *f* de dépoussiérage
d Entstaubungsanlage *f*
e instalación *f* despolvoreadora
n stofverwijderingsinstallatie *f*

890 dwarf
(sc; virus disease)
f "dwarf"; nanisme *m*
d Zwergwuchs *m*; viröse Verzwergung *f*
e enanismo *m*
n dwerggroei *m*

* **dwarf** → 707

* **dwarf mallow** → 577

E

* EAGGF → 917

891 early forget-me-not
(sb; weed)
f myosotis *m* hérissé; myosotis *m* des coteaux
d Hügel-Vergißmeinnicht *n*; rauhes Vergiß-meinnicht *n*
n ruw vergeet-mij-nietje *n*
l Myosotis collina auct.; Myosotis hispida Schl.

* earth almond → 1682

* earth smoke → 573

892 earwig; common earwig; European earwig *(US)*
(sb; pest)
f perce-oreille *m*; forficule *m*
d Ohrwurm *m*
e tijereta *f*
n oorworm *m*; oorwurm *m*
l Forficula auricularia

893 Eastern field wireworm *(US)*
(sb; pest)
l Limonius agronus

894 eccentric worm pump
f pompe *f* à vis excentrique
d Exzenterschneckenpumpe *f*
e bomba *f* helicoidal excéntrica
n excentriekschroefpomp *f*

895 economizer
f économiseur *m*
d Ekonomiser *m*; Ekonomizer *m*
e economizador *m*
n economizer *m*

* EEC-sugar → 587

896 effect-giving procedures
(EEC)
f modalités *fpl* d'application
d Durchführungsbestimmungen *fpl*
e modalidades *fpl* de aplicación
n uitvoeringsmodaliteiten *fpl*

897 effective acidity
f acidité *f* effective
d effektive Acidität *f*
e acidez *f* efectiva
n effectieve aciditeit *f*

898 effective alkalinity
f alcalinité *f* effective
d effektive Alkalität *f*
e alcalinidad *f* efectiva
n effectieve alkaliteit *f*

* effluent → 2119

899 ejector-condenser
f éjecto-condenseur *m*; éjecto *m*
e condensador *m* de eyector; condensador-eyector *m*

900 electrically driven centrifugal
f centrifugeuse *f* électrique; essoreuse *f* électrique
d Zentrifuge *f* mit elektrischem Antrieb
e centrifuga *f* eléctrica
n elektrisch aangedreven centrifuge *f*

901 electrically driven mill
f moulin *m* électrique; moulin *m* à moteur électrique
d Elektromühle *f*; Mühle *f* mit elektrischem Antrieb
e molino *m* eléctrico
n elektrische molen *m*; molen *m* met elektrische motor; elektrisch aangedreven molen *m*

902 electric motor
f moteur *m* électrique
d Elektromotor *m*
e motor *m* eléctrico
n elektromotor *m*; elektrische motor *m*

903 electrodialysis
(Ultrafiltration accelerated by the influence of an electric potential)
f électrodialyse *f*
d Elektrodialyse *f*
e electrodiálisis *f*
n elektrodialyse *f*

* elevating web → 463

904 elevator
f élévateur *m*
d Elevator *m*
e elevador *m*
n elevator *m*

905 eluate
f éluat *m*
d Eluat *n*
e eluato *m*
n eluaat *n*

906 elute *v*
f éluer
d eluieren
e eluir
n elueren

907 elution
f élution *f*
d Elution *f*; Eluierung *f*
e elución *f*
n elutie *f*

* **EMS** → 921

908 end product; final product
f produit *m* final; produit *m* de queue
d Endprodukt *n*; Nachlaufprodukt *n*
e producto *m* final
n eindprodukt *n*

* **endproduct working** → 1513

* **energetic liming** → 2624

* **English daisy** *(US)* → 731

* **English wormseed** → 2536

909 ensilage *(noun)*; **silage; pitting; damping; ensiling**
f ensilage *m*
d Silierung *f*; Silieren *n*; Einmieten *n*; Einsilieren *n*
e silaje *m*; ensilaje *m*
n inkuiling *f*

910 ensilage *v*; **pit** *v*; **clamp** *v*; **ensile** *v*
f ensiler
d einmieten; silieren; einsilieren
e ensilar
n inkuilen; ensileren

* **ensiling** → 909

911 enthalpy
(The sum of the internal and external energies of a fluid system)
f enthalpie *f*
d Enthalpie *f*
e entalpía *f*
n enthalpie *f*

* **entrainment arrestor** → 438

* **entrainment catcher** → 438

* **entrainment separator** → 438

912 entropy
(A measure of the unavailable energy in a thermodynamic system)
f entropie *f*
d Entropie *f*
e entropía *f*
n entropie *f*

913 entropy diagram
(Steam production)
f diagramme *m* entropique
d T-s-Diagramm *n*; Temperatur-Entropie-Diagramm *n*
e diagrama *m* de entropías; diagrama *m* entrópico
n entropiediagram *n*

914 equalizer *(of canes on carrier)*
f égaliseur *m*
d Ebener *m*; Ausgleicher *m*
e nivelador *m*
n laagbegrenzer *m*

915 erect knotweed *(US)*
(sb; weed)
l Polygonum ramosissimum

916 ergots
(sc; fungous disease of inflorescence)
f ergots *mpl*
d Mutterkorn *n*
e cornezuelos *mpl*; enfermedades *fpl* de cornezuelo
n moederkoren *n*; moederkoorn *n*

917 European Agricultural Guidance and Guarantee Fund; EAGGF
(EEC)
f Fonds *m* Européen d'Orientation et de Garantie Agricole; FEOGA *m*
d Europäischer Ausrichtungs- und Garantiefonds *m* für die Landwirtschaft; EAGFL *m*
e Fondo *m* Europeo de Orientación y de Garantía Agrícola; FEOGA *m*
n Europees Oriëntatie- en Garantiefonds *n* voor de Landbouw; EOGFL *n*

918 European Association for Sweeteners
f Association *f* Européenne pour les Edulcorants
d Europäische Vereinigung *f* für Süßstoffe
e Asociación *f* Europea de Edulcorantes
n Europese Vereniging *f* voor Zoetstoffen

* **European cockchafer** *(US)* → 532

919 European Committee of Sugar Manufacturers

f Comité *m* Européen des Fabricants de
  Sucre; CEFS *m*
d Verband *m* der Europäischen Zuckerhersteller
e Comisión *f* Europea de los Fabricantes
  de Azúcar
n Verbond *n* van de Europese
  Suikerfabrikanten

* **European common cockchafer** *(US)*
  → **532**

* **European earwig** *(US)* → **892**

* **European glorybind** → **997**

920 **European heliotrope; heliotrope**
    (sb; weed)
  f tournesol *m*; héliotrope *m*
  d Sonnenwende *f*
  e verrucaria *f*
  n Europese heliotroop *f*
  l Heliotropium europaeum

* **European June beetle** → **2419**

* **European June bug** → **2419**

921 **European Monetary System; EMS**
    (EEC)
  f Système *m* Monétaire Européen; SME *m*
  d Europäisches Währungssystem *n*; EWS *n*
  e Sistema *m* Monetario Europeo; SME *m*
  n Europees Monetair Systeem *n*; Europees
    Muntstelsel *n*; EMS *n*

* **European tarnished plant bug** → **2471**

* **evaporating capacity** → **924**

* **evaporating station** → **933**

* **evaporating surface** *(of evaporator
  vessel)* → **83**

922 **evaporation**
  f évaporation *f*
  d Verdampfung *f*
  e evaporación *f*
  n verdamping *f*

923 **evaporation apparatus**
  f appareil *m* d'évaporation
  d Verdampfapparat *m*
  e aparato *m* de evaporación
  n verdampingsapparaat *n*;
    verdampapparaat *n*; verdamptoestel *n*

924 **evaporation capacity; evaporating
    capacity; evaporative capacity** *(of
    multiple effect)*
  f capacité *f* d'évaporation
  d Verdampfleistung *f*
  e capacidad *f* de evaporación
  n verdampingsvermogen *n*

* **evaporation coefficient** → **928**

925 **evaporation heat**
  f chaleur *f* d'évaporation
  d Verdampfungswärme *f*
  e calor *m* de evaporación
  n verdampingswarmte *f*

926 **evaporation in single effect**
  f évaporation *f* en simple effet;
    évaporation *f* à simple effet
  d einstufige Verdampfung *f*
  e evaporación *f* en simple efecto;
    evaporación *f* en un solo efecto
  n eentrapsverdamping *f*

927 **evaporation plant**
  f installation *f* d'évaporation
  d Verdampfanlage *f*; Verdampfstation *f*
  e planta *f* de evaporación
  n verdampingsinstallatie *f*;
    verdampinstallatie *f*

928 **evaporation rate; evaporation coefficient**
    (Evaporation coefficient of an evaporator
    = weight of vapour furnished by the
    vessel per hour per unit of heating
    surface.
    Evaporation coefficient of a pan =
    weight of vapour evaporated from the
    massecuite per unit heating surface in
    unit of time)
  f taux *m* d'évaporation
  d Verdampfungsrate *f*
  e relación *f* de evaporación; coeficiente *m*
    de evaporación
  n verdampingscijfer *n*

* **evaporation station** → **933**

929 **evaporation under pressure**
  f évaporation *f* sous pression
  d Druckverdampfung *f*
  e evaporación *f* de presión; evaporación *f*
    a presión
  n drukverdamping *f*

* **evaporative capacity** *(of multiple effect)*
  → **924**

930 evaporator
  f évaporateur m
  d Verdampfer m
  e evaporador m
  n verdamper m

**931 evaporator body; evaporator cell;
    evaporator vessel**
  (A single evaporator vessel)
  f caisse f d'évaporation; corps m
    d'évaporation
  d Verdampfkörper m; Verdampfapparat m
  e cuerpo m de evaporación; cuerpo m del
    evaporador
  n verdamplichaam n

932 evaporator effect
  (An evaporator body or group of bodies
  working as one of the series-connected
  units in a multiple-effect evaporator)
  f effet m
  d Körper m
  e efecto m
  n verdamplichaam n

* **evaporator feed tank** → 934

**933 evaporator station; evaporation station;
    evaporating station**
  (The aggregate of the various evaporator
  bodies and their auxiliary equipment)
  f station f d'évaporation; poste m
    évaporatoire; groupe m évaporatoire
  d Verdampfstation f; Eindampfstation f
  e estación f de evaporación; estación f de
    evaporadores; estación f evaporadora;
    departamento m de evaporadores
  n verdampstation n

**934 evaporator supply tank; evaporator feed
    tank**
  f bac m d'alimentation de l'évaporateur
  d Speisebehälter m des Verdampfers
  e tanque m de alimentación del
    evaporador; tanque m alimentador del
    evaporador
  n voedingsbak m van de verdamper

935 evaporator tube
  f tube m d'évaporation
  d Verdampferrohr n
  e tubo m de evaporador
  n verdamperpijp f

* **evaporator vessel** → 931

936 evaporator with forced juice circulation
  f évaporateur m avec circulation de jus
    forcée

  d Verdampfapparat m mit Zwangs-
    zirkulation; Zwangsdurchlauf-
    verdampfer m; Verdampfer m mit
    Zwangsumlauf
  e evaporador m con circulación de guarapo
    forzado
  n verdamplichaam n met gedwongen
    sapcirculatie

* **evening campion** → 2676

937 exchange rate
  (EEC)
  f taux m de change
  d Kassakurs m; Wechselkurs m
  e tipo m de cambio
  n wisselkoers m

* **exhausted beet chips** → 939

938 exhausted cane
  f canne f épuisée
  d erschöpftes Rohr n
  e caña f agotada
  n uitgeput riet n

**939 exhausted cossettes; exhausted slices;
    exhausted beet chips; extracted
    cossettes; beet pulp**
  (Cosettes that have been exhausted by
  diffusion)
  f cossettes fpl épuisées; pulpes fpl (de
    betteraves)
  d ausgelaugte Schnitzel npl; entzuckerte
    Schnitzel npl; erschöpfte Schnitzel npl;
    extrahierte Schnitzel npl
  e cosetas fpl agotadas; cosetas fpl coladas;
    cosetas fpl desazucaradas; pulpa f de
    remolacha
  n uitgeloogde snijdsels npl; bietenpulp f

940 exhausted molasses
  f mélasse f épuisée
  d erschöpfte Melasse f
  e miel f agotada; melaza f agotada
  n uitgeputte melasse f

* **exhausted slices** → 939

**941 exhaustibility of molasses; molasses
    exhaustibility**
  f épuisabilité f de la mélasse
  d Erschöpfbarkeit f der Melasse; Melasse-
    erschöpfung m
  e agotabilidad f de la melaza; agotamiento
    m de la miel
  n uitputbaarheid f van melasse

942 exhaustion
f épuisement *m*
d Erschöpfung *f*
e agotamiento *m*
n uitputting *f*

* **exhaustion of a massecuite** → 700

943 exhaust steam *(of the turbine)*
f vapeur *f* d'échappement; V.E. *f*
d Abdampf *m*; Fabrikdampf *m*; Rück-
   dampf *m*
e vapor *m* de escape; V.E. *m*
n afgewerkte stoom *m*; retourstoom *m*

944 export certificate
   (EEC)
f certificat *m* d'exportation
d Ausfuhrlizenz *f*
e certificado *m* de exportación
n uitvoercertificaat *n*

945 export levy
   (EEC)
f prélèvement *m* à l'exportation
d Ausfuhrabschöpfung *f*
e gravamen *m* a la importación;
   "prélèvement" *m* a la importación
n uitvoerheffing *f*

946 export quota
   (eco)
f quota *m* d'exportation
d Exportquote *f*
e cuota *f* de exportación
n uitvoerquotum *n*

947 express *v*; **press** *v* **out**
f presser; pressurer; exprimer
d auspressen; ausdrücken
e exprimir; estrujar
n uitpersen

* **expressed juice** → 1819

948 external washing
   (Washing of filter cake outside the filter
   press)
f lavage *m* en dehors; lavage *m* hors du
   filtre-presse
d Auswaschen *n* außerhalb der Filterpresse
e lavado *m* exterior
n uitwassen *n* buiten de filterpers

949 extract *v*; **lixiviate** *v*
f extraire; épuiser
d auslaugen; extrahieren

e lixiviar; extraer
n uitlogen; extraheren

* **extracted cossettes** → 939

950 extracted juice
f jus *m* extrait
d extrahierter Saft *m*
e jugo *m* extraido; guarapo *m* extraido
n geëxtraheerd sap *n*

951 extracting liquid
f liquide *m* d'extraction
d Extraktionsflüssigkeit *f*
e líquido *m* de extracción
n extractievloeistof *f*

952 extraction; diffusion
   (The phase of the process in which
   sugar-bearing juice is obtained from the
   cossettes by a combination of leaching
   and cell-membrane dialysis)
f extraction *f*; diffusion *f*
d Extraktion *f*; Diffusion *f*
e extracción *f*; difusión *f*
n extractie *f*; diffusie *f*

953 extraction
   (The net sucrose produced/sucrose
   entered)
f extraction *f*
d Extraktion *f*; Auslaugung *f*
e extracción *f*
n extractie *f*; uitloging *f*

954 extraction apparatus; diffusion apparatus
f appareil *m* d'extraction; appareil *m* de
   diffusion
d Extraktionsapparat *m*; Diffusions-
   apparat *m*
e aparato *m* de difusión; aparato *m* de
   extracción
n extractieapparaat *n*; diffusieapparaat *n*

955 extraction by diffusion
f extraction *f* par diffusion
d Extraktion *f* durch Diffusion;
   Auslaugung *f* durch Diffusion
e extracción *f* por difusión
n extractie *f* door diffusie; uitloging *f* door
   diffusie

* **extraction by dry crushing** → 874

956 extraction drum
f tambour *m* d'extraction
d Extraktionstrommel *f*
e tambor *m* de extracción
n extractietrommel *f*

* **extraction losses** → 807

957 extraction plant; diffusion plant
  f installation *f* d'extraction; installation *f*
    de diffusion
  d Extraktionsanlage *f*; Diffusionsanlage
  e instalación *f* de extracción; instalación *f*
    de difusión
  n extractie-installatie *f*; diffusie-
    installatie *f*

* **extraction process** → 811

* **extraction temperature** → 813

* **extraction time** → 814

**958 extraction tower; diffusion tower; vertical
    diffuser**
  f tour *f* d'extraction; tour *f* de diffusion
  d Diffusionsturm *m*; Extraktionsturm *m*;
    Turmextrakteur *m*
  e torre *f* de extracción; torre *f* de difusión
  n diffusietoren *m*; extractietoren *m*

* **extraction water** → 815

* **extraction water recycling** → 816

* **extraction water return** → 816

* **extraction zone** → 817

* **extractor** → 801

959 eyed click-beetle *(US)*; **snapping beetle**
    (US)
    (sb; pest)
  l Alaus oculatus

960 eye spot; eye spot disease
    (sc; fungous disease)
  f maladie *f* des taches ocellées
  d Augenfleckenkrankheit *f*; Sämlings-
    fäule *f*
  e mancha *f* de ojo
  n oogvlekkenziekte *f*

F

961 factory juice
f jus *m* de fabrication
d Fabriksaft *m*; Betriebssaft *m*
n fabriekssap *n*

962 factory yard
f cour *f* de l'usine
d Fabrikhof *m*
e patio *m* de la fábrica
n fabrieksterrein *n*

963 falling film evaporator
f évaporateur *m* à descendage;
évaporateur *m* à flot tombant;
évaporateur *m* à flux descendant;
évaporateur *m* à flot descendant
d Fallstromverdampfer *m*; Fallstrom-
verdampfapparat *m*; Fallfilmverdampf-
apparat *m*; Fallfilmverdampfer *m*
e evaporador *m* de gravedad; evaporador
m de flujo descendente; evaporador *m*
de película descendente
n valstroomverdamper *m*

964 fall panicum *(US)*
(sb; weed)
l Panicum dichotomiflorum

965 false chinch bug *(US)*
(sb; pest)
f fausse punaise *f* fétide
d kleiner Griesel *m*; Heidekraut-Wanze *f*
e chinche *f* gris
l Nysius ericae

966 false grain; secondary grain
(Undesirably small crystals found among
the crystal population of a massecuite;
the result of unwanted spontaneous
nucleation during the sugar-boiling
operation)
f faux grain *m*; grain *m* secondaire
d Falschkorn *n*; falsches Korn *n*
e falso grano *m*; grano *m* secundario
n vals grein *n*; secundair grein *n*

* false ragweed *(US)* → 1551

967 fan
f ventilateur *m*
d Gebläse *n*; Ventilator *m*
e ventilador *m*
n ventilator *m*

968 fanged beet; sprangled beet
(Having many roots)

f betterave *f* fourchue
d beinige Rübe *f*
e remolacha *f* bifurcada
n vertakte biet *f*

* fanweed → 1003

969 fasciations
(sb; disease)
f fasciations *fpl*
d Verbänderung *f*
e fasciaciones *fpl*
n stengelvergroeiingen *fpl*

970 fat hen; goosefoot; lamb's-quarters *(US)*;
common lamb's-quarters *(US)*; **lamb's-
quarters goosefoot; meld weed**
(sb; weed)
f chénopode *m* blanc; belle dame *f*;
poule *f*; poulette *f*; ansérine *f* blanche;
poule *f* grasse
d weißer Gänsefuß *m*; weiße Melde *f*
e cenizo *m*; cenilgo *m*; ceñiglo *m*; apazote
m blanco; yuyo *m* blanco; apazote *m*
cenizo; cenizo *m* blanco
n melkganzevoet *m*
l Chenopodium album

* fat hen → 2236

* feed channel → 1371

971 feed compartment; flocculating chamber
(of clarifier)
f compartiment *m* d'alimentation;
compartiment *m* de floculation
d Flockungsraum *m*
e compartimiento *m* de floculación
n vlokkingsruimte *f*

972 feeder *(of mills)*
f alimentateur *m*
d Zuführer *m*; Speisevorrichtung *f*;
Aufgeber *m*
e alimentador *m*
n toevoerinrichting *f*; voedingsinrichting *f*

973 feed hopper; feed funnel
f trémie *f* d'alimentation
d Aufgabetrichter *m*; Einfülltrichter *m*;
Einlauftrichter *m*
e tolva *f* de alimentación; tolva *f* de carga
n vultrechter *m*

974 feeding chute
f coulotte *f* d'alimentation; couloir *m*
d'alimentation
d Aufgabeschurre *f*; Aufgaberutsche *f*

e canal *m* de alimentación
n voedingsgoot *f*

* **feeding roll(er)** → **980**

975 **feeding side** *(cane mill; crusher; shredder)*
  f côté *m* alimentation
  d Zuführungsseite *f*
  e lado *m* de alimentación
  n voedingszijde *f*

976 **feeding table**
  (Cane-handling equipment)
  f table *f* d'alimentation
  d Aufgabetisch *m*
  e mesa *f* alimentadora
  n toevoertafel *f*

977 **feed line**
  f tuyau *m* d'alimentation
  d Speiseleitung *f*
  e tubo *m* de alimentación
  n voedingsleiding *f*

978 **feed opening; cane opening** *(between front roller and top roller)*
  f ouverture *f* d'entrée; ouverture *f* avant
  d Zuführungsöffnung *f*; Zuführungsspaltweite *f*; Öffnung *f* der Mühle an der Zuführungsseite
  e abertura *f* de entrada; abertura *f* delantera
  n voedingsopening *f*; toevoeropening *f*

979 **feed pump**
  f pompe *f* alimentaire; pompe *f* d'alimentation
  d Speisepumpe *f*
  e bomba *f* de alimentación
  n voedingspomp *f*

980 **feed roll(er); feeding roll(er); cane roll(er); front roll(er)** *(of mill)*
  f cylindre *m* d'entrée; cylindre *m* alimentateur
  d Zuführungsroller *m*; Zuführungswalze *f*; Speisewalze *f*; Vorderwalze *f*
  e rodillo *m* alimentador; rodillo *m* cañero; cañero *m*; maza *f* cañera; cilindro *m* de entrada
  n rietcilinder *m*; voedingscilinder *m*; voorste onderste rol *f*

981 **feed water**
  f eau *f* d'alimentation
  d Speisewasser *n*
  e agua *f* de alimentación
  n voedingswater *n*

982 **feed water conditioning**
  f conditionnement *m* de l'eau d'alimentation
  d Speisewasseraufbereitung *f*
  e acondicionamiento *m* de agua de alimentación
  n voedingswaterconditionering *f*

983 **feed water pump**
  f pompe *f* alimentaire; pompe *f* d'alimentation (en eau)
  d Speisewasserpumpe *f*
  e bomba *f* de agua de alimentación
  n voedingswaterpomp *f*

984 **feed water purification**
  f purification *f* de l'eau d'alimentation; épuration *f* de l'eau d'alimentation
  d Speisewasserreinigung *f*
  e purificación *f* de agua de alimentación
  n voedingswaterreiniging *f*

985 **feed water tank**
  f bâche *f* d'alimentation; bâche *f* alimentaire
  d Speisewasserbehälter *m*
  e tanque *m* de agua de alimentación; tanque *m* de alimentación
  n voedingswaterbak *m*; voedingswaterballon *m*

986 **feeler; sensing device; finder** *(US)*; **top feeler**
  (agr)
  f tâteur *m*
  d Taster *m*
  e palpador *m*
  n taster *m*

987 **feeler carrier arm; feeler arm; finder arm** *(US)*
  (agr)
  f longeron *m* de tâteur
  d Tasterarm *m*
  e varilla *f* de palpador
  n tasterarm *m*

988 **Fehling's solution**
  f liqueur *f* de Fehling
  d Fehlingsche Lösung *f*
  e solución *f* de Fehling
  n reagens *n* van Fehling

989 **fermentable sugar**
  f sucre *m* fermentescible
  d fermentierbarer Zucker *m*; vergärbarer Zucker *m*
  e azúcar *m* fermentable
  n fermenteerbare suiker *m*

990 fermentation
f fermentation f
d Gärung f; Fermentation f
e fermentación f
n gisting f; fermentatie f

991 fibre; fiber *(US)*
(The dry, water-insoluble matter in the cane)
f fibre f
d Fiber f; Faser f
e fibra f
n vezel f

* **fibre content of cane** → 993

992 fibre in bagasse; fibre content of bagasse
f ligneux m de la bagasse
d Fibergehalt m der Bagasse
e fibra f del bagazo
n vezelstof f in bagasse; vezelstof f in ampas; vezelstofcijfer n van de bagasse; vezelstofcijfer n van de ampas

993 fibre in cane; fibre content of cane
f ligneux m de la canne
d Fibergehalt m des Zuckerrohrs
e fibra f de la caña
n vezelstof f in riet; vezelstofcijfer n van het riet

994 fibrous beet
f betterave f fibreuse
d faserige Rübe f
e remolacha f fibrosa
n vezelachtige biet f; vezelige biet f

995 fibrous mass
f masse f fibreuse
d Fasermasse f
e masa f fibrosa
n vezelachtige massa f; vezelmassa f

* **field beet** → 1543

996 field betony; field woundwort; field-nettle betony
(sb; weed)
f épiaire f des champs
d Ackerziest m
e ortiga f hedienda; hierba f del gato
n akkerandoorn m; akkerandoren m
l Stachys arvensis

997 field bindweed; bearbine; small bindweed; wild morning glory; bearbind; lesser bindweed; cornbirne; European glorybind
(sb; weed)

f liseron m des champs; vrillée f; petit liseron m; petite vrillée f; bédille f
d Ackerwinde f; Drehwurz f; Drehähre f; Feldwinde f; Kornwinde f
e campanilla f; corregüela f menor; correhuela f menor; corregüela f silvestre; correhuela f silvestre
n akkerwinde f; klokjeswinde f; slingerroos f
l Convolvulus arvensis

* **field camomile** → 645

* **field chamomile** → 645

* **field cress** → 1004

998 field cricket *(US)*
(sb; pest)
f grillon m
l Acheta assimilis

999 field emergence
(agr)
f levée f en terre; levée f en champ
d Feldaufgang m
e emergencia f en el campo
n veldopkomst f

1000 field forget-me-not; common forget-me-not
(sb; weed)
f myosotis m des champs
d Ackervergißmeinnicht n
e miosotis f de los campos; nomeolvides f común
n middelst vergeet-mij-nietje n
l Myosotis arvensis

* **field gromwell** → 649

* **field horsetail** → 576

* **field lady's-mantle** → 1730

* **field marigold** → 650

* **field melilot** → 2728

* **field mint** → 651

1001 field mouse-ear chickweed; field mouse-ear; starry cerastium
(sb; weed)
f céraiste m des champs
d Ackerhornkraut n
e cerasto m de los campos; milhojas m
n akkerhoornbloem f; akkermanskruid n
l Cerastium arvense

* **field-nettle betony** → **996**

1002 field pansy; wild pansy; heartsease; Johnny-jump-up
(sb; weed)
f pensée f sauvage; violette f des champs
d Stiefmütterchen n; gewöhnliches Stiefmütterchen n; dreifarbiges Veilchen n
e pensamiento m silvestre; trinitaria f
n driekleurig viooltje n
l Viola tricolor

1003 field pennycress; pennycress; fanweed; French weed; penny grass; stinkweed
(sb; weed)
f tabouret m des champs; monnoyère f
d Ackerhellerkraut n; Ackertäschelkraut n; Bauernkresse f; Bauernsenf m; Pfennigkraut n
e canaspique m; telaspios mpl de los campos; hierba f del ochavo; telaspio m
n witte krodde f; boerenkers f; gemene boerenkers f
l Thlaspi arvense

1004 field pepperwort; field peppergrass; field cress; cow cress; crowdweed; field pepperweed
(sb; weed)
f passerage m des champs; bourse f de Judas; passerage m champêtre
d Feldkresse f
e mostaza f silvestre
n veldkruidkers f
l Lepidium campestre

* **field poppy** → **652**

1005 field slug
(sb; pest)
f limace f grise
d Ackerschnecke f
e babosa f Deroceras reticulatum; babosa f Agriolimax reticulatus
l Deroceras reticulatum; Agriolimax reticulatus

* **field sorrel** → **2130**

* **field sowthistle** → **653**

* **field vole** → **613**

* **field woundwort** → **996**

1006 Fiji disease
(sc; virus disease)
f maladie f de Fidji
d Fidji-Krankheit f; Fidschi-Krankheit f
e enfermedad f de Fiji
n Fidji-ziekte f

1007 filling machine
f remplisseuse f
d Abfüllmaschine f
e máquina f de llenar; llenadora f
n vulmachine f

* **fillmass** *(US)* → **1554**

1008 film effect; climbing film effect *(in evaporator)*
f grimpage m
d Hochsteigen n; Klettern n
e ascensión f; efecto m de ascensión
n stijging f

1009 filter *(noun)*
f filtre m
d Filter nm
e filtro m
n filter nm

1010 filter *v*; **filtrate** *v*
f filtrer
d filtrieren; filtern
e filtrar
n filtreren; filteren

1011 filterability
f filtrabilité f
d Filtrierbarkeit f; Filtrationsfähigkeit f; Filtrierfähigkeit f
e filtrabilidad f
n filtreerbaarheid f

1012 filterable
f filtrable
d filtrierbar; filtrierfähig
e filtrable
n filtreerbaar

1013 filter aid
(A material which is either precoated on a filter membrane, or slurred in the liquid to be filtered, which, because of its porous nature, permits more solids to be filtered out before the filter cake becomes impermeable, or improves the clarity of the filtered liquid)
f adjuvant m de filtration; améliorant m
d Filterhilfsmittel n
e ayuda m de filtración; mejorador m de filtración; auxiliar m de filtración
n filtreerhulpmiddel n

1014 filter area; filtering area; filtration area
f surface f filtrante; surface f de filtration

d Filterfläche f
e superficie f de filtración; superficie f filtrante
n filtrerend oppervlak n; filteroppervlak n

1015 filter bag
f sac m filtrant
d Filterbeutel m
e bolsa f filtrante
n filterzak m

1016 filter cake; cake; press cake
(Solid material deposited on the filter septum during filtration)
f tourteau m; tourteau m de filtre; gâteau m; gâteau m de filtre
d Filterkuchen m; Filterschlamm m
e torta f; torta f de cachaza; torta f de filtro; cachaza f
n filterkoek m

1017 filter cake conveyor
f transporteur m de tourteaux
d Filterkuchentransporteur m
e transportador m de las tortas
n filterkoektransporteur m

1018 filter cake mixing vessel
f bac m de malaxage des tourteaux
d Filterkuchenmaischgefäß n
e depósito m para dilución de las tortas
n filterkoekmengbak m

1019 filter candle
f bougie f filtrante; bougie f de filtre
d Filterkerze f
e bujía f filtrante
n filterkaars f

1020 filter cartridge
f cartouche f de filtration; cartouche f filtrante
d Filterpatrone f
e cartucho m de filtro
n filterpatroon f

1021 filter cell
f cellule f de filtre
d Filterzelle f
e célula f de filtro
n filtercel f

1022 filter cloth; filtering cloth
f toile f filtrante; toile f à filtrer; tissu m filtrant; serviette f
d Filtertuch n
e tela f filtrante; tela f para filtrar; paño m

n filterdoek n; schuimdoek n; filtreerdoek n

1023 filter drum
f tambour m de filtre
d Filtertrommel f
e tambor m de filtro
n filtertrommel f

1024 filtered juice
f jus m filtré
d filtrierter Saft m
e guarapo m filtrado; jugo m filtrado
n gefiltreerd sap n

1025 filter frame
f cadre m filtrant; cadre m de filtre
d Filterrahmen m
e marco m filtrante
n filterraam n

1026 filter gravel
f gravier m pour filtrage; gravier m filtrant
d Filterkies m
e gravilla f filtrante
n filtergrind n; filtergrint n

* **filtering area** → 1014

* **filtering centrifugal** → 1040

* **filtering cloth** → 1022

1027 filtering fabrics
f tissu m filtrant
d Filtergewebe n
e tejido m filtrante
n filtergaas n

1028 filter medium; filter septum
(A layer of cloth, wire, paper, etc., providing actual initial filter surface)
f matériau m filtrant
d Filtermittel n; Filtermedium n; Filtermaterial n
e material m filtrante
n filtermedium n

1029 filter v off
f filtrer
d abfiltrieren; abfiltern
e filtrar
n affiltreren

1030 filter paper
f papier m filtre; papier m à filtrer; papier m filtrant
d Filterpapier n; Filtrierpapier n

e papel *m* filtrante
n filterpapier *n*; filtreerpapier *n*

1031 filter plate
  f plaque *f* filtrante
  d Filterplatte *f*
  e placa *f* filtrante
  n filterplaat *f*

1032 filter pocket
  f poche *f* filtrante
  d Filtertasche *f*
  e bolsillo *m* de filtro
  n filterzak *m*

1033 filter press
  f filtre-presse *m*
  d Filterpresse *f*
  e filtro-prensa *m*
  n filterpers *f*; schuimpers *f*; filtreerpers *f*

1034 filterpress cake
  f tourteau *m* de filtre-presse; gâteau *m* de filtre-presse
  d Filterpreßkuchen *m*
  e torta *f* de filtro-prensa; cachaza *f* de filtro-prensa
  n filterperskoek *m*

1035 filterpress with central feed (channel)
  f filtre-presse *m* à canal central d'alimentation; filtre-presse *m* à alimentation centrale
  d Filterpresse *f* mit zentraler Schlammsaft-zuführung
  e filtro-prensa *m* con alimentación central
  n filterpers *f* met centrale saptoevoer

1036 filterpress with side feed (channel)
  f filtre-presse *m* à canal latéral d'alimentation; filtre-presse *m* à alimentation latérale
  d Filterpresse *f* mit seitlicher Schlammsaft-zuführung
  e filtro-prensa *m* con alimentación lateral
  n filterpers *m* met zijdelingse saptoevoer

* **filter septum** → **1028**

1037 filter station
  f atelier *m* de °iltration
  d Filterstation *f*
  e departamento *m* de filtración
  n filterstation *n*

1038 filter thickener
  f filtre *m* épaississeur
  d Eindickfilter *n*; Eindickungsfilter *n*

e filtro *m* concentrador
n indikfilter *n*

1039 filter tower
  f tour *f* à filtres
  d Filterturm *m*
  e torre *m* de filtros
  n filtertoren *m*

1040 filter-type centrifugal; filtering centrifugal
  f centrifugeuse *f* de filtration
  d Filterzentrifuge *f*
  e centrífuga *f* de filtración
  n filtercentrifuge *f*

1041 filter water
  f eau *f* de filtres
  d Filterwasser *n*
  e agua *f* de filtros
  n filterwater *n*

1042 filtrate *(noun)*
  (Liquid after passing through a filter)
  f filtrat *m*
  d Filtrat *n*
  e filtrado *m*
  n filtraat *n*

* **filtrate** *v* → **1010**

1043 filtrate pump
  f pompe *f* à filtrat
  d Filtratpumpe *f*
  e bomba *f* de filtrado
  n filtraatpomp *f*

1044 filtrate separator; juice separator
  (Juice purification)
  f séparateur *m* de filtrat
  d Filtratabscheider *m*; Saftabscheider *m*
  e separador *m* de filtrado
  n filtraatafscheider *m*; sapafscheider *m*

1045 filtrate tank
  f bac *m* à filtrat
  d Filtratbehälter *m*
  e depósito *m* colector de los filtrados
  n filtraatbak *m*

1046 filtration
  (Removal of solid particles from liquid or particles from an air or gas stream by passing the liquid or gas stream through a permeable membrane)
  f filtration *f*
  d Filtration *f*
  e filtración *f*
  n filtratie *f*

* **filtration area** → 1014

1047 filtration pressure
f pression f de filtration
d Filtrationsdruck m
e presión f de filtración
n filtratiedruk m

1048 filtration rate
f vitesse f de filtration
d Filtrationsgeschwindigkeit f
e velocidad f de filtración
n filtratiesnelheid f

1049 final acidity
f acidité f finale
d Endacidität f
e acidez f final
n eindaciditeit f

1050 final alkalinity *(of juice)*
f alcalinité f finale
d Endalkalität f
e alcalinidad f final
n eindalkaliteit f

* **final bagasse** → 119

1051 final extraction
f extraction f finale
d Endextraktion f
e extracción f final
n eindextractie f

1052 final massecuite
f masse f cuite finale
d Endfüllmasse f
e masa f cocida final
n eind-masse cuite f; eindvulmassa f

1053 final molasses; molasses; blackstrap
(The syrup from the final crystallization
stage in the Recovery House from which
it is impossible or uneconomic to extract
any more sucrose by a crystallization
process)
f mélasse f finale; mélasse f
d Endmelasse f; Melasse f
e melaza f final; melaza f; miel f final;
miel f incristalizable
n eindmelasse f; melasse f

* **final product** → 908

1054 final strike
f cuite f finale
d Endsud m
e templa f final
n eindkooksel n

1055 final sulphitation
f sulfitation f finale
d Endsulfitation f
e sulfitación f final
n eindsulfitatie f

1056 final temperature
f température f finale
d Endtemperatur f
e temperatura f final
n eindtemperatuur f

* **finder** *(US)* → 986

* **finder arm** *(US)* → 987

* **fine bagasse** → 713

* **fine bagasse elevator** → 714

1057 fine bagasse juice
f jus m de folle bagasse
d Feinbagassesaft m
e jugo m de bagacillo; guarapo m de
bagacillo
n fijne-bagassesap n; fijne-ampassap n

* **fine bagasse screen** → 715

* **fine bagasse separator** → 716

1058 fine grain; fines *(US)*
f grain m fin
d Feinkorn n; Feinkristall m
e grano m fino
n fijn grein n

1059 fine pulp
f pulpe f fine; pulpe f folle
d Feinpülpe f
e pulpa f fina; pulpilla f
n fijne pulp f

1060 fine pulp press
f presse f à pulpes folles
d Feinpülpepresse f
e prensa f para la pulpa fina; prensa f
para la pulpilla
n fijne-pulppers f

1061 fine pulp separator
f tamiseur m de pulpes ʒnes; épulpeur m
d Feinpülpeabscheider m
e separador m de la pulpa fina; separador
m de la pulpilla
n fijne-pulpafscheider m

1062 fine pulp silo
f silo m à pulpes folles

d Feinpülpesilo *m*
e silo *m* para la pulpa fina; silo *m* para la
   pulpilla
n fijne-pulpsilo *m*

* **fines** *(US)* → **1058**

1063 finger knife
f couteau *m* à doigts
d Fingermesser *n*
e cuchillo *m* a dedos
n vingermes *n*

1064 finishing centrifugal *(in double purging)*
f finisseuse *f*
d Nachzentrifuge *f*; Affinationszentrifuge *f*
e centrífuga *f* de afinado
n affinagecentrifuge *f*

1065 fire tube boiler
f chaudière *f* à foyer intérieur
d Flammrohrkessel *m*
e caldera *f* tubular de hogar interior
n vlampijpketel *m*

* **fireweed** *(US)* → **1411**

1066 first carbona(ta)tion; first carb.
f première carbonatation *f*
d erste Carbonatation *f*; erste Saturation *f*
e primera carbonatación *f*
n eerste carbonatatie *f*

1067 first carbona(ta)tion juice
f jus *m* de première carbonatation
d erster Carbonatationssaft *m*
e jugo *m* de primera carbonatación;
   guarapo *m* de primera carbonatación
n eerste carbonatatiesap *n*

1068 first centrifugal *(in double purging)*
f ébaucheuse *f*; essoreuse *f* préliminaire
d Vorzentrifuge *f*; erste Zentrifuge *f*
e primera centrífuga *f*
n eerste centrifuge *f*

1069 first-expressed juice
   (The juice expressed by the first two
   rollers of a mill tandem. This is juice to
   which no water has been added. It was
   formerly defined either as the crusher
   juice alone or as the combined crusher
   and first-mill juice or, when there is no
   crusher, the juice from the first mill)
f jus *m* de première pression
d erster Preßsaft *m*
e guarapo *m* de primera expresión; jugo *m*
   de primera expresión
n eerstgeperst sap *n*; voorperssap *n*

* **first massecuite** → **58**

1070 first mill bagasse
f bagasse *f* du premier moulin
d Bagasse *f* der ersten Mühle
e bagazo *m* del primer molino
n bagasse *m* van de eerste molen

1071 first mills
f moulins *mpl* de tête
d erste Mühlen *fpl*
e primeros molinos *mpl*
n eerste molens *mpl*

* **first molasses** → **63**

1072 first product; first runnings product
f produit *m* initial; produit *m* de tête
d Erstprodukt *n*; Anfangsprodukt *n*;
   Vorlaufprodukt *n*
e primero producto *m*
n aanvangsprodukt *n*

* **first sugar** → **94**

1073 fivehook bassia *(US)*
   (sb; weed)
l Bassia hyssopifolia

1074 fixed-bed
   (Referring specifically to adsorbents or
   ion-exchangers, using a non-moving bed
   in a column of the active material)
f lit *m* fixe
d Festbett *n*
e lecho *m* fijo
n vast bed *n*

1075 fixed calandria; fixed type calandria
f faisceau *m* fixe
d feste Heizkammer *f*; festeingebauter
   Heizkörper *m*; Heizkammer *f* mit festen
   Rohrböden
e calandria *f* fija
n vaste stoomtrommel *f*; vaste
   stoomkamer *f*

1076 fixed carrier
   (An intermediate carrier)
f conducteur *m* fixe
d stationärer Transporteur *m*; feststehender Transporteur *m*
e conductor *m* fijo
n vaststaande transporteur *m*

1077 fixed self-aligning share
   (agr)
f soc *m* fixe autoguidé
d selbstführendes festangebrachtes

Rodeschar *n*
e reja *f* fija autolineadora
n zich zelf richtende vaste rooischaar *f*

1078 fixed share; lifting squeeze
(agr)
f soc *f* fixe
d festes Schar *n*
e reja *f* fija
n vaste schaar *f*

* **fixed type calandria** → 1075

* **FL and DH** → 1118

1079 flash chamber *(of decanter)*
f compartiment *m* de détente;
compartiment *m* d'apaisement
d Entspannungskammer *f*
e cámara *f* de expansión
n ontspanningskamer *f*

1080 flash tank; flash vessel; flash pot
f vase *m* de détente; ballon *m* de détente
d Kondensatentspanner *m*
e tanque *m* de expansión; vaso *m* de
expansión; tanque *m* de flash
n expansievat *n*

1081 flash vapo(u)r lines
f conduites *fpl* à vapeur de détente
d Entspannungsbrüdenleitungen *fpl*
e tuberías *fpl* de vahos de expansión
n pijpleidingen *fpl* voor ontspannings-
stoom

* **flash vessel** → 1080

**1082 flat bottomed centrifugal basket;
centrifugal basket with level bottom**
f panier *m* de centrifugeuse à fond plat;
panier *m* d'essoreuse à fond plat
d Flachboden-Zentrifugentrommel *f*
e canasto *m* de centrífuga con fondo plano
n centrifugetrommel *f* met vlakke bodem

1083 flat conveyor-belt
f courroie *f* transporteuse plate
d flaches Förderband *n*
e correa *f* de transporte plana
n vlakke transportband *m*

1084 flat fixed calandria
f faisceau *m* fixe plat
d festeingebaute Heizkammer *f* mit
geraden Rohrplattenböden
e calandria *f* plana y fija
n vaste stoomtrommel *f* met vlakke
pijpplaten

1085 flat fixed calandria pan
f appareil *m* à faisceau fixe plat
d Kochapparat *m* mit festeingebauter Heizkammer und geraden Böden;
Verdampfungskristallisator *m* mit festeingebauter Heizkammer und geraden
Böden
e tacho *m* al vacío de calandria plana y
fija
n kookpan *f* met vaste stoomtrommel en
vlakke pijpplaten

1086 flat millipede
(sb; pest)
l Brachydesmus superus

1087 flat millipede
(sb; pest)
f polydesme *m* étroit
l Polydesmus angustus

1088 flavo(u)red sugar
f sucre *m* aromatisé
d aromatisierter Zucker *m*
e azúcar *m* aromatizado
n gearomatiseerde suiker *m*

1089 flea beetle
(sb; pest)
f altise *f*
d Erdfloh *m*
e altisa *f*
n aardvlo *f*
l Chaetocnema sp.

* **flixweed** → 2461

* **flixweed tansy mustard** → 2461

**1090 floating calandria; suspended calandria;
basket calandria**
f faisceau *m* flottant; faisceau *m*
suspendu; faisceau *m* amovible
d eingehängte Heizkammer *f*; hängende
Heizkammer *f*
e calandria *f* flotante; calandria *f*
suspendida; calandria *f* de canasta
n hangende stoomkamer *f*; hangende
stoomtrommel *f*

**1091 floating calandria pan; suspended
calandria pan; basket calandria pan**
f appareil *m* à cuire à faisceau flottant;
appareil *m* à cuire à faisceau suspendu;
appareil *m* à cuire à faisceau amovible
d Kochapparat *m* mit eingehängter Heizkammer; Verdampfungskristallisator *m*
mit eingehängter Heizkammer
e tacho *m* al vacío de calandria flotante;

85

tacho *m* al vacío de calandria
suspendida; tacho *m* al vacío de
calandria de canasto
n kookpan *f* met hangende stoomkamer

1092 floating chain
(agr)
f chaîne *f* flottante
d schwingende Siebkette *f*
e cadena *f* oscilante
n zwevende zeefketting *f*

1093 floc
f floculat *m*
d Flocke *f*
e floculado *m*; precipitado *m*; flóculo *m*
n vloksel *n*

1094 flocculant
f floculant *m*
d Flockungsmittel *n*; Ausflockungsmittel *n*
e agente *m* de floculación
n uitvlokkingsmiddel *n*; uitvlokker *m*

1095 flocculate *v*
f floculer
d ausflocken; flocken
e separarse en flóculos
n vlokken; uitvlokken

* **flocculating chamber** *(of clarifier)* → **971**

* **flocculation** → **1097**

1096 flocculation aid
f adjuvant *m* de floculation
d Flockungshilfsmittel *n*
e auxiliar *m* para la floculación
n uitvlokkingshulpmiddel *n*

1097 floc formation; flocculation
f floculation *f*
d Ausflockung *f*; Floc-Bildung *f*; Flocken-bildung *f*
e floculación *f*
n uitvlokking *f*; vlokvorming *f*

* **flowerfly** *(US)* → **1276**

* **flower-of-an-hour** → **2614**

1098 flower thrips
(sb; pest)
l Frankliniella intonsa; Frankliniella tritici; Frankliniella brevistylus

1099 flume *v*
(To transport beets by means of a beet flume)

f emporter dans le caniveau
d schwemmen
e transportar en el canal de flotación
n in de zwemgoot vervoeren; in de spoelgoot vervoeren

1100 flume gate valve
f vanne *f* du caniveau
d Schwemmrinnenschieber *m*
e compuerta *f* del canal de flotación
n zwemgootschuif *f*; spoelgootschuif *f*

1101 flume water
(The water used to transport beets in a beet flume)
f eau *f* de transport
d Schwemmwasser *n*
e agua *f* de flotación; agua *f* para saetines; agua *f* de conducción
n transportwater *n*

1102 flushing the beets; water-jet discharge of beets
f abattage *m* hydraulique des betteraves; déchargement *m* hydraulique des betteraves
d Abspritzen *n* der Rüben; Entladen *n* der Rüben mittels Wasserstrahls
e descarga *f* de las remolachas por chorro de agua; descarga *f* de las remolachas por eyección de agua
n lossen *n* van de bieten met waterstraal; lossen *n* van de bieten met het waterkanon

* **foalfoot** → **559**

* **foam killer** → **72**

1103 fodder beet; fodder sugar beet; half mangel beet
f betterave *f* demi-sucrière
d Futterzuckerrübe *f*; Halbzuckerrübe *f*; Gehaltsrübe *f*
e remolacha *f* semiazucarera
n voederbiet *f*

* **fodder beet** → **1543**

* **follower** → **1645**

1104 following complete harvesters
(agr)
f machines *fpl* combinées en série
d hintereinander fahrende Vollernte-maschinen *fpl*
e cosechadoras *fpl* de remolacha de reata; máquinas *fpl* combinadas en serie

n achter elkaar rijdende verzamelrooiers
m pl

1105 footing; pied de cuite
(In pan boiling, a quantity of high-purity
massecuite or syrup which is used to
furnish crystal surface for crystallization
from lower purity liquors)
f pied-de-cuite *m*; pied *m* de cuite
d Kornfuß *m*; Kristallfuß *m*
e pie *m* de templa; pie *m* de cuite; pie *m*
de cocida; pie *m* de semilla
n pied-de-cuite *m*; voet *m*

1106 footing of second massecuite
f pied *m* de cuite de deuxième jet
d Mittelproduktkornfuß *m*; Mittelprodukt-
kristallfuß *m*
e pie *m* de semilla de segundo producto
n voet *m* van middenprodukt

1107 footing volume; seed volume
f volume *m* de pied de cuite
d Kornfußvolumen *n*; Kristallfuß-
volumen *n*
e volumen *m* de pie de templa; volumen *m*
del pie de cocida
n volume *n* van pied-de-cuite

1108 footing work
f travail *m* avec pied de cuite
d Kristallfußarbeit *f*; Kornfußarbeit *f*
e trabajo *m* con pie de semilla
n werken *n* met pied-de-cuite

1109 forced circulation
f circulation *f* forcée
d Zwangszirkulation *f*; Zwangsumlauf *m*
e circulación *f* forzada
n gedwongen circulatie *f*; geforceerde
circulatie *f*

1110 forced draught *(of boiler)*
f tirage *m* soufflé
d Saugzug *m*
e tirado *m* forzado
n geforceerde trek *m*

1111 forced feeding
f alimentation *f* forcée
d Zwangszuführung *f*
e alimentación *f* forzada
n gedwongen voeding *f*; gedwongen
toevoer *f*

1112 forced ventilation *(of beet silo)*
f ventilation *f* forcée
d verstärkte Belüftung *f*; Zwangs-
belüftung *f*

e ventilación *f* forzada
n gedwongen ventilatie *f*; geforceerde
ventilatie *f*

* **forelayer of knife-box** → 1407

1113 formaldehyde
f formaldéhyde *m*; formal *m*; aldéhyde *m*
formique
d Formaldehyd *m*
e formaldehído *m*; aldehído *m* fórmico
n formaldehyde *n*

1114 formalin
(Disinfectant)
f formol *m*; formaline *f*
d Formalin *n*; Formol *n*
e formol *m*; formalina *f*
n formaline *n*; formol *n*

* **forward market** → 1150

* **forward price** → 1151

**1115 four-product boiling scheme; four-
product boiling system; four-massecuite
system**
f schéma *m* de cristallisation en quatre
jets; schéma *m* de cuisson en quatre jets;
travail *m* en quatre jets; procédé *m* des
quatre masses cuites
d Vierproduktenkristallisationsschema *n*;
vierstufiges Kristallisationsschema *n*;
Vier-Produkte-Schema *n*; Vierprodukten-
schema *n*
e esquema *m* de cristalización de cuatro
productos; esquema *m* de cuatro
productos; sistema *m* de cuatro templas;
sistema *m* de cuatro cocciones; sistema
m de cristalización de cuatro productos;
esquema *m* de cuatro templas
n kristallisatieschema *n* met vier
produkten; vierproduktenkookschema *n*

* **fourth disease** *(Java)* → 480

**1116 foxtail barley; squirreltail; squirreltail
grass; squirreltail barley**
(sb; weed)
f orge *f* à crinière; orge *f* à épis à crinière
d Mähnengerste *f*
n kwispelgerst
l Hordeum jubatum

1117 fractional liming
f chaulage *m* fractionné
d fraktionierte Kalkung *f*; fraktionierte
Kalkzugabe *f*; fraktionierte Scheidung *f*
e alcalinización *f* fraccionada; encalado *m*

fraccionario; alcalización f fraccionada
n gefractioneerde kalking f;
gefractioneerde scheiding f

**1118 fractional liming with double heating;
fractional liming and double heating; FL
and DH**
f chaulage m fractionné et double
réchauffage
d fraktionierte Kalkung f mit zweimaliger
Erwärmung; fraktionierte Kalkzugabe f
mit zweimaliger Anwärmung
e encalado m fraccionario y doble
calentamiento; alcalinización f
fraccionada y doble calefacción;
alcalización f fraccionada y doble
calentamiento
n gefractioneerde kalking f met dubbele
verwarming

1119 fractional preliming
f préchaulage m fractionné
d unterbrochene Vorkalkung f; unterteilte
Vorkalkung f; unterbrochene Vorscheidung f; unterteilte Vorscheidung f
e preencalado m fraccionario;
prealcalización f fraccionada;
prealcalinización f fraccionada
n gefractioneerde voorkalking f;
gefractioneerde voorscheiding f

1120 frame *(of filter press)*
f cadre m
d Rahmen m
e marco m
n raam m

1121 frame filter
f filtre m à cadres
d Rahmenfilter n
e filtro m de marcos
n raamfilter n

1122 frame filter press
f filtre-presse m à cadres
d Rahmenfilterpresse f
e filtro-prensa m de marcos
n raamfilterpers f

1123 free circulation
(EEC)
f libre circulation f
d freier Verkehr m
e libre circulación f
n vrij verkeer n

1124 free-flow Krajewski
(A crusher)
f Krajewski m à écoulement libre;

défibreur m Krajewski à écoulement
libre
e desmenuzadora f Krajewski con
escurrimiento libre; molino m
desmenuzador Krajewski con
escurrimiento libre

1125 free sugar market; open sugar market
f marché m libre du sucre
d freier Zuckermarkt m
e mercado m libre del azúcar; mercado m
azucarero libre
n vrije suikermarkt f

1126 freetrade area
(eco)
f zone f de libre-échange
d Freihandelszone f
e zona f de libre intercambio
n vrijhandelszone f

* **freeze-damaged cane** → **1139**

* **French mercury** *(US)* → **70**

* **French spinach** *(US)* → **1909**

* **French weed** → **1003**

1127 fresh bagasse
f bagasse f fraîche
d frische Bagasse f
e bagazo m fresco
n verse bagasse m; verse ampas m

1128 fresh beet
f betterave f fraîche
d frische Rübe f
e remolacha f fresca
n verse biet f

1129 fresh cossettes
f cossettes fpl fraîches
d frische Schnitzel npl
e cosetas fpl frescas
n verse snijdsels npl

1130 fresh water
f eau f fraîche
d Frischwasser n
e agua f fresca
n vers water n

1131 fresh water heater
f réchauffeur m d'eau fraîche
d Frischwasser-Wärmer m
e calentador m de agua fresca
n vers-waterverwarmer m

1132 **fresh water pump**
  f pompe *f* à eau fraîche
  d Frischwasser-Pumpe *f*
  e bomba *f* de agua fresca
  n vers-waterpomp *f*

1133 **fresh water tank**
  f caisse *f* à eau fraîche
  d Frischwasser-Kasten *m*
  e depósito *m* de agua fresca
  n vers-waterkist *f*

1134 **froghoppers; cuckoo spit (insects);**
  spittlebugs; spittle insects; cuckoospittles
  (sc; pest)
  f cercopes *mpl*
  d Schaumzikaden *fpl*; Schaumzirpen *fpl*;
     Schildzirpen *fpl*; Stirnzirpen *fpl*
  e cercópidos *mpl*
  n schuimcicaden *fpl*; schuimbeestjes *npl*;
     schuimdiertjes *npl*
  l Cercopoidae

  * **front roll(er)** *(of mill)* → 980

1135 **frosted beet**
  f betterave *f* gelée (partiellement)
  d angefrorene Rübe *f*
  e remolacha *f* helada (parcialmente)
  n (gedeeltelijk) bevroren biet

1136 **froth fermentation**
  f fermentation *f* des écumes
  d Schaumgärung *f*
  e fermentación *f* de espuma
  n schuimgisting *f*

1137 **froth pressure**
  f pression *f* des écumes
  d Schaumdruck *m*
  e' presión *f* de espuma
  n schuimdruk *m*

1138 **frozen beet**
  f betterave *f* gelée (complètement)
  d erfrorene Rübe *f*
  e remolacha *f* helada (completamente)
  n geheel bevroren biet *f*

1139 **frozen cane; freeze-damaged cane**
  f canne *f* gelée
  d gefrorenes Zuckerrohr *n*
  e caña *f* helada
  n bevroren riet *n*

1140 **fructose; levulose; laevulose**
  f fructose *m*; lévulose *m*
  d Fructose *f*; Fruktose *f*; Fruchtzucker *m*;
     Lävulose *f*

  e fructosa *f*; levulosa *f*
  n fructose *f*; vruchtesuiker *m*; levulose *f*

  * **fugal** → 449

1141 **fugalability; centrifugalability**
  f turbinabilité *f*
  d Schleuderfähigkeit *f*
  e centrifugabilidad *f*
  n centrifugeerbaarheid *f*

  * **fugalling** → 454

  * **fugal operator** → 456

1142 **fulgoroids**
  (sc; pest)
  f fulgoroïdes *mpl*
  d Langkopfzirpen *fpl*; Langkopfzikaden
     fpl; Spitzköpfe *mpl*
  e fulgoroidos *mpl*
  n lantaarndragers *mpl*
  l Fulgoroidae

  * **fuller's thistle** → 2215

  * **fumitory** → 573

1143 **fungal disease; fungous disease; mycosis**
  f maladie *f* fongique; mycose *f*; maladie *f*
     cryptogamique
  d Pilzkrankheit *f*
  e enfermedad *f* fungosa; micosis *f*;
     enfermedad *f* criptogámica
  n schimmelziekte *f*; mycose *f*; door
     schimmels veroorzaakte ziekte *f*

1144 **funnel**
  f entonnoir *m*
  d Trichter *m*
  e embudo *m*
  n trechter *m*

1145 **furfural**
  (By-product)
  f furfurol *m*; furfural *m*
  d Furfural *n*; Fural *n*
  e furfurol *m*; furfural *m*
  n furfural *n*; furfurol *n*

1146 **fusarium**
  (sb; disease)
  f fusariose *f*
  d Fusarium-Welke *f*
  e fusariosis *f*
  n "Stalk Blight"

1147 **fusarium rot**
  (sb; disease)

  f fusariose *f*
  d Fusarium-Kopffäule *f*
  e fusariosis *f*
  n fusarium-rot *n*

**1148 fusarium sett; stem rot; stem rot disease;
     knife cut**
     (sc; fungous disease)
  f pourriture *f* fusarienne de la tige
  d Stengelfäule *f*
  e ataque *m* del fusarium; pudrición *f* del
     tallo
  n fusarium-stengelziekte *f*

1149 futures contract
     (eco)
  f contrat *m* à terme
  d Terminkontrakt *m*; Terminvertrag *m*
  e contrato *m* a plazo
  n termijncontract *n*

1150 futures market; forward market
     (eco)
  f marché *m* à terme; marché *m* à livrer
  d Terminmarkt *m*
  e mercado *m* a plazo; mercado *m* de
     futuros
  n termijnmarkt *f*

1151 futures price; forward price
     (eco)
  f prix *m* sur le marché à terme; prix *m* à
     terme
  d Preis *m* für Termingeschäfte; Kurs *m*
     für Termingeschäfte; Terminkurs *m*;
     Terminpreis *m*
  e precio *m* a plazo
  n termijnprijs *m*

G

1152 galactan
f galactane *m*
d Galaktan *n*
e galactana *f*
n galactaan *n*

1153 galactose
f galactose *m*
d Galaktose *f*
e galactosa *f*
n galactose *f*

1154 galinsoga; smallflower galinsoga; kew weed; gallant soldier; little-flower quickweed
(sb; weed)
f galinsoga *m* à petites fleurs; galinsoga *m*; galinsoge *f*
d kleinblütiges Franzosenkraut *n*; Franzosenkraut *n*; kleinblütiges Knopfkraut *n*; Choleradistel *f*; Goldköpfchen *n*
e galinsoga *f* de flor pequeña; soldado *m* galante
n klein knopkruid *n*; knopkruid *n*
l Galinsoga parviflora

1155 galvanometer
f galvanomètre *m*
d Galvanometer *n*
e galvanómetro *m*
n galvanometer *m*

1156 gamma moth; common silver Y moth; silver Y moth; common silver Y
(sb; pest)
f noctuelle *f* gamma; noctuelle *f* grise; vers *f* gris *(larve)*
d Gammaeule *f*; Feldstaudenflur-Silbereule *f*
e noctuido *m* gamma; noctuido *m* plateado
n gamma-uiltje *n*; gammavlinder *m*
l Plusia gamma; Phytometra gamma

* **gantry crane → 324**

* **garden beet → 226**

* **garden centipede *(US)* → 1159**

* **garden daisy → 731**

1157 garden dart moth; garden dart
(sb; pest)
f sombre *f*; noir-âtre *m*
d schwärzliche Ackereule *f*; violettschwarze Erdeule *f*; Geröllkräuterflur-

Erdeule *f*
l Euxoa nigricans

* **garden mercury → 70**

* **garden sorrel → 582**

1158 garden springtail
(sb; pest)
f collembole *m* des jardins
d Gurkenspringschwanz *m*; Gartenspringschwanz *m*; schädlicher Springschwanz *m*; Kugelspringer *m*
l Bourletiella hortensis; Bourletiella signata

1159 garden symphylan *(US)*; garden centipede *(US)*; garden symphilid *(US)*
(sb; pest)
f scutigérelle *f* immaculée; scutigérelle *f* des jardins
d Zwergskolopender *m*; Gewächshaus-Zwergfüßler *m*
l Scutigerella immaculata

1160 garden webworm *(US)*
(sb; pest)
l Loxostege similalis

1161 gas washer
(Apparatus used to remove entrained solids and tarry substances from carbon dioxide gas from a lime kiln)
f laveur *m* de gaz
d Gaswascher *m*
e lavador *m* de gas
n gaswasser *m*

1162 gauze *(of centrifugal)*
f toile *f*
d Siebgewebe *n*; Siebtuch *n*
e malla *f*
n gaas *n*

* **G.C.V. → 1201**

1163 genetic monogerm seed
(agr)
f graine *f* monogerme génétique; semence *f* monogerme génétique
d genetisch einkeimiges Saatgut *n*; genetisch monogermes Saatgut *n*; Monogermsaatgut *n*
e semilla *f* monogermen genética
n genetisch eenkiemig zaad *n*

* **German camomile → 2694**

* **German chamomile → 2694**

1164 **germination capacity; germinative capacity**
(agr)
f faculté f germinative; capacité f germinative
d Keimfähigkeit f; Keimvermögen n
e facultad f germinativa; capacidad f germinativa
n kiemkracht f; kiemvermogen n

1165 **germination energy; germination power**
(agr)
f énergie f germinative; pouvoir m germinatif
d Keimenergie f; Keimkraft f
e energía f germinativa; poder m germinativo
n kiemenergie f

* **germinative capacity** → 1164

* **giant coccids** → 1548

1166 **giant foxtail**
(sb; weed)
l Setaria faberii

* **giant ragweed** *(US)* → 1194

1167 **girth scab; girdle scab**
(sb; disease)
f gale f en ceinture
d Gürtelschorf m
e sarna f de la cintura; sarna f en cintura; roña f en cintura
n gordelschurft f

* **glasswort** → 2025

1168 **glassy cutworm** *(US)*
(sb; pest)
l Crymodes devastator

1169 **globe-podded whitetop**
(sb; weed)
l Cardaria pubescens

1170 **glucose; dextrose; grape-sugar**
f glucose m; dextrose m; cérélose m; glycose m; sucre m d'amidon; sucre m de chiffons; sucre m de raisins; sucre m de miel
d Glucose f; Traubenzucker m; Glukose f; Dextrose f
e glucosa f; dextrosa f
n glucose f; dextrose f; druivesuiker m; glycose f; bloedsuiker m

* **glucose quotient** → 1929

* **glucose ratio** → 1929

* **goatsbeard** → 1559

* **goat's beard** → 1559

* **goldcup** → 361

1171 **golden ragwort**
(sb; weed)
l Senecio aureus

* **goldlocks** → 647

1172 **Goller knife**
f couteau m Goller
d Gollermesser n; gewalztes Messer n
e cuchilla f Goller
n Gollermes n

1173 **Good-King-Henry; allgood; Good-King-Henry goosefoot**
(sb; weed)
f chénopode m Bon-Henri; ansérine f Bon-Henri; épinard m sauvage; Bon-Henri m
d Guter Heinrich m
e pie m de ánade; buen Enrique m; buen armuelle m
n Brave Hendrik m; Goede Hendrik m; algoede ganzevoet m
l Chenopodium bonus-henricus

* **goosefoot** → 970

* **goose grass** → 509

* **gosmore** → 441

* **go** *v* **to seed** → 310

* **G. Pur.** → 1192

* **grab** *(of cane crane)* → 1175

1174 **grabbing crane; grab crane**
f grue f preneuse; grue f à grappin
d Greiferkran m
e grúa f con cuchara; grúa f con mordazas
n grijperkraan f

1175 **grab loader; grab** *(of cane crane)*
f grappin m
d Greiferlader m
e araña f
n grijper m

1176 **graded seed**
(agr)

f graine f calibrée; semence f calibrée
d kalibriertes Saatgut n
e semilla f calibrada
n gecalibreerd zaad n

1177 grain *(noun)*
(The aggregate of the crystals in a massecuite)
f grain m
d Korn n
e grano m
n grein n

1178 grain v
(To introduce or initiate a crop of crystal nuclei in a graining charge)
f grainer
d Korn bilden
e granar; cristalizar
n greineren

1179 graining; grain formation
f grainage m; formation f de germes
d Kornmachen n; Kornholen n; Keim-bildung f; Kornbildung f
e graneación f; formación f de grano; cristalización f; granulación f; fabricación f del grano; formación f de nuevos cristales
n greinvorming f

1180 graining by introduction of powdered sugar
f grainage m par introduction d'amorce; grainage m par amorce
d Kornbildung f durch Pudereinzug; Impfung f mit Puderzucker
e graneación f con azúcar pulverizado; cristalización f con azúcar pulverizado; fabricación f del grano por inyección de cristales finos
n enten n met poedersuiker

1181 graining by seeding
f grainage m par ensemencement
d Kornbildung f durch Impfen; Kornbildung f durch Saatimpfen
e graneación f por semilla; cristalización f por semilla; fabricación f del grano por semillamiento; granulación f por vacuna
n greinvorming f door enting

1182 graining by shocking; graining by shock seeding
f grainage m par choc
d Kornbildung f durch Schockimpfen; Kornbildung f durch Schock
e graneación f por choque; cristalización f por choque; granulación f por choque;

semillamiento m por choque
n schokgreinen n

1183 graining by waiting method; graining by waiting
f grainage m par attente
d normale Kornbildung f
e graneación f por espera; cristalización f por espera; granulación f por espera
n greinvorming f zonder enting

1184 graining volume
f volume m de grainage
d Kornvolumen n
e volumen m del grano
n greinvolume n

1185 grain size
f grosseur f du grain
d Körnung f; Korngröße f
e tamaño m del grano
n greingrootte f

1186 granulate cutworm
(sb; pest)
l Feltia subterranea

1187 granulated sugar
(Crystalline sugar of market quality)
f sucre m granulé
d "Granulated Sugar"
e azúcar m granulado
n "granulated sugar"

1188 granulator
f granulateur m; sécheur m à tambour
d Granulator m; Trommeltrockner m
e granulador m; tambor m secadero de azúcar
n trommeldroger m

* **grape-sugar** → 1170

* **grass catcher** → 2532

1189 grasshopper conveyor; shaker conveyor; grasshopper
f transporteur m à secousses; transporteur m à secousses Kreiss; Kreiss m
d Schüttelrinne f; Schwingförderer m
e conductor m saltamontes; conductor m tembladero; conductor m de sacudidas; conductor m de chapulín
n schudgoot f

1190 grassy shoot
(sc; virus disease)
f "grassy shoot"
d Grassy shoot-Krankheit f

e mata f zacatosa
n grassy shoot-ziekte f

1191 gravimetric solids
f matières fpl sèches gravimétriques
d gravimetrischer Trockengehalt m;
gravimetrisch bestimmter Trockengehalt m; Trockengehalt m
(gravimetrisch)
e sólidos mpl gravimétricos
n gravimetrisch bepaalde droge stof f

1192 gravity purity; G. Pur.
(The percentage of (true) sucrose (e.g., by
Clerget), percent Brix)
f pureté f Clerget
e pureza f por gravedad; pureza f en peso
n gravimetrische reinheid f

* **greasy cutworm** → 278

* **greater nettle** → 2283

**1193 greater plantain; broad-leaved plantain;
common plantain; ripple grass; great
plantain; ripple-seed plantain**
(sb; weed)
f plantain m majeur; grand plantain m
d großer Wegerich m; Breitwegerich m
e llantén m major; llantén m común;
llantén m grande
n grote weegbree f
l Plantago major

1194 great ragweed (US); giant ragweed (US)
(sb; weed)
d dreilappiges Traubenkraut n
l Ambrosia trifida

* **green amaranth** → 2191

* **greenfly** → 1735

**1195 green foxtail; green pigeongrass (US);
pigeongrass; green panic grass;
bristlegrass (US); green bristlegrass**
(sb; weed)
f sétaire f verte; panic m vert
d grüne Borstenhirse f; grüner Fennich m
e almorejo m
n groene naaldaar f
l Setaria viridis

**1196 green leafhopper; Southern garden
leafhopper (US)**
(sb; pest)
f cicadelle f de la pomme de terre
d hellgrüne Zwergikade f; grasgrüne

Zikade f
l Empoasca solani; Empoasca flavescens

* **green molasses** → 1198

* **green panic grass** → 1195

* **green-peach aphid (US)** → 1735

* **green pea louse** → 1734

* **green pigeongrass (US)** → 1195

1197 green rate
(EEC)
f taux m vert
d grüner Kurs m
e tasa f verde; cambio m verde; tipo m
verde; paridad f verde
n groene koers m

**1198 green syrup; raw washings; green
molasses; initial molasses; poor running
(US); green run off; poor molasses;
heavy molasses**
(The filtrate leaving a centrifugal before
initiation of the wash, consisting of
mother liquor and whatever fine crystals
pass the screen)
f égout m pauvre; sirop m vert
d Grünablauf m; grüner Ablauf m
e miel f pobre; miel f inicial; miel f verde;
derrame m verde; miel f pesada
n groene stroop f; groene afloop m

* **grinding season** → 1596

1199 grip of the rollers (of a crusher, mill)
f prise f des cylindres
e toma f de los cilindros
n knijpen n van de cilinders

* **grizzly** → 152

* **grooved cossettes** → 1981

1200 grooves of a roller (of a roller mill)
f rainures fpl d'un cylindre
d Riffelung f eines Rollers; Riffeln fpl
eines Rollers; Rillung f eines Rollers
e ranurado m de un cilindro; ranuras fpl
de un cilindro
n groeven fpl van een cilinder

**1201 gross calorific value; G.C.V.; higher
calorific value (of bagasse)**
f pouvoir m calorifique supérieur;
P.C.S. m
d spezifischer Brennwert m; oberer Heiz-

wert *m*
e valor *m* calorífico superior
n bovenste verbrandingswaarde *f*;
   calorische bovenwaarde *f*

* **ground almond** → **1682**

1202 ground flea
   (sb; pest)
f collembole *m*
d Springschwanz *m*; Blindspringer *m*
n springstaart *m*
l Onychiurus sp.

* **ground lime** → **1862**

1203 groundsel; common groundsel
   (sb; weed)
f séneçon *m* vulgaire; herbe *f* aux
   charpentiers
d gemeines Kreuzkraut *n*; gemeines Greis-
   kraut *n*; Goldkraut *n*; Grindkraut *n*
e hierba *f* cana; hierba *f* de las
   quemaduras; senecio *m*; pan *m* de
   pájaros
n klein kruiskruid *n*
l Senecio vulgaris

1204 ground sugar
f sucre *m* semoule
d Sandzucker *m*
n castorsuiker *m*

1205 grow *v (grain)*
f grossir
d wachsen
e crecer
n groeien; aangroeien

1206 guaranteed price
   (EEC)
f prix *m* garanti
d Garantiepreis *m*; garantierter Preis *m*
e precio *m* garantizado; precio *m* de
   garantía
n gegarandeerde prijs *m*

1207 guaranteed quantity
   (EEC)
f quantité *f* garantie
d Garantiemenge *f*
e cantidad *f* garantizada
n gegarandeerde hoeveelheid *f*

1208 guide finger
   (agr)
f guide *m*
d Leitfinger *m*

e dedo *m* guiador; guía *f*
n leivinger *m*

1209 guide price
   (EEC)
f prix *m* d'orientation
d Orientierungspreis *m*
e precio *m* de orientación
n oriëntatieprijs *m*

* **Guinea grass** → **1366**

1210 gum
f gomme *f*
d Gummi *n*
e goma *f*
n gom *mn*

**1211 gumming disease; gumming; gummosis;
Cobb's disease of sugar cane**
   (sc; bacterial disease)
f maladie *f* de la gomme; gommose *f*
d Gummikrankheit *f*; Gummosis *f*;
   bakterieller Gummifluß *m*; Schleim-
   krankheit *f*; Cobb'sche Krankheit *f*
e gomosis *f*; enfermedad *f* de la gomosis
n gomziekte *f*

1212 gur
   (Type of sugar)
f gur *m*
d Gur *m*
e gur *m*
n gur *m*

* **gyrating sieve** → **2010**

H

* **hairy crabgrass** → 1426

* **hairy finger-grass** → 1426

1213 **hairy gall nematode** *(US)*
(sb; pest)
l Nacobbus batatiformis

* **hairy gold-of-pleasure** → 2188

* **halberd-leaved orache** → 1223

1214 **half lean beet**
(EEC)
f betterave *f* demi-maigre
d halbmagere Rübe *f*
e remolacha *f* semimagra
n halfmagere biet *f*

* **half mangel beet** → 1103

1215 **hammer; beater** *(of shredder)*
f marteau *m*
d Hammer *m*
e martillo *m*
n hamer *m*

* **hammer mill of Searby type** → 2077

1216 **hammer-type shredder**
f shredder *m* à marteaux
d Hammershredder *m*
e desfibradora *f* de martillos
n hamershredder *m*

1217 **hand singling**
(agr)
f démariage *m* à la main
d Handvereinzelung *f*; Handausdünnung *f*
e aclareo *m* a mano
n opeenzetten *n* met de hand

* **hard-black beetles** → 1975

1218 **hardboard**
(By-product)
f panneau *m* dur
d Hartfaserplatte *f*
e tablero *m* duro
n hardboard *m*

1219 **hard crystal**
f cristal *m* dur
d harter Kristall *m*
e cristal *m* duro
n hard kristal *n*

1220 **hard grain**
f grain *m* dur
d hartes Korn *n*
e grano *m* duro
n hard grein *n*

1221 **hare's-ear mustard; treacle hare's-ear**
(sb; weed)
f roquette *f* d'Orient
d weißer Ackerkohl *m*; weißer Schöterich *m*; morgenländischer Ackerkohl *m*
e collejón *m*
n witte steenraket *f*
l Conringia orientalis

* **hare's lettuce** → 583

* **hariff** → 509

1222 **harmful nitrogen; noxious nitrogen; noxious N**
f azote *m* nuisible
d schädlicher Stickstoff *m*
e nitrógeno *m* perjudicial; nitrógeno *m* nocivo
n schadelijke stikstof *f*

* **harvest fly** → 486

1223 **hastate orache; halberd-leaved orache; spear-leaved orache; spear-leaved fat hen saltbush**
(sb; weed)
f arroche *f* hastée
d spießblättrige Melde *f*; Spießmelde *f*
e armuelle *m* astado
n spiesmelde *f*; spiesbladige melde *f*
l Atriplex hastata

1224 **Hawaii ratio**
(Brix of absolute juice/Brix of first-expressed juice)
f coefficient *m* Hawaii
d Hawaii-Verhältnis *n*; Hawaii-Quotient *m*
e relación *f* Hawaii
n Hawaii-coëfficiënt *m*; Hawaii-verhouding *f*; Hawaii-quotiënt *n*

* **health speedwell** → 584

* **heart rot** → 315

* **heartsease** → 1002

* **heat balance** → 2476

1225 **heater**
f réchauffeur *m*

d Wärmer *m*
e calentador *m*
n verwarmer *m*

1226 **heat exchanger**
f échangeur *m* thermique; échangeur *m*
de température; échangeur *m* de chaleur
d Wärmetauscher *m*; Wärme-
austauscher *m*; Wärmeübertrager *m*
e intercambiador *m* de calor; cambiador *m*
de calor
n warmtewisselaar *m*

1227 **heat-exchange surface; heat-exchanging
surface**
f surface *f* d'échange
d Wärmeaustauschfläche *f*; Wärme-
übertragungsfläche *f*
e superficie *f* de intercambio
n warmteuitwisselingsoppervlak *n*

1228 **heating calandria; heating element**
f corps *m* de chauffe; élément *m*
chauffant
d Heizkörper *m*
e elemento *m* de calefacción
n verwarmingslichaam *n*;
verwarmingselement *n*

1229 **heating chamber** *(of evaporator)*
f chambre *f* de chauffe
d Heizkammer *f*
e cámara *f* de calefacción
n verwarmingskamer *f*;
verwarmingsruimte *f*

1230 **heating coil**
f serpentin *m* de chauffage
d Heizschlange *f*
e serpentín *m* de calefacción
n verwarmingsspiraal *f*;
verwarmingsslang *f*;
verwarmingsserpentijn *n*

* **heating element** → 1228

1231 **heating medium**
f fluide *m* de chauffage
d Heizmittel *n*
e fluido *m* de calefacción
n verwarmingsmiddel *n*

1232 **heating mixer**
f malaxeur *m* de réchauffage
d Anwärmmaische *f*
e cristalizador-calentador *m*
n verwarmingskristallisoir *m*

1233 **heating stage** *(of juice heater)*
f étage *m* de réchauffage
d Heizstufe *f*
e etapa *f* de calentamiento
n verwarmingstrap *m*

1234 **heating steam**
f vapeur *f* de chauffage
d Heizdampf *m*
e vapor *m* de calentamiento
n verwarmingsstoom *m*

1235 **heating surface**
f surface *f* de chauffe; surface *f*
chauffante
d Heizfläche *f*
e superficie *f* de calentamiento; superficie
f de calefacción
n verwarmend oppervlak *n*

1236 **heating tube**
f tube *m* de chauffe
d Heizrohr *n*; Siederohr *n*
e tubo *m* de calefacción
n verwarmingsbuis *f*

1237 **heat losses**
f pertes *fpl* de chaleur
d Wärmeverluste *mpl*
e pérdidas *fpl* de calor
n warmteverliezen *npl*

1238 **heat of crystallization**
f chaleur *f* de cristallisation
d Kristallisationswärme *f*
e calor *m* de cristalización
n kristallisatiewarmte *f*

1239 **heavy boiling**
f cristallisation *f* difficile; cuisson *f*
difficile
d Schwerkochen *n*
e cocción *f* difícil
n traag koken *n*

* **heavy liming** → 2624

* **heavy molasses** → 1198

1240 **hedge bindweed; wild morning glory;
hedge glorybind**
(sb; weed)
f liseron *m* des haies; grand liseron *m*
d Zaunwinde *f*; gemeine Zaunwinde *f*;
deutsche Purgierwinde *f*; deutsche
Skammonie *f*; Uferzaunwinde *f*
e corregüela *f* mayor; correhuela *f* mayor
n haagwinde *f*
l Convolvulus sepium

* **hedge mustard** → 2461

* **heliotrope** → 920

* **hemlock stork's bill** → 586

* **hemp dogbane** → 1305

* **hemp nettle** → 575

* **henbane fly** → 1545

1241 **henbit; bee nettle; henbit dead nettle**
(sb; weed)
f lamier *m* amplexicaule; lamier *m* à
feuilles embrassantes
d stengelumfassende Taubnessel *f*; Ackertaubnessel *f*
e ortiga *f* muerta
n hoenderbeet *m*
l Lamium amplexicaule

* **herb mercury** → 70

* **hercules beetles** → 1975

1242 **hexose**
f hexose *m*
d Hexose *f*
e hexosa *f*
n hexose *f*

* **higher calorific value** *(of bagasse)*
→ 1201

* **high-grade massecuite** → 58

1243 **high-grade sugar**
f sucre *m* de premier jet
d Erstproduktzucker *m*
e azúcar *m* de alta calidad
n ruwe suiker *m* eerste produkt

* **high gravity factor centrifugal** → 1247

* **high mallow** → 578

1244 **high purity juice**
f jus *m* de pureté élevée
d Saft *m* hoher Reinheit
e jugo *m* de alta pureza; guarapo *m* de
alta pureza
n sap *n* van hoge reinheid

1245 **high purity magma**
f magma *m* de pureté élevée
d Magma *n* hoher Reinheit
e magma *m* de alta pureza
n magma *n* van hoge reinheid

1246 **high purity massecuite**
f masse *f* cuite de haute pureté
d Füllmasse *f* hoher Reinheit
e masa *f* cocida de alta pureza
n vulmassa *f* van hoge reinheid; masse
cuite *f* van hoge reinheid

1247 **high speed centrifugal; high gravity
factor centrifugal**
f centrifugeuse *f* rapide; centrifuge *f*
rapide; essoreuse *f* rapide
d hochtourige Zentrifuge *f*; schnellaufende
Zentrifuge *f*
e centrífuga *f* de alta velocidad
n centrifuge *f* met hoog toerental;
sneldraaiende centrifuge *f*

* **high syrup** → 2638

1248 **high test molasses; HTM; invert syrup**
(Heavy partially inverted cane syrup
having a Brix about 85)
f égout *m* riche en sucre inverti
d High-Test-Melasse *f*; an Invertzucker
reicher Ablauf *m*
e melaza *f* rica invertida; miel *f* rica;
melaza *f* de alta capacidad de prueba;
melaza *f* de alta calidad; jarabe *m*
invertido
n high-test-melasse *f*

1249 **high vacuum**
f grand vide *m*; vide *m* profond
d Hochvakuum *n*
e alto vacío *m*
n hard vacuüm *n*

1250 **high-velocity evaporator**
f évaporateur *m* à courant rapide
d Schnellstromverdampfer *m*
e evaporador *m* de flujo rápido
n snelstroomverdamper *m*

1251 **high-velocity preheater**
f réchauffeur *m* à courant rapide
d Schnellstromwärmer *m*
e calentador *m* continuo rápido
n snelstroomverwarmer *m*

* **high wash syrup** → 2638

1252 **hoary pepperwort; hoary cress; hoary
peppergrass; whitlow peppergrass;
pepperweed whitetop**
(sb; weed)
f passerage *m* drave; passerage *m* lépidier
d Pfeilkresse *f*; gemeine Pfeilkresse *f*
e mastuerzo *m* oriental

n pijlkruidkers *f*
l Cardaria draba; Lepidium draba

1253 hoary plantain; lamb's tongue; sweet plantain
(sb; weed)
f plantain *m* moyen
d mittlerer Wegerich *m*; Weiden-
wegerich *m*
e llantén *m* blanquecino; llantén *m*
mediano
n ruige weegbree *f*
l Plantago media

1254 hoe singling
(agr)
f démariage *m* à la rasette
d Verkrehlen *n*
e aclareo *m* con escardillo
n opeenzetten *n* met de hark

* **honey clover** → **2679**

* **hop clover** → **281**

1255 hop flea beetle *(US)*
(sb; pest)
l Psylliodes punctalata

* **hop flea (beetle)** → **1544**

1256 hopper
f trémie *f*
d Trichter *m*
e tolva *f*
n trechter *m*

* **horizontal beet slicer** → **237**

* **horizontal beet slicing machine** → **237**

1257 horizontal drier
f sécheur *m* horizontal
d Horizontaltrockner *m*
e secador *m* horizontal
n horizontale droger *m*

1258 horizontal evaporator
f évaporateur *m* horizontal
d liegender Verdampfapparat *m*
e evaporador *m* horizontal
n horizontaal verdamplichaam *n*;
kofferapparaat *n*

1259 horizontally fed shredder
f shredder *m* à alimentation horizontale
d Shredder *m* mit horizontaler Zuckerrohr-
einspeisung
e desfibradora *f* con alimentación

horizontal de la caña
n shredder *m* met horizontale riettoevoer

1260 horse nettle *(US)*; **ball nettle** *(US)*; **bull nettle** *(US)*
(sb; weed)
l Solanum carolinense

1261 horse purslane *(US)*
(sb; weed)
l Trianthema portulacastrum

* **horsy daisy** → **2059**

1262 hot digestion
f digestion *f* à chaud
d heiße Digestion *f*
e digestión *f* en caliente
n warme digestie *f*

1263 hot filtrate
(Liquid effluent from the filters, if any,
that separate the hot saccharate per se)
f filtrat *m* chaud
d heißes Filtrat *n*
e filtrado *m* caliente
n warm filtraat *n*

1264 hot imbibition
f imbibition *f* à chaud
d warme Imbibition *f*
e imbibición *f* caliente
n warme imbibitie *f*

1265 hot liming
f chaulage *m* à chaud
d heiße Kalkung *f*
e encalado *m* en caliente; alcalización *f* en
caliente; alcalinización *f* en caliente
n warme kalking *f*

1266 hot maceration *(of bagasse)*
f imbibition *f* à haute température
d heiße Mazeration *f*
e maceración *f* caliente
n warme maceratie *f*

1267 hot preliming
f préchaulage *m* à chaud
d heiße Vorkalkung *f*; heiße
Vorscheidung *f*
e preencalado *m* caliente; prealcalización *f*
caliente; prealcalinización *f* caliente
n warme voorkalking *f*; warme
voorscheiding *f*

1268 hot raw juice
f jus *m* brut chaud
d warmer Rohsaft *m*

e jugo *m* crudo caliente; guarapo *m* crudo
  caliente
n warm ruwsap *m*

1269 hot saccharate
  f saccharate *m* chaud
  d heißes Saccharat *n*
  e sacarato *m* caliente
  n warm saccharaat *n*

1270 hot sulphitation
  f sulfitation *f* à chaud
  d heiße Sulfitation *f*
  e sulfitación *f* caliente; sulfitación *f* en
    caliente
  n warme sulfitatie *f*

1271 hot water pump
  f pompe *f* à eau chaude
  d Warmwasserpumpe *f*
  e bomba *f* de agua caliente
  n warmwaterpomp *f*

  * **hot well → 2076**

1272 housing; mill cheek
  f chapelle *f*
  e virgen *f*; castillejo *m*

**1273 housing with inclined headstocks;
       housing with inclined mounting**
  f chapelle *f* à chapeaux inclinés
  e virgen *f* de cabezotes inclinados

**1274 housing with king-bolts and horizontal
       bolts**
  f chapelle *f* à chandeliers et boulons
    horizontaux

1275 housing with vertical headstock
  f chapelle *f* à axe de chapeau vertical
  e virgen *f* de cabezotes verticales; virgen *f*
    con el eje de los cabezotes vertical

1276 hover fly; flowerfly *(US)*
  (sb)
  f syrphe *m*
  d Schwebfliege *f*; Schwirrfliege *f*; Blattlaus-
    fliege *f*
  e sirfido *m*; sirfida *f*; mosca *f* de las flores
  n zweefvlieg *f*
  l Syrphus sp.

  * **HTM → 1248**

1277 hulled seed
  (agr)
  f semence *f* nue; graine *f* nue
  d unpilliertes Saatgut *n*

e semilla *f* sin cáscara
n naakt zaad *n*

1278 humidity content; moisture content
  f teneur *f* en humidité; degré *m*
    d'humidité
  d Feuchtigkeitsgehalt *m*
  e contenido *m* de humedad
  n vochtgehalte *n*

1279 Hungarian beet moth
  (sb; pest)
  f noctuelle *f* hongroise
  d ungarische Rübenmotte *f*
  l Scrobipalpa ocellata

  * **hunger grass → 279**

  * **hurt-sickle → 648**

  * **hydrated lime → 2169**

1280 hydraulic cap *(of cane mill)*
  f chapeau *m* hydraulique
  d hydraulischer Deckel *m*
  e cabezote *m* hidráulico
  n hydraulisch deksel *n*

1281 hydraulic head
  f tête *f* hydraulique
  d Hydraulikkolben *m*
  e cabezal *m* hidráulico
  n hydraulische zuiger *m*

**1282 hydraulic plunger; hydraulic piston;
       cheese** *(slang) (of cane mill)*
  f piston *m* hydraulique; fromage *m*
  d Hydraulikkolben *m*
  e pistón *m* hidráulico
  n hydraulische zuiger *m*

1283 hydraulic pressure
  f pression *f* hydraulique
  d hydraulischer Druck *m*
  e presión *f* hidráulica
  n hydraulische druk *m*

1284 hydrochloric acid
  f acide *m* chlorhydrique
  d Salzsäure *f*; Chlorwasserstoffsäure *f*
  e ácido *m* clorhídrico
  n zoutzuur *n*; chloorwaterstofzuur *n*

1285 hydrometer; areometer
  (An instrument by which the specific
  gravity or density of a liquid may be
  determined through the depth of
  immersion of its stem, when it floats in
  the liquid)

f densimètre *m*; hydromètre *m*;
aréomètre *m*
d Hydrometer *n*; Aräometer *n*
e hidrómetro *m*; areómetro *m*
n areometer *m*

1286 hydrostatic pressure
f pression *f* hydrostatique
d hydrostatischer Druck *m*
e presión *f* hidrostática
n hydrostatische druk *m*

1287 hygrometer
f hygromètre *m*
d Hygrometer *n*
e higrómetro *m*
n hygrometer *m*

* **hygroscopic water** → **2580**

I

1288 icing sugar
f sucre *m* glace
d "Icing Sugar" *(sehr fein gemahlener Zucker)*
e azúcar *m* "icing"
n "icing sugar" *(zeer fijne poedersuiker)*

* ICUMSA → 1336

1289 iliau
(sc; fungous disease)
f iliau *m*
d Iliaukrankheit *f*
e enfermedad *f* de iliau
n iliauziekte *f*

1290 imbibition; saturation
(Adding of water or juice to the partially exhausted cane)
f imbibition *f*
d Imbibition *f*
e imbibición *f*
n imbibitie *f*

1291 imbibition juice
f jus *m* d'imbibition; jus *m* diluant
d Imbibitionssaft *m*
e jugo *m* de imbibición; guarapo *m* de imbibición
n imbibitiesap *n*

1292 imbibition trough
f cuvette *f* d'imbibition
d Imbibitionstrog *m*
e tolva *f* de imbibición
n imbibitietrog *m*

1293 imbibition water; saturation water
f eau *f* d'imbibition
d Imbibitionswasser *n*
e agua *f* de imbibición
n imbibitiewater *n*

1294 imbibition water spray
f rampe *f* d'imbibition
d Imbibitionswasserzerstäuber *m*
e pulverizador *m* de agua de imbibición
n imbibitiewaterverstuiver *m*

1295 imbricated snout beetle *(US)*
(sb; pest)
l Epicaerus imbricatus

1296 import certificate
(EEC)
f certificat *m* d'importation

d Einfuhrlizenz *f*
e certificado *m* de importación
n invoercertificaat *n*

1297 import subsidy
(EEC)
f subvention *f* à l'importation
d Einfuhrsubvention *f*
e subvención *f* a la importación
n invoersubsidie *f*

1298 in bulk
f en vrac
d lose
e a granel
n in bulk

1299 inclined-plate calandria
f faisceau *m* à plaques inclinées
d Heizkammer *f* mit schrägen Rohrböden
e calandria *f* de placas inclinadas
n stoomtrommel *f* met hellende pijpplaten

1300 inclined-plate calandria pan
f appareil *m* à cuire à faisceau à plaques inclinées
d Kochapparat *m* mit Heizkammer mit schrägen Rohrböden; Verdampfungs-kristallisator *m* mit Heizkammer mit schrägen Rohrböden
e tacho *m* al vacío con calandria de placas inclinadas
n kookpan *f* met stoomtrommel met hellende pijpplaten

1301 incondensable gases; incondensables
f gaz *mpl* incondensables; incondensables *mpl*
d nichtkondensierbare Gase *npl*; unkondensierbare Gase *npl*
e gases *mpl* incondensables; incondensables *mpl*
n oncondenseerbare gassen *npl*

1302 incondensable gas pipe
f tuyau *m* des incondensables
d Rohr *n* für nichtkondensierbare Gase
e tubo *m* de los incondensables
n buis *f* voor oncondenseerbare gassen

* incondensables → 1301

1303 incrustations; scalings; scales
f incrustations *fpl*
d Inkrustationen *fpl*; Belag *m*; Ablagerungen *fpl*
e incrustaciones *fpl*
n incrustaties *fpl*; afzetsels *npl*

1304 indanthrene blue
f bleu *m* d'indanthrène
d Indanthrenblau *n*
e azul *m* de indantreno
n indantreenblauw *n*

1305 Indian hemp *(US)*; **hemp dogbane**
(sb; weed)
f chanvre *m* du Canada; apocyn *m*
chanvrin
d hanfartige Hundswolle *f*; Hanfhundsgift *n*
n Amerikaanse hennep *m*
l Apocynum cannabinum

* **Indian mallow** *(US)* → **2612**

1306 indicator
(pH control)
f indicateur *m*
d Indikator *m*
e indicador *m*
n indicator *m*

1307 indicator paper
(pH control)
f papier *m* indicateur
d Indikatorpapier *n*
e papel *m* indicador
n indicatorpapier *n*

1308 individual condenser
f condenseur *m* individuel
d Einzelkondensator *m*
e condensador *m* independiente
n individuele condensor *m*

1309 individual vacuum
f vide *m* individuel
d Einzelvakuum *n*
e vacío *m* individual
n individueel vacuüm *n*

1310 industrial beet
f betterave *f* industrielle
d Fabrikrübe *f*
e remolacha *f* industrial
n fabrieksbiet *f*

1311 inflorescence binding
(sc; fungous disease of inflorescence)
d Blütenverknäulung *f*
e adherencia *f* de la inflorescencia

* **infusorial earth** → **797**

1312 ingots of sugar
f lingots *mpl* de sucre
d Zuckerstangen *fpl*; Zuckerstreifen *mpl*

e lingotes *mpl* de azúcar
n suikerstaven *fpl*

1313 inhibitor
f inhibiteur *m*
d Hemmstoff *m*; Inhibitor *m*
e inhibidor *m*
n inhibitor *m*

* **initial molasses** → **1198**

1314 initial temperature
f température *f* initiale
d Anfangstemperatur *f*
e temperatura *f* inicial
n begintemperatuur *f*;
aanvangstemperatuur *f*

1315 injection water
f eau *f* d'injection
d Einspritzwasser *n*; Injektionswasser *n*
e agua *f* de inyección
n injectiewater *n*

1316 injection water pump
f pompe *f* à eau froide
d Einspritzwasserpumpe *f*
e bomba *f* de inyección de agua fría
n injectiewaterpomp *f*

1317 inorganic ash
f cendres *fpl* minérales
d anorganische Asche *f*
e ceniza *f* inorgánica
n anorganische as *f*

1318 inorganic nonsugar
f non-sucre *m* minéral
d anorganischer Nichtzucker *m*
e no-azúcar *m* inorgánico
n anorganische nietsuiker *m*

1319 inorganic salt
f sel *m* minéral
d Mineralsalz *n*
e sal *f* mineral
n anorganisch zout *n*

**1320 insoluble lime salts; non-soluble lime
salts**
f sels *mpl* de chaux insolubles
d unlösliche Kalksalze *npl*
e sales *fpl* de cal insolubles
n onoplosbare kalkzouten *npl*

1321 insufficient liming
f chaulage *m* insuffisant
d ungenügende Kalkung *f*
e encalado *m* insuficiente; alcalización *f*

insuficiente; alcalinización f insuficiente
n onvoldoende kalking f

1322 insulating board
(By-product of bagasse)
f panneau m isolant
d Dämmplatte f
e plancha f aislante
n isolerende plaat f

1323 intake of syrup; syrup intake; drink of syrup
f alimentation f en sirop; addition f de sirop; charge f de sirop
d Sirupeinzug m; Dicksafteinzug m
e introducción f de jarabe
n intrekken n van diksap

* interconnecting pipe *(between evaporator vessels)* → **608**

1324 intermediate bagasse
f bagasse f intermédiaire
d Zwischenbagasse f
e bagazo m intermedio
n tussenbagasse m; tussenampas m

1325 intermediate carrier
f conducteur m intermédiaire
d Zwischentransporteur m
e conductor m intermedio
n tussentransporteur m; tussencarrier m

1326 intermediate centrifugal
f centrifugeuse f de jet intermédiaire; essoreuse f de jet intermédiaire
d Mittelproduktzentrifuge f
e centrífuga f intermedia
n tussenproduktcentrifuge f

1327 intermediate juice
f jus m intermédiaire
d Zwischensaft m
e jugo m intermedio; guarapo m intermedio
n tussensap n

1328 intermediate liming
f chaulage m intermédiaire
d Zwischenkalkung f
e encalado m intermedio; alcalización f intermedia; alcalinización f intermedia
n tussenkalking f

1329 intermediate mill
f moulin m intermédiaire
d Zwischenmühle f
e molino m intermedio
n tussenmolen m

1330 intermediate pump
f pompe f intermédiaire
d Zwischenpumpe f
e bomba f intermedia
n tussenpomp f

1331 intermediate vacuum pan
f appareil m à cuire pour jet intermédiaire
d Mittelproduktkochapparat m
e tacho m al vacío intermedio
n tussenproduktkookpan f

1332 intermediate zone
(One of the three zones in the supersaturated phase)
f zone f intermédiaire
d intermediäre Zone f; intermediäres Gebiet n
e zona f intermedia
n intermediaire zone f

1333 intermittent liming; batch liming
f chaulage m discontinu
d diskontinuierliche Kalkung f
e encalado m intermitente; alcalización f intermitente; alcalinización f intermitente
n discontinue kalking f

* intermittent settler → **167**

* intermittent subsider → **167**

1334 International Association of the European Sugar Beet Growers; CIBE
f Confédération f Internationale des Betteraviers Européens; CIBE f
d Internationale Vereinigung f Europäischer Zuckerrübenanbauer; CIBE f
e Confederación f Internacional de Remolacheros Europeos; CIBE f
n Internationale Vereniging f van Europese Bietentelers; CIBE f

1335 International Commission for Sugar Technology
f Commission f Internationale Technique de Sucrerie; CITS f
d Internationale Kommission f für Zuckertechnologie
e Comisión f Internacional para la Tecnología del Azúcar
n Internationale Commissie f voor Suikertechnologie

1336 International Commission for Uniform Methods of Sugar Analysis; ICUMSA
f Commission f Internationale pour

l'Uniformisation des Méthodes d'Analyse
du Sucre; ICUMSA *f*
d Internationale Kommission *f* für Einheit-
liche Methoden der Zuckeruntersuchung;
ICUMSA *f*
e ICUMSA *f*
n ICUMSA *f*

**1337 International Institute für Sugar Beet
Research**
f Institut *m* International de Recherches
Betteravières; IIRB *m*
d Internationales Institut *f* für Zuckerrübenforschung
e Instituto *m* Internacional de
Investigaciones sobre la Remolacha
n Internationaal Instituut *n* voor Bieten-
onderzoek

**1338 International Society of Sugar Cane
Technologists; ISSCT**
f ISSCT *f*
d ISSCT *f*
e Sociedad *f* Internacional de Técnicos de
la Caña de Azúcar; ISSCT *f*
n ISSCT *f*

1339 International Sugar Agreement; ISA
f Accord *m* International sur le Sucre;
AIS *m*
d Internationales Zuckerabkommen *n*;
Internationales Zuckerüber-
einkommen *n*; IZÜ *n*
e Acuerdo *m* Internacional Azucarero;
Acuerdo *m* Internacional del Azúcar;
Acuerdo *m* Internacional sobre el Azúcar
n Internationale Suikerovereenkomst *f*

1340 International Sugar Council
f Conseil *m* International du Sucre
d Internationaler Zuckerrat *m*;
Intersugar *m*
e Consejo *m* Internacional del Azúcar
n Internationale Suikerraad *m*

1341 International Sugar Organization; ISO
f Organisation *f* Internationale du Sucre;
O.I.S. *f*
d Internationale Zuckerorganisation *f*
e Organización *f* Internacional del Azúcar
n Internationale Suikerorganisatie *f*

1342 intervention
(EEC)
f intervention *f*
d Intervention *f*
e intervención *f*
n interventie *f*

1343 intervention agency
(EEC)
f organisme *m* d'intervention
d Interventionsstelle *f*
e organismo *m* de intervención
n interventiebureau *n*

1344 intervention price
(EEC)
f prix *m* d'intervention
d Interventionspreis *m*
e precio *m* de intervención
n interventieprijs *m*

1345 inversion *(of sugar)*
f inversion *f*
d Inversion *f*
e inversión *f*
n inversie *f*

1346 inversion losses; losses by inversion
f pertes *fpl* par inversion
d Inversionsverluste *mpl*
e pérdidas *fpl* por inversión
n inversieverliezen *npl*

* **invert → 1350**

1347 invert *v*
f invertir
d invertieren
e invertir
n inverteren

1348 invertase; invertin; saccharase; sucrase
(An enzyme which catalyzes the splitting
of sucrose into glucose and fructose)
f invertase *f*; invertine *f*
d Invertase *f*; Invertin *n*
e invertasa *f*; invertina *f*
n invertase *n*

* **invert/ash ratio → 1928**

1349 invert destruction
f destruction *f* du sucre inverti;
destruction *f* de l'inverti
d Invertzuckerzerstörung *f*
e destrucción *f* del azúcar invertido
n vernietiging *f* van invertsuiker

* **invertin → 1348**

* **invert ratio → 1929**

1350 invert sugar; invert
(A mixture of glucose and fructose,
formed in equal quantities by the
hydrolysis of sucrose)

f sucre m inverti; inverti m
d Invertzucker m
e azúcar m invertido
n invertsuiker m

1351 invert-sugar balance
f bilan m du sucre inverti
d Invertzuckerbilanz f
e balance m del azúcar invertido
n invertsuikerbalans f

1352 invert-sugar content
(Content of an equal-part mixture of
glucose and fructose)
f teneur f en sucre inverti
d Invertzuckergehalt m
e contenido m de azúcar invertido
n invertsuikergehalte n

* **invert syrup** → 1248

1353 inward processing arrangements
(EEC)
f trafic m de perfectionnement actif;
TPA m
d aktiver Veredelungsverkehr m
e tráfico m de perfeccionamiento activo
n actief veredelingsverkeer n

**1354 ion exchange column; ion exchanger
column**
f colonne f échangeuse d'ions
d Austauscherkolonne f
e columna f de intercambio iónico
n ionenuitwisselaar m

1355 ion exchanger
f échangeur m d'ions
d Ionenaustauscher m
e intercambiador m iónico; intercambiador
m de iones; permutador m de iones;
cambiador m de iones
n ionenuitwisselaar m; ionenwisselaar m

* **ion exchanger column** → 1354

1356 ion exchange resin
f résine f échangeuse d'ions
d Ionenaustauscher m; Austauscherharz n;
Ionenaustauscherharz n
e resina f de intercambio iónico; resina f
intercambiadora de iones; resina f
permutadora de iones; resina f
cambiadora de iones
n ionenuitwisselaarhars n;
ionenuitwisselend hars n

* **IPA** → 1359

1357 iron deficiency
(sb; disease)
f carence f en fer
d Eisenmangel m
e deficiencia f de hierro; carencia f de
hierro
n ijzergebrek n

* **ISA** → 1339

* **ISO** → 1341

1358 isoglucose
(EEC)
f isoglucose m
d Isoglukose f
e isoglucosa f
n isoglucose f

1359 isopropyl alcohol; isopropanol; IPA
f alcool m isopropylique; isopropanol m
d Isopropylalkohol m; Isopropanol n
e isopropanol m; alcohol m isopropílico
n isopropylalcohol m; isopropanol n

* **ISSCT** → 1338

J

1360 jaggery
(Type of sugar)
f jaggery *m*
d Jaggery *m*
e jaggery *m*
n jaggery *m*

* **jamestown weed** → 2497

1361 Java ratio
(Ratio expressing the relationship
between sucrose (Pol) in the first-
expressed juice and the sucrose (Pol) in
the cane)
f coefficient *m* Java
d Saftquotient *m*; "Java ratio" *m*
e relación *f* de Java; razón *m* de Java;
Java ratio *f*
n Javaquotiënt *n*

1362 jet condenser
f condenseur *m* à jet; condenseur *m* à
mélange
d Strahlkondensator *m*; Einspritz-
kondensator *m*; Mischkondensator *m*
e condensador *m* de chorro; condensador
m de mezcla
n straalcondensor *m*; mengcondensor *m*

1363 jet of water; water jet
f jet *m* d'eau
d Wasserstrahl *m*
e chorro *m* de agua
n waterstraal *m*

1364 jet pump
f pompe *f* à jet; pompe *f* trompe
d Strahlpumpe *f*
e bomba *f* de chorro
n straalpomp *f*

**1365 jet-type washer; nozzle washer; washer
with water nozzles**
(A beet washer)
f lavoir *m* à jet; laveur *m* à jet; lavoir *m* à
gicleurs; laveur *m* à gicleurs
d Düsenwaschmaschine *f*; Düsenwäsche *f*
e lavador *m* de tobera tipo "jet"; lavador
m de toberas
n straalwasmolen *m*

* **jimpson(weed)** → 2497

* **jimpson-weed datura** → 2497

* **jimson(weed)** → 2497

* **jimson-weed datura** → 2497

* **Johnny-jump-up** → 1002

**1366 Johnson grass; Cuba grass; Guinea
grass; Aleppo grass**
(sb; weed)
f sorgho *m* d'Alep; herbe *f* de Cuba;
houlque *f* d'Alep
d Aleppohirse *f*; Johnsongras *n*; wilde
Mohrenhirse *f*; Aleppobartgras *n*
e maicillo *m*; hierba *f* Johnson; sorgo *m*
de Alepo
n johnsongras *n*
l Sorghum halepense

* **jointed charlock** → 2697

**1367 Joint Working Party of the Advisory
Committee on Sugar**
(EEC)
f Groupe *m* Paritaire de Comité
Consultatif du Sucre
d Paritätische Gruppe *f* des Beratenden
Ausschusses für Zucker
e Grupo *m* Paritario del Comité Consultivo
del Azúcar
n Paritaire Groep *f* van het Adviescomité
voor Suiker

1368 juice
f jus *m*
d Saft *m*
e guarapo *m*; jugo *m*
n sap *n*

1369 juice box *(of clarifier)*
f boîte *f* à jus
d Saftsammler *m*
e caja *f* de jugo; caja *f* de guarapo
n sapverzamelkist *f*; sapkist *f*

1370 juice chamber; juice space
f chambre *f* à jus
d Saftraum *m*
e espacio *m* de jugo; espacio *m* de guarapo
n sapruimte *f*

1371 juice channel; feed channel
f canal *m* à jus
d Saftkanal *m*
e canal *m* de jugo; canal *m* de guarapo
n sapkanaal *n*

1372 juice circulation
f circulation *f* du jus
d Saftzirkulation *f*; Saftumlauf *m*
e circulación *f* del jugo; circulación *f* del

guarapo
n sapcirculatie *f*

1373 juice collecting box
  f caisse *f* collectrice de jus
  d Saftsammelkasten *m*
  e depósito *m* colector del jugo; depósito *m*
  colector del guarapo
  n sapverzamelkist *f*

1374 juice colo(u)ration
  f coloration *f* du jus
  d Färbung *f* des Saftes
  e coloración *f* del jugo; coloración *f* del
  guarapo
  n kleuring *f* van het sap

1375 juice density
  f densité *f* du jus
  d Saftdichte *f*
  e densidad *f* del jugo
  n densiteit *f* van het sap

1376 juice distributor
  f répartiteur *m* de jus
  d Saftverteiler *m*
  e distribuidor *m* de jugo; distribuidor *m*
  de guarapo; repartidor *m* de jugo;
  repartidor *m* de guarapo
  n sapverdeler *m*

1377 juice draw-off
  f soutirage *m* du jus
  d Saftabzug *m*
  n sapaftapping *f*; sapaftrek *m*

* **juice epuration** → 1390

1378 juice evaporation
  f évaporation *f* du jus
  d Safteindampfung *f*
  e evaporación *f* del jugo; evaporación *f* del
  guarapo
  n indampen *n* van het sap

1379 juice extraction
  f extraction *f* du jus
  d Saftgewinnung *f*; Saftextraktion *f*
  e extracción *f* del jugo
  n sapwinning *f*; sapextractie *f*

1380 juice flange (of cane mill)
  f bride *f* à jus
  d Saftflansch *m*
  e brida *f* guardajugo
  n sapflens *m*

1381 juice flow
  f schéma *m* du jus

  d Saftschema *n*
  e esquema *m* de jugo
  n sapschema *n*

1382 juice flow
  f flot *m* de jus
  d Saftstrom *m*
  e flujo *m* de jugo; flujo *m* de guarapo
  n sapstroom *m*

* **juice groove** → 1578

1383 juice heater
  f réchauffeur *m* de jus
  d Safterhitzer *m*
  e calentador *m* de jugo; calentador *m* de
  guarapo
  n sapverwarmer *m*

1384 juice heating
  f réchauffage *m* du jus
  d Saftwärmung *f*; Saftanwärmung *f*; Saft-
  erhitzung *f*
  e calentamiento *m* del jugo; calentamiento
  m del guarapo
  n sapverwarming *f*

1385 juice level
  f niveau *m* du jus
  d Saftstand *m*
  e nivel *m* del jugo; nivel *m* del guarapo
  n sapstand *m*

**1386 juice level ga(u)ge glass; juice level
   indicator tube**
  f tuyau *m* d'indicateur de niveau du jus
  d Saftstandrohr *n*
  e tubo *m* del indicador de nivel del jugo;
  tubo *m* del indicador de nivel del
  guarapo
  n peilglas *n* voor sapstand

1387 juice level regulator
  f régulateur *m* de niveau de jus
  d Saftstandregler *m*
  e regulador *m* del nivel de jugo; regulador
  m del nivel de guarapo
  n sapstandregelaar *m*

1388 juice pipe
  f tuyau *m* de jus
  d Saftrohr *n*
  e tubo *m* del jugo; tubo *m* del guarapo
  n sapleiding *f*

1389 juice pump
  f pompe *f* à jus
  d Saftpumpe *f*

e bomba *f* de jugo; bomba *f* de guarapo
n sappomp *f*

1390 juice purification; juice epuration
f épuration *f* du jus
d Saftreinigung *f*
e purificación *f* del jugo; purificación *f* del guarapo
n sapzuivering *f*; sapreiniging *f*

1391 juice purification station
f station *f* d'épuration du jus
d Saftreinigungsstation *f*
e estación *f* de purificación de jugo; estación *f* de purificación de guarapo
n sapzuiveringsstation *n*; sapreinigingsstation *n*

1392 juice purity
f pureté *f* du jus
d Saftreinheit *f*; Reinheit *f* des Saftes
e pureza *f* del jugo; pureza *f* del guarapo
n sapreinheid *f*

1393 juice sample boy
f échantillonneur *m* de jus
d Saftprobenehmer *m*
e muestreador *m* de jugo; muestreador *m* de guarapo
n sapmonsternemer *m*

1394 juice sampler
f échantillonneur *m* de jus
d Saftprobenehmer *m*
e muestreador *m* de guarapo; tomador *m* de muestras de guarapo; muestreador *m* de jugo; tomador *m* de muestras de jugo
n sapmonstertrekker *m*

1395 juice scale
f balance *f* à jus
d Saftwaage *f*
e balanza *f* de jugo; balanza *f* de guarapo
n sapweegschaal *f*; sapschaal *f*

1396 juice screen; juice strainer
f tamis *m* à jus
d Saftsieb *n*
e tamiz *m* para jugo; tamiz *m* para guarapo
n sapzeef *f*

* **juice screening** → **1398**

* **juice separator** → **1044**

* **juice space** → **1370**

1397 juice stability
f stabilité *f* du jus
d Stabilität *f* des Saftes
e estabilidad *f* del jugo
n stabiliteit *f* van het sap

* **juice strainer** → **1396**

1398 juice straining; juice screening
f tamisage *m* du jus
d Saftabsiebung *f*
e tamizado *m* del jugo; tamizado *m* del guarapo
n zeven *n* van het sap

1399 juice tank
f bac *m* à jus
d Saftkasten *m*; Saftbehälter *m*
e depósito *m* de jugo; depósito *m* de guarapo
n sapbak *m*; sapkist *f*

1400 juice tray
f cuvette *f* à jus
d Safttrog *m*
e cubeta *f* de jugo; cubeta *f* de guarapo
n saptrog *m*

1401 juice vapo(u)r
f vapeur *f* de jus
d Brüdendampf *m*; Brüden *m*
e vapor *m* de jugo; vapor *m* de guarapo
n sapdamp *m*

* **June beetle** → **2419**

* **June bug** → **2419**

K

* **kelpwort** → 2025

* **Kestner evaporator** → 1499

* **kew weed** → 1154

* **kieselguhr** → 797

1402 kinematic viscosity *(of the juice)*
f viscosité f cinématique
d kinematische Viskosität f
e viscosidad f cinemática
n kinematische viscositeit f

1403 kingbolt *(of cane mill)*
f chandelier m; boulon m royal
d Königsbolzen m
e perno m real
n hoofdbout m

1404 kingboltless headstock *(of cane mill)*
f bâti m sans boulons royaux
d königsbolzenloser Ständer m
e virgen f sin pernos reales
n frame n zonder hoofdbouten

* **kingcup** → 361

* **Klamath weed** → 585

1405 knife blade
f lame f de couteau
d Messerblatt n
e hoja f de cuchilla
n meslemmet n

**1406 knife-block; knife-box; knife-frame;
     knife-holder**
     (Steel frames for supporting the beet
     knives in the slicers)
f porte-couteaux m
d Messerkasten m
e caja f portacuchillos; caja f de cuchillos;
   portacuchillas m
n messenkast f; mesblok n

**1407 knife-box counterblade; forelayer of
     knife-box**
f contre-plaque f de porte-couteaux
d Messerkastenvorlage f
e contraplaca f de la caja de cuchillos;
   contrarregla f de portacuchillas
n voorlegstuk n van messenkast

* **knife cut** → 1148

1408 knife disc
f disque m de coupe-racines; plateau m de
   coupe-racines
d Schneidscheibe f
e disco m portacuchillos; disco m cortador
n messchijf f

* **knife-frame** → 1406

* **knife-holder** → 1406

* **knife set with horizontal blades** → 392

1409 knotgrass; knotweed *(US)*; **prostrate
     knotweed**
     (sb; weed)
f renouée f des oiseaux; traînasse f; herbe
   f à cochon
d Vogelknöterich m; Knicker m
e corregüela f de los caminos; sanguinaria
   f mayor; lengua f de pájaro; cien nudos
   m pl; correhuela f de los caminos;
   centinodio m; hierba f de las calenturas;
   pico m de gorrión
n varkensgras n; bargegras n;
   zwijnegras n; mottegras n; weggras n;
   kreupelgras n
l Polygonum aviculare

1410 knotweed
     (sb; weed)
f renouée f
d Knöterich m
e polígono m
n duizendknoop m
l Polygonum sp.

* **knotweed** *(US)* → 1409

* **known losses** → 5

1411 kochia *(US)*; **fireweed** *(US)*; **tumbleweed**
     (US); **mock cypress; summer cypress;
     belvedere summer cypress**
     (sb; weed)
f ansérine f à balais; ansérine f belvédère;
   kochie f à balais
d Besenradmelde f; Studentenkraut n;
   Besenkraut n
e ceniglo m de jardín
n studentenkruid n; zomercipres n
l Kochia scoparia

**1412 Königsfeld knife; Königsfelder-type
     knife; Königsfelder knife**
f couteau m fraisé Königsfeld; couteau m
   Königsfeld
d Königsfelder Schnitzelmesser n
e cuchilla f Königsfeld; cuchilla f tipo

Königsfeld
n Königsfelder bietenmes n

1413 Krajewski
(A two-roller crusher)
f Krajewski m; défibreur m Krajewski
d Krajewski-Crusher m
e desmenuzadora f Krajewski; Krajewski f
n Krajewski-crusher m; Krajewski-
kneuzer m

L

1414 labile zone
(The region of supersaturation in which nucleation will readily occur)
f zone f labile
d labile Zone f; labiles Gebiet n
e zona f lábil
n labiele zone f

1415 laboratory mill juice
f jus m de moulin de laboratoire
d Saft m der Labormühle; Labormühlen-saft m
e jugo m de molino de laboratorio; guarapo m de molino de laboratorio
n sap n van laboratoriummolen

1416 lace bug; beet leaf bug; leaf bug; beet bug
(sb; pest)
f punaise f de la betterave; punaise f de la feuille de betterave
d Rübenblattwanze f
e chinche f de la remolacha
n bietenbladwants f
l Piesma quadratum

1417 lace bug *(US)*
(sb; pest)
l Piesma cinerea

* **lace flower** → 2693

1418 lactate
f lactate m
d Lactat n; Laktat n
e lactato m
n lactaat n

1419 lactic acid
f acide m lactique
d Milchsäure f
e ácido m láctico
n melkzuur n

1420 lactose
f lactose m; lactobiose m; sucre m de lait; sel m de lait
d Laktose f; Milchzucker m; Lactose f
e lactosa f
n lactose f; melksuiker m

1421 ladybird (beetle); lady beetle; ladybug
f coccinelle f
d Marienkäfer m
e mariquita f

n lieveheersbeestje n
l Coccinella sp.

* **lady's-comb** → 2131

* **lady's thumb** *(US)* → 1750

1422 laevorotation; levorotation
f rotation f lévogyre
d Linksdrehung f
e rotación f levogira
n linksdraaiing f

1423 laevorotatory sugar; levorotatory sugar
f sucre m lévogyre
d linksdrehender Zucker m
e azúcar m levógiro; azúcar m levo-rotatorio
n linksdraaiende suiker m

* **laevulose** → 1140

1424 Lafeuille crystallizer; crystallizer pan
f cuite f rotative Lafeuille
e cristalizador m Lafeuille; tacho m cristalizador
n Lafeuille-kristallisoir m

* **lamb's-quarters** *(US)* → 970

* **lamb's-quarters goosefoot** → 970

* **lamb's tongue** → 1253

1425 lance nematodes
(sc; pest)
l Hoplolaimus spp.

1426 large crabgrass *(US)*; **hairy crabgrass; hairy finger-grass**
(sb; weed)
f digitaire f sanguine; panic m sanguin; millet m sanguin; sanguinette f; manne f terrestre; manne f sanguinale
d Bluthirse f; Blutfennich m; Blutfinger-hirse f; Blutfingergras n; blutrote Finger-hirse f; rotes Fingergras n
e pata f de gallina; garrachuelo m; pasto m de cuaresma; pasto m de sembrados
n harig vingergras n; bloedgierst f; bloedgerst f
l Digitaria sanguinalis (L.) Scop.; Panicum sanguinale

1427 large-flowered hemp nettle; nice hemp nettle
(sb; weed)
f galéopsis m à fleurs variées; ortie f ornée; galéope m bigarré

d bunter Hohlzahn *m*
n dauwnetel *f*
l Galeopsis speciosa Mill.; Galeopsis versicolor Curtis

1428 last-expressed juice
(The juice expressed by the last two rollers of a tandem)
f jus *m* de dernière pression
d letzter Preßsaft *m*
e guarapo *m* de última expresión; jugo *m* de última expresión
n laatstgeperst sap *n*; naperssap *n*

* **last-mill bagasse** → 119

1429 last-mill juice
(The juice expressed by the last mill of a tandem)
f jus *m* du dernier moulin
d Saft *m* der letzten Mühle
e guarapo *m* del último molino; jugo *m* del último molino
n sap *n* van de laatste molen; naperssap *n*

1430 latent heat
f chaleur *f* latente
d latente Wärme *f*; gebundene Wärme *f*
e calor *m* latente
n latente warmte *f*

1431 lateral downtake; lateral well *(of evaporator, vacuum pan)*
f puits *m* latéral
d seitliches Rohr *n*
e tubo *m* lateral; vía *f* de descenso lateral
n zijbuis *f*; zijpijp *f*

1432 lateral feeding table; lateral feed table
(Cane-handling equipment)
f table *f* latérale d'alimentation; table *f* d'alimentation latérale
d seitlicher Aufgabetisch *m*
e mesa *f* alimentadora lateral
n laterale toevoertafel *f*

* **lateral well** *(of evaporator, vacuum pan)* → 1431

1433 layer of sugar; sugar layer; sugar cake; wall of sugar *(in the centrifugal)*
f couche *f* de sucre; mur *m* de sucre
d Zuckerschicht *f*; Zuckerkuchen *m*
e capa *f* de azúcar; pared *f* de azúcar
n suikerlaag *f*

* **LDP** → 1496

1434 LDP(R); London Daily Price for Raw Sugar
(eco)
f LDP(R)
d Londoner Tageskurs *m* für Rohzucker; LDP(R)
e LDP(R)
n LDP(R)

1435 LDP(W); London Daily Price for White Sugar
(eco)
f LDP(W)
d Londoner Tageskurs *m* für Weißzucker; LDP(W)
e LDP(W)
n LDP(W)

1436 lead acetate
f acétate *m* de plomb
d Bleiacetat *n*
e acetato *m* de plomo
n loodacetaat *n*; loodijzer *n*

1437 leaf blast
(sc; fungous disease)
d Blattbräune *f*
e flameado *m* de la hoja
n bladbruinheid *f*

1438 leaf blight
(sc; fungous disease)
d Leptosphaeria-Blattfleckenkrankheit *f*
e tizón *m* de la hoja
n Leptosphaeria-bladvlekkenziekte *f*

* **leaf bug** → 1416

1439 leaf catcher; leaf separator
f épailleur *m*
d Krautabscheider *m*
e separador *m* de hoja; retenedor *m* de hojarasca
n bladvanger *m*

1440 leaf curl
(sb; virus disease)
f frisolée *f* de la betterave
d Wanzen-Kräuselkrankheit *f*
e rizado *m*
n krulziekte *f*

1441 leaf filter
f filtre *m* à disques
d Blattfilter *n*; Scheibenfilter *n*
e filtro *m* de láminas
n schijffilter *n*

1442 **leafhopper**
      (sb; pest)
   f cicadelle f
   d Zwergzikade f
   n dwergcicade f
   l Empoasca sp.

1443 **leaf scald**
      (sc; bacterial disease)
   f échauffement m des feuilles
   d bakterieller Blattbrand m; bakterielle
      Blattstreifigkeit f
   e escaldadura f de la hoja
   n bacteriële bladbrand m

1444 **leaf scorch**
      (sc; fungous disease)
   f brûlure f de la feuille
   d Stagonospora-Blattbrand m
   e chamuscado m de la hoja
   n bladbrand m

   * **leaf separator → 1439**

1445 **leaf-splitting disease**
      (sc; fungous disease)
   d Blattrissigkeit f
   e rajadura f de la hoja

1446 **leaf stripper**
      (agr)
   f effeuilleuse f
   d Rübenblattschläger m
   e deshojadora f de remolacha
   n ontbladermachine f

1447 **leaf stripper-lifter**
      (agr)
   f effeuilleuse-arracheuse f
   d Blattschläger m und Rübenroder
   e deshojadora-arrancadora f
   n ontbladermachine f en bietenrooier

1448 **leaf stripper-topper**
      (agr)
   f effeuilleuse-décolleteuse f
   d Blattschläger m und Köpfer
   e deshojadora-descoronadora f de
      remolacha
   n ontbladermachine f en kopper

1449 **leafy spurge leafy euphorbia**
      (sb; weed)
   f euphorbe f ésule
   d Eselswolfsmilch f; Rutenwolfsmilch f;
      scharfe Wolfsmilch f
   e lechetrezna f
   n heksenmelk f
   l Euphorbia esula; Euphorbia virgata

1450 **lean beet**
      (EEC)
   f betterave f maigre
   d magere Rübe f
   e remolacha f magra
   n magere biet f

1451 **leatherjacket**
      (sb; pest)
   f tipule f; tipule f des prairies
   d Wiesen-Schnaken-Larve f; Wiesen-Erd-
      Schnaken-Larve f; Wiesenwurm m
   l Tipula paludosa (larva)

   * **leatherjacket → 571**

   * **leatherjacket (larva) → 674**

1452 **lens-podded whitetop**
      (sb; weed)
   l Cardaria repens

1453 **lesser armyworm; beet armyworm**
      (sb; pest)
   f noctuelle f de la betterave
   d Zuckerrübeneule f
   e gardama f
   l Laphygma exigua

   * **lesser bindweed → 997**

1454 **lesser burdock; common burdock (US);
      rough cocklebur; smaller burdock**
      (sb; weed)
   f petite bardane f; grappille f; petit
      glouteron m
   d kleine Klette f; gemeine Spitzklette f
   e lampazo m menor; cadillo m común
   n kleine klis f; ongedoornde stekelnoot m
   l Arctium minus; Xanthium strumarium

1455 **lesser migratory grasshopper (US)**
      (sb; pest)
   l Melanoplus mexicanus

1456 **lesser snapdragon; calfs snout;
      weaselsnout; corn snapdragon**
      (sb; weed)
   f muflier m; muflier m rubicond; tête-de-
      mort f; muflier m oronce
   d Ackerlöwenmaul n; Feldlöwenmaul m;
      kleiner Dorant m
   e cabeza f de muerto; hierba f becerra
   n akkerleeuwebek m; akkerleeuwebekje n
   l Antirrhinum orontium

1457 **level feeler** (of cane layer on the carrier)
   f tâteur m de niveau; palpeur m de niveau
   d Niveautaster m

    e palpador *m* de nivel
    n niveautaster *m*

1458 leveller; leveller knives
    f niveleur *m*; coupe-cannes *m* niveleur
    d Ausgleicher *m* (mit Messern)
    e nivelador *m*; cuchillas *fpl* niveladoras
    n laagbegrenzer *m* (met messen)

    * **levorotation** → **1422**

    * **levorotatory sugar** → **1423**

    * **levulose** → **1140**

1459 levy
    (EEC)
    f prélèvement *m*
    d Abschöpfung *f*
    e gravamen *m*; "prélèvement" *m*
    n heffing *f*

1460 lift *v*; **dig** *v (US; beet)*
    f arracher
    d roden; ausmachen
    e arrancar
    n rooien

1461 lifter-loader; digger-loader *(US)*
    (agr)
    f arracheuse-chargeuse *f*
    d Rüben-Rodelader *m*; Wagenroder *m*
    e arrancadora-cargadora *f* de remolacha
    n rooier-lader *m*

1462 lifter-windrower; beet lifter and collector
    (agr)
    f arracheuse-groupeuse *f*; arracheuse-
      aligneuse *f*
    d Sammelroder *m*; Querschadroder *m*;
      Rübenvorratsroder *m*; Rübenroder *m*
      und Schwadleger
    e arrancadora-amontonadora *f* de
      remolacha; arrancadora-hileradora *f* de
      remolacha
    n verzamelrooier *m*; voorraadrooier *m* op
      dwarszwad; rooier-zwadlegger *m*; bieten-
      bunkerrooier *m*

1463 lifting belt
    (agr)
    f courroie *f* arracheuse
    d Klappgreifer *m*; Greiferband *n*
    e correa *f* extractora de remolacha

1464 lifting disc; digging disc
    (agr)
    f disque *m* souleveur
    d Rodescheibe *f*

    e disco *m* desarraigador; disco *m* extractor
    n rooischijf *f*

1465 lifting share; digging share
    (agr)
    f soc *m*
    d Rodeschar *n*
    e reja *f*
    n rooischaar *f*

    * **lifting squeeze** → **1078**

    * **lifting wheel** → **2005**

    * **lifting wheel digger** → **2005**

    * **light molasses** → **2638**

1466 ligneous beet
    (sb; disease)
    f betterave *f* ligneuse
    d verholzte Rübe *f*
    e remolacha *f* lignificada
    n verhoute biet *f*

1467 lime *(noun)*
    f chaux *f*
    d Kalk *m*
    e cal *f*
    n kalk *m*

1468 lime *v*
    f chauler
    d kalken
    e encalar; alcalizar; alcalinizar
    n kalken; chauleren

1469 lime addition
    f addition *f* de chaux
    d Kalkzugabe *f*; Kalkzusatz *m*
    e adición *f* de cal
    n kalkgift *f*; kalktoevoeging *f*

    * **limed juice** → **749**

1470 limed juice pump
    f pompe *f* à jus chaulé
    d Kalkungssaftpumpe *f*; Scheidesaft-
      pumpe *f*
    e bomba *f* para jugo encalado; bomba *f*
      para guarapo encalado
    n pomp *f* voor gekalkt sap

    * **lime house** → **1481**

1471 lime kiln
    f four *m* à chaux
    d Kalkofen *m*

e horno *m* de cal
n kalkoven *m*

1472 lime milk; milk of lime
f lait *m* de chaux
d Kalkmilch *f*
e lechada *f* de cal
n kalkmelk *f*

* **lime-milk classifier → 1590**

1473 lime-milk preparation
f préparation *f* du lait de chaux
d Kalkmilchaufbereitung *f*; Kalkmilch-
herstellung *f*
e preparación *f* de la lechada de cal
n kalkmelkbereiding *f*

1474 lime powder
f poudre *f* de chaux
d Kalkmehl *n*
e polvo *m* de cal
n kalkmeel *n*

* **limer → 1484**

1475 lime salts
f sels *mpl* de chaux
d Kalksalze *npl*
e sales *fpl* de calcio; sales *fpl* cálcicas
n kalkzouten *pl*

1476 lime-salts content
(The content of dissolved calcium salts,
expressed as percent CaO on dry
substance)
f teneur *f* en sels de chaux
d Kalksalzgehalt *m*
e contenido *m* de sales de calcio
n gehalte *n* aan kalkzouten

1477 lime slaker
f appareil *m* à éteindre la chaux; appareil
m d'extinction de la chaux
d Kalklöscher *m*; Kalklöschapparat *m*
e aparato *m* para apagar la cal
n kalkblusapparaat *n*

1478 lime slaking
f extinction *f* de la chaux
d Kalklöschen *n*
e apagamiento *m* de la cal
n blussen *n* van kalk

1479 lime slaking drum; rotary lime slaker
f appareil *m* Mick; tambour *m* Mick
d Kalklöschtrommel *f*
e tambor *m* rotativo para apagar la cal
n kalkblustrommel *f*

1480 lime slaking installation
f installation *f* d'extinction de la chaux
d Kalklöschanlage *f*
e instalación *f* para apagar cal
n kalkblusinstallatie *f*

1481 lime station; lime house
f atelier *m* de la chaux; atelier *m* de
chaulage
d Kalkhaus *n*; Kalkstation *f*
e estación *f* de la cal; estación *f* para
encalado; estación *f* para alcalización;
estación *f* para alcalinización
n kalkstation *n*

1482 limestone
f pierre *f* à chaux
d Kalkstein *m*
e piedra *f* de cal; piedra *f* caliza
n kalksteen *nm*

* **lime sucrate → 2030**

1483 liming
f chaulage *m*
d Kalkung *f*; Scheidung *f*
e encalado *m*; alcalinización *f*;
alcalización *f*
n kalking *f*; scheiding *f*

1484 liming tank; limer
f bac *m* chauleur; chauleur *m*; bac *m* de
chaulage
d Kalkungspfanne *f*; Kalkungsgefäß *n*;
Scheidepfanne *f*
e tanque *m* de encalar; tanque *m* de
alcalinizar
n chauleur *m*

1485 linear vibrator
(A water separator)
f vibrateur *m* linéaire
d Linearschwinger *m*
e vibrador *m* rectilíneo
n lineaire vibrator *m*

1486 liquid raffinade
f sucre *m* raffiné liquide
d flüssige Raffinade *f*
e azúcar *m* refinado líquido
n vloeibare suikerstroop *f*

1487 liquid ring pump
f pompe *f* à anneau liquide
d Flüssigkeitsringpumpe *f*
e bomba *f* hidrorrotativa
n waterringpomp *f*

1488 **liquid sugar**
(Marketable syrup, usually prepared by dissolving granulated sugar in water)
f sucre *m* liquide
d Flüssigzucker *m*
e azúcar *m* líquido
n vloeibare suiker *m*

1489 **liquid sugar factory**
f fabrique *f* de sucre liquide; usine *f* de sucre liquide
d Flüssigzuckerfabrik *f*
e fábrica *f* de azúcar líquido
n vloeibare-suikerfabriek *f*; fabriek *f* van vloeibare suiker

* **little-flower quickweed** → 1154

* **little mallow** *(US)* → 474

* **little-pod false flax** → 2188

1490 **live steam**
f vapeur *f* directe; vapeur *f* vive; VD *f*
d Frischdampf *m*; Direktdampf *m*
e vapor *m* vivo; vapor *m* directo
n verse stoom *m*; directe stoom *m*

* **lixiviate** *v* → 949

1491 **lixiviation**
f lixiviation *f*
d Auslaugung *f*; Extraktion *f*
e lixiviación *f*
n uitloging *f*; extractie *f*

1492 **loader-cleaner**
(agr)
f chargeur-décrotteur *m*
d Rübenreiniger *m* und Lader
e limpiadora *f* cargadora de remolacha
n bietenreiniger *m* en lader

1493 **loading beet harvester**
(agr)
f décolleteuse-arracheuse-chargeuse *f*; arracheuse *f* combinée de betteraves
d Wagenköpfroder *m*; Rübenvollernter *m*; Rübenvollerntegerät *n*; Vollerntegerät *n*
e arrancadora-cargadora *f* de remolachas; descoronadora-arrancadora-cargadora *f*
n bietenoogstmachine *f*

1494 **loaf sugar**
f sucre *m* en pain
d Hutzucker *m*
e azúcar *m* de pilón; azúcar *m* en pan
n broodsuiker *m*

1495 **local overheating**
f surchauffe *f* locale
d örtliche Überhitzung *f*; lokale Überhitzung *f*
e sobrecalentamiento *m* local
n plaatselijke oververhitting *f*; lokale oververhitting *f*

* **locust** → 486

1496 **London Daily Price; LDP**
(eco)
f LDP
d Londoner Tageskurs *m*; LDP
e LDP
n LDP

* **London Daily Price for Raw Sugar**
→ 1434

* **London Daily Price for White Sugar**
→ 1435

* **long condenser** → 151

1497 **longleaf groundcherry** *(US)*
(sb; weed)
l Physalis longifolia

1498 **long tube vertical falling film evaporator; LTVFF evaporator**
f évaporateur *m* à descendage à longs tubes verticaux
d Fallstromverdampfer *m* mit senkrechten langen Heizrohren; Fallfilmverdampfer *m* mit senkrechten langen Heizrohren
e evaporador *m* de película descendente de tubo largo
n valstroomverdamper *m* met verticale lange verwarmingsbuizen

1499 **long tube vertical rising film evaporator; LTVRF evaporator; Kestner evaporator**
f évaporateur *m* Kestner
d Kestner-Verdampfer *m*; Kletterfilmverdampfer *m* mit senkrechten langen Heizrohren
e evaporador *m* Kestner de película ascendente; evaporador *m* vertical de película elevada de tubo largo
n Kestner-verdamper *m*

* **lopside oat** → 327

1500 **losses by caramelization**
f pertes *fpl* par caramélisation
d Verluste *mpl* durch Karamelisierung
e pérdidas *fpl* por caramelización

n verliezen *npl* door caramelisatie;
verliezen *npl* door caramelvorming

1501 losses by convection *(of heat)*
f pertes *fpl* par convection
d Verluste *mpl* durch Konvektion
e pérdidas *fpl* por convección
n verliezen *npl* door convectie

1502 losses by entrainment
f pertes *fpl* par entraînements
d Verluste *mpl* durch Mitreißen
e pérdidas *fpl* por arrastre
n verliezen *npl* door meeslepen; verliezen
npl door overspatten

* **losses by inversion** → 1346

1503 losses by radiation *(of heat)*
f pertes *fpl* par rayonnement
d Strahlungsverluste *mpl*; Abstrahlungs-
verluste *mpl*
e pérdidas *fpl* por radiación
n stralingsverliezen *npl*

1504 lost juice
f jus *m* perdu
d verlorener Saft *m*
e jugo *m* perdido; guarapo *m* perdido
n verloren sap *n*

* **lovage weevil** → 47

* **love vine** → 843

1505 lower grade sugar
f sucre *m* de qualité courante
d Verbrauchszucker *m*
e azúcar *m* de consumo
n consumptiesuiker *m*

* **lower roll(er)** *(of crusher, cane mill)*
→ 319

1506 lower tube plate
f plaque *f* tubulaire inférieure
d unterer Rohrboden *m*; unterer
Rohrplattenboden *m*
e placa *f* tubular inferior; plancha *f* de
tubos inferior; fondo *m* de tubos inferior;
fondo *m* tubular inferior
n onderpijpenplaat *f*; onderpijpplaat *f*;
ondertubenplaat *f*

* **low-grade boiling** → 1512

1507 low-grade centrifugal; after worker
f centrifuge *f* bas-produits; centrifugeuse *f*
bas-produits; essoreuse *f* bas-produits;

centrifuge *f* troisième jet; essoreuse *f*
troisième jet
d Nachproduktzentrifuge *f*
e centrífuga *f* de productos de bajo grado;
centrífuga *f* para azúcar de bajo grado
n naproduktcentrifuge *f*

1508 low-grade magma
f magma *m* de bas-produit
d Nachproduktmagma *n*
e magma *m* de bajo grado
n naproduktmagma *n*

**1509 low-grade massecuite; low raw
massecuite; afterproduct massecuite**
f masse *f* cuite du dernier jet; masse *f*
cuite bas-produits; masse *f* cuite de bas-
produits; masse *f* cuite arrière-produit
d Nachproduktfüllmasse *f*; Nachprodukt-
kochmasse *f*
e masa *f* cocida de bajo grado; masa *f*
cocida de baja pureza; masa *f* cocida de
baja calidad; masa *f* cocida de productos
de bajo grado; masa *f* cocida de tercer
producto
n naprodukt-masse cuite *f*;
naproduktvulmassa *f*

* **low-grade materials** → 1511

1510 low-grade pan
f appareil *m* bas-produits
d Nachproduktkochapparat *m*
e tacho *m* de templas de baja calidad
n naproduktkookpan *f*

1511 low grades; low-grade materials
f bas-produits *mpl*; arrière-produits *mpl*
d Nachprodukte *npl*
e materiales *mpl* de bajo grado; materiales
mpl de baja calidad
n naprodukten *npl*

1512 low-grade strike; low-grade boiling
f cuite *f* de bas-produit; cuite *f* bas-
produits
d Nachproduktsud *m*
e templa *f* de baja calidad; cocción *f* de
templas de baja calidad; templa *f* de
tercer producto; cocción *f* de templas de
tercer producto
n naproduktkooksel *n*

* **low-grade sugar** → 38

* **low mallow** → 577

**1513 low product recovery; endproduct
working**

f traitement *m* des bas-produits;
traitement *m* des arrière-produits;
traitement *m* des égouts et des masses
cuites du dernier jet
d Nachproduktarbeit *f*
e elaboración *f* de productos de bajo grado;
elaboración *f* de producto derivado
n naproduktverwerking *f*

1514 low purity juice
f jus *m* de faible pureté
d Saft *m* geringer Reinheit
e jugo *m* de baja pureza; guarapo *m* de
baja pureza
n sap *n* van lage reinheid

1515 low purity massecuite
f masse *f* cuite de basse pureté
d Füllmasse *f* geringer Reinheit
e masa *f* cocida de baja pureza
n masse cuite *f* van lage reinheid

1516 low purity molasses
f mélasse *f* de basse pureté
d Melasse *f* geringer Reinheit
e melaza *f* de baja pureza
n melasse *f* van lage reinheid

* **low raw massecuite** → 1509

* **low raw sugar** → 38

1517 low test molasses
f mélasse *f* pauvre; mélasse *f* de faible
polarisation
d Low-Test-Melasse *f*
e melaza *f* de baja polarización; melaza *f*
de baja calidad
n low-test-melasse *f*

1518 low vacuum
f petit vide *m*
d geringes Vakuum *n*
e bajo vacío *m*
n klein vacuüm *n*

* **LTVFF evaporator** → 1498

* **LTVRF evaporator** → 1499

* **lump quicklime** → 1520

1519 lumps *(of sugar)*
f grumeaux *mpl*; agglomérats *mpl*
d Zuckerknoten *mpl*; Knollen *mpl*
e grumos *mpl*
n klonters *mpl*

**1520 lumps of quicklime; rock lime; lump
quicklime**
f chaux *f* vive en roche concassée;
morceaux *mpl* de chaux vive; chaux *f*
vive en morceaux
d Stückkalk *m*
e trozos *mpl* de cal viva; cal *f* viva en
trozos
n ongebluste kalk *m* in stukken

1521 lygus bug *(US)*
(sb; pest)
l Lygus hesperus

1522 lygus bug *(US)*
(sb; pest)
l Lygus elisus

M

* MA → 1560

1523 maceration; bath maceration
(A form of imbibition in which the
bagasse is steeped in an excess of fluid)
f macération f
d Mazeration f
e maceración f; maceración f en baño
n maceratie f

1524 maceration juice
f jus m de macération
d Mazerationssaft m
e jugo m de maceración; guarapo m de
maceración
n maceratiesap n

1525 maceration pump
f pompe f à jus de macération
d Mazerationssaftpumpe f; Nachpreß-
saftpumpe f
e bomba f para jugo de maceración; bomba
f para guarapo de maceración
n maceratiesappomp f; naperssappomp f

* machine → 449, 460

* machine man → 456

* machine syrup → 2451

1526 magma
(A mechanical mixture of crystals and
molasses or heavy syrup. In beet work,
the term is synonymous with
"massecuite")
f magma m
d Magma m; Kristall-Sirup-Gemisch n;
Kristallbrei m; Kristallsuspension f
e magma m
n magma n; aangepapte suiker m

1527 magma footing
f pied m de magma; magma m de pied de
cuite
d Kornfußmagma n; Kristallfußmagma n
e pie m de magma; magma m de cristales
para semilla
n pied-de-cuite m van magma

1528 magma mixer
f malaxeur-réempâteur m; réempâteur m
d Aufmaische f; Magmamaische f
e mezclador m de magma
n magmamenger m

1529 magnesium deficiency
(sb; disease)
f carence f en magnésium
d Magnesiummangel m
e deficiencia f de magnesio; carencia f de
magnesio
n magnesiumgebrek n

1530 magnetic eliminator
f éliminateur m magnétique
d Magnetabscheider m; Magnetscheider m
e eliminador m magnético
n magneetafscheider m; magnetische
afscheider m

* **magnetic separator** → 2530

1531 main filtrate
f filtrat m principal
d Hauptfiltrat n
e filtrado m principal
n hoofdfiltraat n

1532 main liming
(The treatment of the juice with
relatively large amounts of lime, e.g., 1%
CaO or more, to precipitate certain
nonsucroses and to promote the
decomposition reactions between lime
and certain nonsucroses)
f chaulage m principal
d Hauptscheidung f; Hauptkalkung f
e encalado m principal; alcalización f
principal; alcalinización f principal
n hoofdkalking f

1533 main surplus area
(EEC)
f région f la plus excédentaire
d Hauptüberschußgebiet n
e región f más excedentaria
n gebied n met het grootste overschot

1534 make-up water
f eau f de complément; eau f d'appoint;
eau f additionnelle
d Zusatzwasser n
e agua f complementaria
n toegevoegd water n

1535 malate
f malate m
d Malat n
e malato m
n malaat n

1536 malic acid
f acide m malique
d Apfelsäure f

e ácido *m* málico
n appelzuur *n*

1537 mallow
(sb; weed)
f mauve *f*
d Malve *f*; Käsepappel *f*; Hasenpappel *f*;
Gänsepappel *f*
e malva *f*
n kaasjeskruid *n*; maluwe *f*; malve *f*;
kaasjesblad *n*; kaasjesbloem *f*
l Malva sp.

1538 maltose
f maltose *m*; sucre *m* de malt;
maltobiose *m*; ptyalose *m*
d Maltose *f*; Maltzucker *m*
e maltosa *f*
n maltose *f*; moutsuiker *m*

1539 malva yellows
(sb; disease)
f jaunisse *f* de la mauve
d Vergilbungskrankheit *f* der Malve
e amarillez *f* de la malva
n vergelingsziekte *f* van het kaasjeskruid

1540 mammoth pump; airlift beet pump
f pompe *f* mammouth; élévateur *m* à
émulsion
d Mammutpumpe *f*
e bomba *f* mamut; bomba *f* elevadora
n mammoetpomp *f*

1541 Management Committee for Sugar
(EEC)
f Comité *m* de Gestion du Sucre
d Verwaltungsausschuß *m* für Zucker
e Comité *m* de Gestión del Azúcar
n Beheerscomité *n* voor Suiker

1542 manganese deficiency
(sb; disease)
f carence *f* en manganèse
d Manganmangel *m*
e deficiencia *f* de manganeso; carencia *f*
de manganeso
n mangaangebrek *n*

**1543 mangel; mangel wurzel (US); fodder
beet; common beet; field beet**
f betterave *f* fourragère
d Futterrübe *f*; Runkelrübe *f*
e remolacha *f* forrajera
n voederbiet *f*
l Beta vulgaris ssp. rapacea

**1544 mangold flea beetle; hop flea (beetle);
brassy(-toothed) flea beetle; toothed-**

legged turnip beetle
(sb; pest)
f altise *f* de la betterave
d Rübenerdfloh *m*; Erdfloh *m*; Erdfloh-
käfer *m*; nordeuropäischer Rübenerdfloh *m*; erzfarbiger Erdfloh *m*
e pulguilla *f* de la remolacha
n aardvlo *f*
l Chaetocnema concinna

**1545 mangold fly; beet leaf miner (US);
spinach leaf miner (US); henbane fly;
beet and mangold fly**
(sb; pest)
f mouche *f* de la betterave; pégomyie *f* de
la betterave
d Rübenfliege *f*; Nachtschattenfliege *f*;
Bilsenkrautfliege *f*
e mosca *f* de la remolacha
n bietenvlieg *f*
l Pegomya hyoscyami; Pegomya betae

* **many-seeded goosefoot → 56**

1546 marc content
f teneur *f* en marc
d Markgehalt *m*; Rübenmarkgehalt *m*
e contenido *m* de marco
n merggehalte *n*; bietenmerggehalte *n*

1547 march fly (US); bibionid fly
(sb; pest)
f bibion *m* des jardins; bibion *m* horticole;
bibion *m* maraîcher
d Gartenhaarmücke *f*
e bibio *m* de las huertas
n rouwvlieg *f*
l Bibio hortulanus

**1548 margarodid scales; giant coccids;
marsupial coccids (US)**
(sc; pest)
d Erdperlen *fpl*; große Schildläuse *fpl*
e perlas *fpl* de tierra; cochinillas *fpl*
subterráneas
l Margarodidae

1549 marketing price
(EEC)
f prix *m* d'écoulement
d Absatzpreis *m*
e precio *m* de venta
n verkoopprijs *m*

1550 marketing year
(EEC)
f campagne *f* de commercialisation
d Wirtschaftsjahr *n*

e campaña *f* de comercialización
n verkoopseizoen *n*

* **marsh betony** → 1552

* **marsh crane fly** → 571

1551 **marsh elder** *(US)*; **burweed marsh elder**
(US); **false ragweed** *(US)*
(sb; weed)
l Iva xanthifolia

1552 **marsh woundwort; marsh betony**
(sb; weed)
f épiaire *f* des marais
d Sumpf-Ziest *m*
e salvia *f* de pantano
n moerasandoorn *m*; moerasandoren *m*
l Stachys palustris

* **marsupial coccids** *(US)* → 1548

1553 **mash; mush**
f râpure *f*
d Mus *n*; Musanteile *mpl*; Brei *m*
e pasta *f*; masa *f*
n moes *n*

1554 **massecuite; fillmass** *(US)*
(The concentrated syrup or molasses in
which the sugar has been crystallized or
the material has been concentrated to a
point at which it will crystallize.
Massecuites are designated by names,
numbers, or letters, indicating their
relative purity or the number of crops of
crystals of sugar that are to be removed)
f masse *f* cuite; m c *f*
d Füllmasse *f*; Kochmasse *f*; FM *f*
e masa *f* cocida; masacocida *f*; masa *f* de
relleno
n masse cuite *f*; vulmassa *f*

1555 **massecuite charging valve; massecuite
gate** *(US)*; **massecuite mixer gate** *(US)*
f trappe *f* à masse cuite; trappe *f* du
malaxeur distributeur
d Füllmasseschieber *m*
e compuerta *f* de la masa cocida
n masse cuiteschuif *f*; vulmassaschuif *f*

1556 **massecuite pump**
f pompe *f* à masse cuite
d Füllmassepumpe *f*
e bomba *f* de masa cocida
n masse cuitepomp *f*; vulmassapomp *f*

1557 **maximum quota**
(EEC)

f quota *m* maximum
d Höchstquote *f*
e cuota *f* máxima
n maximumquotum *n*

* **mayweed camomile** → 2286

* **mayweed chamomile** → 2286

1558 **meadow froghopper; cuckoo-spit insect;
common froghopper** *(adult)*; **meadow
spittlebug** *(US)*
(sb; pest)
f aphrophore *f* écumeuse; cigale *f*
bedeaude; cicadelle *f* écumeuse; cercope
m écument
d Schaumzikade *f*; gemeine Schaumzikade *f*; Wiesenschaumzikade *f*;
Weidenschaumzikade *f*
e espumadora *f*; cigarra *f* de la espuma;
cigarra *f* espumosa; cercópido *m*
espumoso; cicádula *f* espumosa
n schuimbeestje *n*; schuimcicade *f*
l Philaenus leucophthalmus; Philaenus
spumarius

* **meadow moth** → 257

1559 **meadow salsify; goatsbeard; yellow
goatsbeard; yellow goat's-beard** *(US)*;
goat's beard; meadow goat's-beard
(sb; weed)
f salsifis *m* des prés; barbe *f* de bouc
d Wiesenbocksbart *m*
e barba *f* cabruna; barbón *m*; salsifí *m*
común
n gele morgenster *f*
l Tragopogon pratensis

* **meadow soft grass** → 2737

* **meadow spittlebug** *(US)* → 1558

1560 **mean aperture; MA**
(The average dimension of a screen
opening, through which granulated sugar
will just pass. Used to characterize grain
size)
f ouverture *f* moyenne
d mittlere Maschenweite *f*; durch-
schnittliche lichte Maschenweite *f*
e apertura *f* media; A.M. *f*
n gemiddelde maaswijdte *f*

1561 **measure *v* the polarization**
f mesurer la polarisation
d die Polarisation messen; die Polarisation
bestimmen
e medir la polarización

n de polarisatie meten; de polarisatie bepalen

1562 measuring tank
f bac *m* de mesurage; bac *m* jaugeur; bac *m* mesureur
d Meßgefäß *n*
e tanque *m* de medición
n meetbak *m*

1563 mechanical circulation
f circulation *f* mécanique
d mechanische Zirkulation *f*
e circulación *f* mecánica
n mechanische circulatie *f*

1564 mechanical feeder
f alimentateur *m* mécanique
d mechanischer Zuführer *m*
e alimentador *m* mecánico
n mechanische toevoerinrichting *f*

1565 mechanical filter
f filtre *m* mécanique
d mechanisches Filter *n*
e filtro *m* mecánico
n mechanisch filter *n*

1566 mechanical losses
f pertes *fpl* mécaniques
d mechanische Verluste *mpl*
e pérdidas *fpl* mecánicas
n mechanische verliezen *npl*

1567 mechanical thickener
f décanteur *m* mécanique
d mechanischer Eindicker *m*
e decantador *m* mecánico
n mechanische indikker *m*

1568 mechanical unloading
f déchargement *m* mécanique
d mechanische Entladung *f*
e descarga *f* mecánica
n mechanisch lossen *n*

* **megass(e)** *(British Commonwealth)*
→ 119

1569 melanoidin
(Colo(u)ring material formed by complex reactions of reducing sugars with amino compounds, especially with amino acids and peptides)
f mélanoïdine *f*
d Melanoidin *n*
e melanoidina *f*
n melanoïdine *f*

1570 melassigenic; molasses forming
(Tends to increase amount of molasses produced)
f mélassigène
d melassebildend
e melasigénico
n melassevormend

* **melassigenic matter** → 1626

* **meld weed** → 970

* **melitriose** → 1883

* **melon and cotton aphid** → 662

* **melon aphid** → 662

1571 melt *v*
(To dissolve)
f refondre
d auflösen
e fundir; diluir; refundir
n oplossen

1572 melter; dissolver; melting pan
(The apparatus for dissolving crystalline sugar in juice, syrup, or water)
f bac *m* de refonte
d Auflösegefäß *n*; Auflösepfanne *f*
e tanque *m* para refundición
n oplosbak *m*

1573 melter-type centrifuge
f centrifugeuse *f* de dissolution
d Auflösezentrifuge *f*
e centrífuga *f* de refundición
n oploscentrifuge *f*

1574 melt-house
f atelier *m* de refonte
d Auflösestation *f*
e estación *f* de refundición
n oplosstation *n*

* **melting pan** → 1572

1575 melting point
f point *m* de fusion
d Schmelzpunkt *m*
e punto *m* de fusión
n smeltpunt *n*

1576 melt liquor; washed sugar liquor
f clairce *f*; refonte *f* de sucre affiné; refonte *f* de sucre claircé
d Kläre *f*
e licor *m* clarificado; jugo *m* clarificado;

licor *m* de azúcar lavado
n klaarsel *n*

1577 mesh *(of screen)*
f maille *f*
d Masche *f*
e malla *f*
n maas *f*

**1578 Messchaert groove; juice groove;
Messchaert**
f rainure *f* Messchaert; messchaert *f*;
mécharte *f*
d Messchaert *f*
e ranura *f* messchaert; Messchaert *f*;
mecharte *f*
n Messchaert-groef *f*

1579 Messchaert scraper
f peigne *m* à messchaert
d Messchaert-Abstreifer *m*
e raspador *m* Messchaert
n Messchaert-schraper *m*

1580 metastable zone
(The region of supersaturation in which
crystals already present will grow, but
new-crystal formation (nucleation) will
not readily occur)
f zone *f* métastable
d metastabile Zone *f*; metastabiles
Gebiet *n*
e zona *f* metaestable; zona *f* metastable
n metastabiele zone *f*

1581 methyl orange
(pH control)
f méthylorange *m*
d Methylorange *n*
e naranja *f* de metilo
n methyloranje *n*

1582 methyl red
(pH control)
f rouge *m* de méthyle
d Methylrot *n*
e rojo *m* de metilo
n methylrood *n*

* **michaelis** → 100

1583 micro-organism
f micro-organisme *m*
d Mikroorganismus *m*
e microorganismo *m*
n micro-organisme *n*

1584 middle juice
f jus *m* d'effet du milieu

d Mittelsaft *m*
e jugo *m* medio; guarapo *m* medio
n middensap *n*

1585 middle juice carbona(ta)tion
f carbonatation *f* de demi-sirop
d Mittelsaft-Carbonatation *f*
e carbonatación *f* del jugo medio;
carbonatación *f* del guarapo medio
n carbonatatie *f* van het middensap

1586 middle product massecuite
f masse *f* cuite de deuxième jet
d Mittelproduktfüllmasse *f*
n tussenprodukt-masse cuite *f*;
tussenproduktvulmassa *f*

1587 migratory locust
(sc; pest)
f criquet *m* migrateur; criquet *m* voyageur
d europäische Wanderheuschrecke *f*;
Wanderheuschrecke *f*
e langosta *f* migratoria europea; langosta *f*
migratoria; langosta *f* emigrante
l Locusta migratoria

* **milk gowan** → 733

* **milk of lime** → 1472

1588 milk of lime distributor
f doseur *m* de lait de chaux
d Kalkmilchverteiler *m*
e dosificador *m* de lechada de cal
n kalkmelkverdeler *m*

1589 milk of lime pump
f pompe *f* à lait de chaux
d Kalkmilchpumpe *f*
e bomba *f* para lechada de cal
n kalkmelkpomp *f*

1590 milk of lime screen; lime-milk classifier
f tamiseur *m* à lait de chaux
d Kalkmilchsieb *n*; Kalkmilchklassierer *m*;
Kalkmilchgrießabscheider *m*
e tamisador *m* de lechada de cal
n kalkmelkzeef *f*

* **milk sowthistle** → 583

* **mill** → 405

* **mill cheek** → 1272

1591 mill choke
f engorgement *m* du moulin
d Mühlenverstopfung *f*
e atasco *m* en el molino; atascamiento *m*

en el molino
n molenverstopping f

1592 mill control
f contrôle m des moulins
d Mühlenkontrolle f
e control m de molinos
n molencontrole f

* **mill-crusher** → 2499

1593 milling *(of cane)*
f broyage m
d Vermahlung f
e molienda f
n vermalen n

1594 milling capacity *(of cane mill)*
f capacité f de broyage
d Vermahlungskapazität f
e capacidad f de molienda
n vermaalcapaciteit f

1595 milling loss; sucrose/fibre content
f coefficient m saccharose/ligneux; pertes
fpl aux moulins
d Saccharose/Fiber-Verhältnis n
e pérdidas fpl en los molinos
n saccharose/vezelstof-verhouding f

* **milling plant** → 1601

**1596 milling season; grinding season;
crushing season**
f campagne f
d Vermahlungssaison f; Verarbeitungs-
saison f
e zafra f; campaña f
n campagne f

1597 mill juice
f jus m de moulin
d Mühlensaft m
e jugo m de molino; guarapo m de molino
n molensap n

1598 mill roll(er) *(of cane mill)*
f cylindre m de moulin
d Mühlenwalze f; Mühlenroller m; Zucker-
rohrmühlenroller m
e cilindro m de molino; rodillo m de
molino; maza f de molino
n molencilinder m

1599 mill speed
f vitesse f du moulin
d Mühlenrollerdrehzahl f; Mühlenwalzen-
drehzahl f

e velocidad f del molino
n toerental n van de molencilinders

1600 mill station
f atelier m de broyage
d Mühlenstation f
e estación f de molinos
n molenstation n

**1601 mill tandem; milling plant; train of
mills; mill train**
f train m de moulins; batterie f de
moulins
d Mühlenzug m; Mühlenstraße f; Mühlen-
tandem n
e batería f de molinos; tren m de molinos;
tándem m de molinos
n molenbatterij f; moleninstallatie f;
molensysteem n

1602 mill tray
f cuvette f de moulin
d Mühlentrog m
e cubeta f de molino
n molentrog m

* **mill with fixed ratio** → 609

1603 mingle v
f malaxer
d einmaischen
e mezclar
n mengen

1604 mingler
(Apparatus for receiving the massecuite
discharged from a crystallizer, adjusting
or controlling its temperature, and
distributing it among a set of
centrifugals; usually a trough or tube
equipped with a rotating coil of pipe
through which water at a controlled
temperature is circulated)
f malaxeur m
d Maische f; Maischtrog m; Mischtrog m
e mezclador m
n malaxeur m; mengbak m; mengtrog m

1605 minimum premium
(EEC)
f bonification f minimum
d Mindestzuschlag m
e bonificación f mínima
n minimumbonificatie f

1606 minimum price
(EEC)
f prix m minimum
d Mindestpreis m

e precio *m* mínimo
n minimumprijs *m*

1607 minimum stock
(EEC)
f stock *m* minimum
d Mindestlagermenge *f*; Mindestbestand *m*
e stock *m* mínimo
n minimumvoorraad *m*

1608 mix *v*
f mélanger; malaxer
d mischen; maischen; aufmaischen;
einmaischen
e mezclar
n mengen

1609 mixed juice; dilute juice
(The mixture of primary and secondary
juices which enters the boiling house)
f jus *m* mélangé
d Mischsaft *m*
e guarapo *m* mezclado; guarapo *m* diluido;
jugo *m* mezclado; jugo *m* diluido
n gemengd sap *n*

1610 mixed price
(EEC)
f prix *m* mixte
d Mischpreis *m*
e precio *m* mixto
n mengprijs *m*

1611 mixer
(A surge vessel for receiving massecuite
batchwise from a vacuum pan and
discharging it more or less continuously
to a centrifugal station or continuous
crystallizer)
f malaxeur *m*
d Füllmasseverteiler *m*
e mezclador *m*
n menger *m*

1612 mixer-type centrifuge
f centrifugeuse *f* d'empâtage
d Einmaischzentrifuge *f*
e centrífuga *f* para mezclar
n mengcentrifuge *f*

1613 mixing
f malaxage *m*
d Maische *f*; Maischenarbeit *f*; Maisch-
arbeit *f*
e mezclado *m*
n malaxeren *n*

1614 mixing machine
f mélangeur *m*

d Mischmaschine *f*
e mezclador *m*
n mengmachine *f*

**1615 mixing plant for molasses and dried
pulp**
f installation *f* pour fourrages mélassés
d Melassieranlage *f*
e instalación *f* de melazar
n installatie *f* voor melassevoer

1616 mixing tank
f bac *m* mélangeur
d Mischkessel *m*
e tanque *m* mezclador
n mengbak *m*

1617 mixture
(Cane molasses containing 26.5% or more
water and 42.5% or less total sugars)
f "Mixture"
d "Mixture"
e "Mixture"
n "Mixture"

* **mock cypress** → **1411**

* **moisture content** → **1278**

1618 molasse *v*
f mélasser
d melassieren
e melazar
n melasseren

1619 molassed dried pulp
f pulpes *fpl* séchées mélassées; pulpes *fpl*
sèches mélassées
d melassierte Trockenschnitzel *npl*
e pulpa *f* seca melazada
n gemelasseerde droge pulp *f*; droge
melassepulp *f*

1620 molassed pulp; molasses pulp
(By-product)
f pulpes *fpl* mélassées
d Melassepülpe *f*; Melasseschnitzel *npl*
e pulpa *f* de melaza; pulpa *f* melazada
n melassepulp *f*

* **molasses** → **1053**

1621 molasses balance (*of a country*)
f bilan *m* de mélasse
d Melassebilanz *f*
e balance *m* de melaza
n melassebalans *f*

1622 molasses desugarizing plant; molasses desugaring plant
f sucraterie f; installation f de désucrage de la mélasse; installation f de dessucrage de la mélasse
d Melasseentzuckerungsanlage f
e planta f para desazucaración de melaza; fábrica f de extracción de azúcar de las melazas
n melasseontsuikeringsinstallatie f

* **molasses exhaustibility** → 941

1623 molasses exhaustion
f épuisement m de la mélasse
d Melasseerschöpfung f
e agotamiento m de la melaza
n melasseuitputting f

1624 molasses film *(surrounding the crystals)*
f film m d'égout
d Ablauffilm m
e película f de miel
n stroopfilm m

1625 molasses fodder
f mélapaille f; fourrage m mélassé
d Melassemischfutter n; Melassefutter n
e forraje m melazado
n melassevoeder n; melassevoer n

* **molasses forming** → 1570

1626 molasses-forming matter; melassigenic matter
f substances fpl mélassigènes
d Melassebildner mpl
e sustancias fpl melasigénicas; sustancias fpl capaces de formar melaza
n melassevormende stoffen fpl

* **molasses pulp** → 1620

* **molasses slop** → 1630

1627 molasses sugar
f sucre m de mélasse; sucre-mélasse m
d Melassezucker m
e azúcar m de melaza; azúcar m melado
n melassesuiker m

1628 molasses tank
f réservoir m à mélasse
d Melassebehälter m
e depósito m para melaza
n melassetank m

1629 molasses utilization plant
f installation f d'utilisation de la mélasse

d Melasseverwertungsanlage f
e planta f de utilización de la melaza
n melasseverwerkingsinstallatie f

1630 molasses vinasses; molasses slop
f vinasse f de mélasse
d Melasseschlempe f
e melaza f residual; vinaza f de melaza
n melassevinasse f

1631 molasses yield
f rendement m en mélasse
d Melasseausbeute f
e rendimiento m de melaza
n melasseopbrengst f

1632 mole cricket
(sb; pest)
f taupe-grillon f; courtilière f commune
d Maulwurfsgrille f; europäische Werre f; Werre f
e alacrán m cebollero; grillotalpa m; grillo m topo; grillo m real
n veenmol m
l Gryllotalpa vulgaris; Gryllotalpa gryllotalpa

1633 molybdene deficiency
(sb; disease)
f carence f en molybdène
d Molybdänmangel m
e deficiencia f de molibdeno
n molybdeengebrek n

1634 monetary compensating amount
(EEC)
f montant m compensatoire monétaire; MCM m
d Währungsausgleichsbetrag m; WAB m
e importe m compensatorio monetario; IMC m; montante m compensatorio monetario; MCM m
n monetair compenserend bedrag n

1635 monitor; spray nozzles
(In handling beets, a swiveling hydraulic nozzle used to flush the stored beets into the flumes of a beet slab)
f mitrailleuse f
d Spritzköpfe mpl
e cabezales mpl de extrusión
n waterkanon n

1636 monogermity; singleness *(US)*
(agr)
f monogermie f
d Einkeimigkeit f; Monogermität f
e monogermia f
n eenkiemigheid f

1637 monogerm seed
(agr)
f graine f monogerme; semence f
   monogerme
d einkeimiges Saatgut n; monogermes
   Saatgut n
e semilla f monogermen
n eenkiemig zaad n

1638 monosaccharide
f monosaccharide m
d Monosaccharid n
e monosacárido m
n monosaccharide f

1639 monte jus
(A juice-pumping device without moving
parts except for valves, which operates
by steam or air pressure)
f monte-jus m
d Saftpumpe f; Saftheber m
e montajugos m
n sappomp f; montejus m

1640 mormon cricket *(US)*
(sb; pest)
l Anabrus simplex

* **mosaic, beet** ~ → 212

* **mosaic, sugar cane** ~ → 2346

1641 mother liquor
(The solution from which crystals are
formed)
f égout-mère m; liqueur-mère f
d Muttersirup m; Mutterlauge f
e licor m madre
n moederloog f

1642 mottled stripe
(sc; disease)
f rayure f bigarrée
d scheckige Streifigkeit f
e estría f moteada
n vlekkerige-strepenziekte f; vlekkerige
   streping f

**1643 mouse-ear(ed) chickweed; mouse-ear;
clammy chickweed; common mouse-ear**
(sb; weed)
f céraiste m vulgaire; mouron m d'alouette
d gemeines Hornkraut n
e hierba f del cuerno
n gewone hoornbloem f
l Cerastium vulgatum

* **mouse foxtail** → 279

1644 movement water
(Sugar boiling)
f eau f de circulation
e agua f de circulación; agua f de
   movimiento
n circulatiewater n

1645 moving end plate; press plate; follower
f sommier m mobile
d Spindelbock m
e cabezal m movible; cabezal m móvil
n losse kop m

1646 mud
(First or second carbonatation
precipitates)
f boue f
d Schlamm m; Carbonatationsschlamm m
e cachaza f
n schuimaarde f

1647 mud box; mud outlet box *(of clarifier)*
f boîte f à boues; boîte f de décharge des
   boues
d Schlammkasten m
e caja f de cachazas
n slibkist f

1648 mud concentrate *(from the decanter)*
f boues fpl épaissies; boues fpl épaisses
d Dickschlammsaft m
e concentrado m; concentrado m fangoso
n ingedikt slib n

1649 mud mixing tank
f récipient m de malaxage des tourteaux
d Schlammaischgefäß n
e depósito m mezclador para suspender la
   torta
n schuimaardemengbak m

* **mud outlet box** *(of clarifier)* → 1647

1650 mud pump
f pompe f à boues
d Schlammpumpe f
e bomba f de cachaza
n slibpomp f

1651 muds
f boues fpl; écumes fpl
d Schlamm m
e lodo m; fango m; cachaza f; espuma f
n schuimaarde f

* **mud settling pond** → 2116

* **mud water** → 2119

* **multi-effect evaporator station** → **1656**

1652 multigerm seed
(agr)
f semence f multigerme; graine f
multigerme
d mehrkeimiges Saatgut n; multigermes
Saatgut n; polykarpes Saatgut n
e semilla f multigermen
n meerkiemig zaad n

1653 multijet condenser
f condenseur m multijet; condenseur m à
multijets; condenseur m à plusieurs
tuyères
d Strahlkondensator m mit mehreren
Strahldüsen
e condensador m de chorros múltiples
n waterstraalcondensor m met meerdere
straalpijpen

* **multiple effect** → **1655**

1654 multiple-effect evaporation
f évaporation f à multiple effet
d mehrstufige Verdampfung f; Mehrfach-
verdampfung f; Mehrkörper-
verdampfung f; Vielkörperverdampfung f
e evaporación f en múltiple efecto;
evaporación f multietapa
n meervoudige verdamping f

**1655 multiple-effect evaporator; multiple
effect**
(A series of evaporator effects operating
at successively reduced vapo(u)r
pressure, and so connected that the first
effect is heated by steam, the second by
vapo(u)r generated by the first effect, and
so on)
f multiple effet m; évaporateur m à
multiple effet
d Mehrfachverdampfer m; Mehrstufen-
verdampfer m; Mehrkörper-
verdampfer m; Mehrkörperverdampf-
apparat m; Vielkörperverdampfer m
e múltiple efecto m
n meervoudige verdamper m

**1656 multiple effect evaporator station; multi-
effect evaporator station**
f station f d'évaporation à multiple effet;
groupe m évaporatoire à multiple effet
d mehrstufige Verdampfanlage f;
Mehrstufenverdampfanlage f;
mehrstufige Verdampfstation f
e instalación f de evaporación de múltiple
efecto
n meervoudige verdampingsinstallatie f

* **mush** → **1553**

1657 musk thistle; musk bristle thistle
(sb; weed)
f chardon m penché; chardon m nu
d nickende Distel f; Bisamdistel f; Esel-
distel f
e cardo m almizclero
n knikkende distel m
l Carduus nutans

* **mycosis** → **1143**

1658 myriapods; centipedes; myriopods
(sc; pest)
f myriapodes mpl; mille-pattes mpl;
mille-pieds mpl
d Tausendfüßler mpl; Tausendfüßler mpl;
Vielfüßler mpl
e miriápodos pl
n duizendpootachtigen mpl
l Myriapoda

1659 myriogenospora leaf binding
(sc; fungous disease)
d Blattverknäulung f
e ligadura f myriogenospora de la hoja

* **myriopods** → **1658**

N

* **narrow-leaved plantain** → 1979

1660 natural alkalinity
 f alcalinité *f* naturelle
 d natürliche Alkalität *f*
 e alcalinidad *f* natural
 n natuurlijke alkaliteit *f*

1661 natural circulation
 f circulation *f* naturelle
 d natürliche Zirkulation *f*
 e circulación *f* natural
 n natuurlijke circulatie *f*

1662 natural circulation pan; pan with natural circulation
 f appareil *m* à cuire à circulation naturelle
 d Kochapparat *m* mit natürlicher Zirkulation; Verdampfungskristallisator *m* mit natürlicher Zirkulation
 e tacho *m* al vacío con circulación natural
 n kookpan *f* met natuurlijke circulatie

1663 natural fibre
 (Fibre plus water of constitution (Java))
 f fibre *f* naturelle
 d natürliche Fiber *f*
 e fibra *f* natural
 n natuurlijke vezel *f*

1664 natural seed; whole seed *(US)*
 (agr)
 f graine *f* ordinaire; graine *f* normale; graine *f* naturelle
 d Normalsamen *m*
 e semilla *f* ordinaria; semilla *f* natural; semilla *f* normal
 n normaal zaad *n*

* **N.C.V.** *(of bagasse)* → **1667**

* **near-white sugar** → **1690**

1665 necrotic tips of heart leaves
 (sb; disease)
 f nécrose *f* terminale sur les feuilles du coeur
 d Spitzennekrosen *fpl* an den Herzblättern
 e necrosis *f* del ápice de las hojas del corazón
 n necrosis *f* van de punten van de hartbladeren

1666 needle nematodes
 (sc; pest)

 d Nadelälchen *npl*
 l Longidorus spp.

1667 net calorific value; net C.V.; N.C.V. *(of bagasse)*
 f pouvoir *m* calorifique inférieur; PCI *m*
 d spezifischer Heizwert *m*; unterer Heizwert *m*
 e valor *m* calorífico inferior
 n stookwaarde *f*; onderste verbrandingswaarde *f*

1668 nettleleaf goosefoot *(US)*; **nettle-leaved goosefoot**
 (sb; weed)
 f chénopode *m* des murs; ansérine *f* des murs
 d Mauergänsefuß *m*
 e cenizo *m*
 n muurganzevoet *m*
 l Chenopodium murale

1669 neutral juice
 f jus *m* neutre
 d neutraler Saft *m*
 e jugo *m* neutral; guarapo *m* neutral
 n neutraal sap *n*

1670 neutral medium
 f milieu *m* neutre
 d neutrales Milieu *n*; neutraler Bereich *m*
 e medio *m* neutro
 n neutraal milieu *n*

* **nice hemp nettle** → **1427**

1671 nitrogen
 f azote *m*
 d Stickstoff *m*
 e nitrógeno *m*
 n stikstof *f*

1672 nitrogen deficiency
 (sb; disease)
 f carence *f* en azote
 d Stickstoffmangel *m*
 e deficiencia *f* de nitrógeno; carencia *f* de nitrógeno
 n stikstofgebrek *n*

1673 nitrogen fertilizer; nitrogenous fertilizer
 (agr)
 f engrais *m* azoté
 d Stickstoffdünger *m*
 e abono *m* nitrogenado
 n stikstofmeststof *f*

1674 noble cane
 f canne *f* noble

d edles Rohr *n*
e caña *f* noble
n edel riet *n*

1675 non-centrifugal sugar
f sucre *m* non centrifugé
d nichtzentrifugierter Zucker *m*; nicht-
abgeschleuderter Zucker *m*
e azúcar *m* no centrífugo
n niet-gecentrifugeerde suiker *m*

1676 non-crystallizable sugar
f sucre *m* non cristallisable; sucre *m*
incristallisable
d nichtkristallisierbarer Zucker *m*
e azúcar *m* incristalizable
n niet-kristalliseerbare suiker *m*

* **nonsaccharide** → **1680**

* **nonsaccharide content** → **1681**

* **non-soluble lime salts** → **1320**

1677 non-Steffen factory
f sucrerie *f* sans sucraterie
d Zuckerfabrik *f* ohne Steffen-Anlage
e azucarera *f* sin instalación de sacarato
n suikerfabriek *f* zonder Steffen-installatie

1678 nonsucrose
(Any water-soluble matter present which
is not sucrose)
f non-saccharose *m*
d Nichtsaccharose *f*
e no-sacarosa *f*
n nietsaccharose *f*

1679 nonsucrose content
(The content of water-soluble matter
other than sucrose)
f teneur *f* en non-saccharose
d Nichtsaccharosegehalt *m*
e contenido *m* de no-sacarosa
n nietsaccharosegehalte *n*

1680 nonsugar; nonsaccharide
(Any water-soluble matter present which
is not a saccharide)
f non-sucre *m*; N.S. *m*
d Nichtzucker *m*; Nichtzuckerstoff *m*;
NZ *m*
e no-azúcar *m*
n nietsuiker *m*

1681 nonsugar content; nonsaccharide content
(The content of water-soluble matter
other than saccharides)
f teneur *f* en non-sucre

d Nichtzuckergehalt *m*
e contenido *m* de no-azúcar
n nietsuikergehalte *n*

* **normal juice** → **2581**

**1682 Northern nut grass *(US)*; yellow nut
sedge *(US)*; yellow nut grass; chufa;
earth almond; ground almond; rush nut
(US); chufa flat sedge**
(sb; weed)
f souchet *m* comestible; amande *f* de
terre; gland *m* de terre
d Erdmandel *f*; Erdmandelzypergras *n*;
Kaffeewurzel *f*
e chufa *f*; juncia *f* avellanada
n aardamandel *f*
l Cyperus esculentus

1683 Northern root-knot nematode
(sb; pest)
d nördliches Wurzelgallenälchen *n*
l Meloidogyne hapla

* **noxious N** → **1222**

* **noxious nitrogen** → **1222**

* **nozzle washer** → **1365**

1684 nucleation
(Formation of small nuclei, such as seed
crystals)
f nucléation *f*; formation *f* des germes
cristallins; formation *f* de centres
cristallins; formation *f* de nuclei
d Kristallkeimbildung *f*; Kristallkern-
bildung *f*
e nucleación *f*; formación *f* de núcleos
n vorming *f* van kristalkernen

**1685 number of the mill; place of the mill in
the set; place of the mill in the tandem;
position of the mill in the tandem**
f rang *m* du moulin
d Mühlennummer *f*
e colocación *f* del molino
n molennummer *n*

**1686 nut grass *(US)*; purple nut sedge; coco
grass; nut-grass flat sedge**
(sb; weed)
f souchet *m* rond; souchet *m* officinal
d Nußgras *n*
e juncia *f*; castañuela *f*; junquilla *f*; juncia
f redonda; tamascán *m*; cipero *m*
l Cyperus rotundus

O

1687 oakleaf goosefoot *(US)*; **oak-leaved goosefoot**
(sb; weed)
f chénopode *m* glauque
d graugrüner Gänsefuß *m*
n zeegroene ganzevoet *m*
l Chenopodium glaucum

* **OCT** → **1711**

1688 off-season; slack season
f intercampagne *f*; période *f* d'intercampagne; entrecoupe *f*
d Stillstandszeit *f*
e tiempo *m* muerto; período *m* fuera de temporada
n intercampagne *f*

1689 offsetting of storage costs
(EEC)
f compensation *f* des frais de stockage
d Ausgleich *m* der Lagerkosten
e compensación *f* de los gastos de almacenamiento
n vereveningsregeling *f* voor opslagkosten

1690 off-white sugar; near-white sugar
f sucre *m* quasi blanc
d beinah weißer Zucker *m*
e azúcar *m* casi blanco
n bijna witte suiker *m*

1691 oil separator
f séparateur *m* d'huile
d Ölabscheider *m*
e separador *m* de aceite
n olieafscheider *m*

1692 oligosaccharide
f oligosaccharide *m*
d Oligosaccharid *n*
e oligosacárido *m*
n oligosaccharide *f*

1693 onion thrips; onion louse
(sb; pest)
f thrips *m* de l'oignon; thrips *m* du tabac (et de l'oignon)
d Zwiebelblasenfuß *m*; Tabakblasenfuß *m*; Blasenfuß *m*; Tabakthrips *m*
e trips *m* del tabaco; tripido *m* del tabaco; trips *m* de la cebolla; piojillo *m* del tabaco
n thrips *m* van de ui; thrips *m* van de tabak
l Thrips tabaci

1694 open pan evaporator
f évaporateur *m* ouvert
d offener Verdampfer *m*
e evaporador *m* abierto
n open verdamper *m*

* **open sugar market** → **1125**

* **open web elevator** → **463**

1695 optical density
f densité *f* optique
d optische Dichte *f*
e densidad *f* óptica
n optische dichtheid *f*

1696 orach(e); saltbush
(sb; weed)
f arroche *f*
d Melde *f*; Graumelde *f*
e armuelle *m*
n melde *f*
l Atriplex sp.

* **orange ear** → **2000**

* **orchard grass** *(US)* → **533**

1697 organic acid
f acide *m* organique
d organische Säure *f*
e ácido *m* orgánico
n organisch zuur *n*

1698 organic nonsugar
f non-sucre *m* organique
d organischer Nichtzucker *m*
e no-azúcar *m* orgánico
n organische nietsuiker *m*

1699 original juice
f jus *m* d'origine
d ursprünglicher Saft *m*
e jugo *m* original; guarapo *m* original
n oorspronkelijk sap *n*

1700 original water *(of syrup)*
f eau *f* originelle; eau *f* d'origine
d ursprüngliches Wasser *n*
e agua *f* original
n oorspronkelijk water *n*

1701 oscillating basket
(agr)
f panier *m* oscillant
d Schwingsieb *n*
e depósito *m* oscilante de rejilla
n slingerzeef *f*

1702 osmotic diffusion
  f diffusion f osmotique
  d osmotische Diffusion f
  e difusión f osmótica
  n osmotische diffusie f

1703 ovariicolous smuts; culmicolous smut
  (sc; fungous disease of inflorescence)
  f charbon m ovarien
  d Stengelbrand m
  e carbón m de los ovarios
  n vruchtbeginselbrand m

1704 overall recovery; total recovery
  f récupération f générale
  d Gesamtausbeute f
  e recuperación f total
  n totale opbrengst f

1705 over-carbonated juice; burned juice *(US)*
  f jus m surcarbonaté; jus m bicarbonaté
  d übercarbonatierter Saft m; über-
    saturierter Saft m
  e jugo m sobrecarbonatado; guarapo m
    sobrecarbonatado
  n overgecarbonateerd sap n

1706 overflow pipe
  f trop-plein m; tuyau m de trop-plein
  d Überlaufrohr n
  e tubo m de derrame
  n overlooppijp f

1707 overlime *v*
  f surchauler
  d überkalken
  e sobreencalar
  n overkalken

1708 overliming
  f surchaulage m
  d Überkalkung f
  e sobreencalado m
  n overkalking f

1709 overnight supply
  f provision f de nuit
  d Nachtvorrat m
  e provisión f nocturna; provisión f de la
    noche
  n nachtvoorraad m

1710 overscalding
  f suréchaudage m
  d Überbrühung f
  e sobreescaldamiento m
  n overbroeiing f

1711 Overseas Countries and Territories; OCT
  (EEC)
  f pays mpl et territoires d'outre-mer;
    PTOM mpl
  d überseeische Länder npl und Gebiete;
    ÜLG npl
  e países mpl y territorios de ultramar
  n overzeese landen npl en gebieden;
    landen npl en gebieden overzee

1712 oxalate
  f oxalate m
  d Oxalat n
  e oxalato m
  n oxalaat n

1713 oxalic acid
  f acide m oxalique
  d Oxalsäure f
  e ácido m oxálico
  n oxaalzuur n

* **ox-tongue** → 566

* **oyster plant** → 2611

P

1714 packaging machine
  f machine f à paqueter; empaqueteuse f
  d Paketiermaschine f
  e empaquetadora f
  n pakketteermachine f

* **packed sugar** → 135

1715 paddles
  f pales fpl d'agitation; pales fpl
    agitatrices
  d Rührarme mpl
  e paletas fpl agitadoras
  n roerarmen mpl

1716 Pahala blight
  (sc; fungous disease)
  f "Pahala"
  d Pahala-Krankheit f
  e enfermedad f de Pahala
  n Pahala-ziekte f

1717 pale persicaria
  (sb; weed)
  f renouée f à feuilles de patience
  d filziger Knöterich m
  e polígono m pata perdíz
  n viltige duizendknoop m
  l Polygonum lapathifolium ssp. pallidum

1718 pale-spotted millipede
  (sb; pest)
  f iule m moucheté; iule m à taches jaunes
  d gelbgefleckter Tausendfüßler m
  l Archiboreoilulus pallidus

1719 pale-striped flea beetle *(US)*
  (sb; pest)
  l Systena blanda

1720 pale Western cutworm *(US)*
  (sb; pest)
  l Agrotis orthogonia

* **pan** → 2595

* **pan boiler** *(US)* → 1723

* **pan boiling** → 2300

1721 pan control
  f contrôle m de la cuite; contrôle m des
    appareils à cuire
  d Kochkontrolle f; Kontrolle f der Koch-
    apparate

  e control m de tachos
  n controle f van de kookpannen

* **pan house** → 1724

**1722 panic grass; barnyard grass; barn grass;
barnyard millet; cockspur grass; prickly
grass; cock's-foot; cockspur**
  (sb; weed)
  f oplismène f; pieds mpl de coq; panic m
    pied-de-coq; millet m pied-de-coq
  d Hühnerhirse f; gemeine Hühnerhirse f;
    Stachelhirse
  e cola f de caballo; pierna f de gallo; pata
    f de gallo; zacate m de agua
  n hanepoot m
  l Echinochloa crusgalli

**1723 pan man; sugar boiler; sugar cook;
boiler *(US)*; pan operator *(US)*; pan
boiler *(US)***
  f cuiseur m; ouvrier m cuiseur
  d Kocher m; Zuckerkocher m
  e tachero m; puntista m; operador m de
    tachos; cocedor m
  n koker m; suikerkoker m

**1724 pan station; pan house; boiling station;
boiling house**
  f chantier m des cuites; station f de
    cristallisation; atelier m de
    cristallisation; atelier m de cuite;
    chantier m de cristallisation;
    concentration f
  d Kochstation f; Vakuumstation f
  e departamento m de tachos; estación f de
    tachos; cuarto m de cocción; casa f de
    cocimientos; instalación f de cocción
  n kookstation n

1725 pan vapo(u)rs
  f vapeurs fpl des appareils à cuire;
    vapeurs fpl de cuite
  d Kochbrüden m
  e vahos mpl desprendidos en los tachos;
    vahos mpl de los tachos
  n stoom m van de kookpannen

1726 pan with mechanical circulation
  f appareil m à cuire à circulation forcée;
    appareil m à cuire à circulation
    mécanique
  d Kochapparat m mit mechanischer
    Zirkulation; Verdampfungskristallisator
    m mit mechanischer Zirkulation
  e tacho m al vacío con circulación
    mecánica
  n kookpan f met gedwongen circulatie;

kookpan *f* met geforceerde circulatie;
kookpan *f* met mechanische circulatie

* **pan with natural circulation** → **1662**

1727 paper pulp
(By-product)
f pâte *f* à papier
d Papierzellstoff *m*
e pulpa *f* de papel
n papierpap *f*

**1728 parallel complete harvesters; cross-wise
complete harvesters**
(agr)
f machines *fpl* combinées en parallèle
d nebeneinander fahrende Vollernte-
maschinen *fpl*
e cosechadoras *fpl* de remolacha en
paralelo; máquinas *fpl* combinadas en
paralelo
n naast elkaar rijdende verzamelrooiers
mpl

* **parallel-current condenser** → **536**

1729 parallel flow
f circulation *f* parallèle
d Gleichstrom *m*; Parallelstrom *m*
e circulación *f* paralela
n gelijkstroom *m*

1730 parsley piert; field lady's-mantle
(sb; weed)
f alchémille *f*; alchémille *f* des champs;
alchimilla *f*
d Ackerfrauenmantel *m*; Feldlöwenklau *f*
e alquimila *f*; pie *m* de león
n akkerleeuweklauw *m*; kleine leeuwe-
voet *m*
l Alchemilla arvensis; Aphanes arvensis

1731 part payment; advance payment
(EEC)
f acompte *m*
d Abschlagszahlung *f*
e pago *m* adelanto
n voorschot *n*

1732 Pauly pan
f Pauly *m*
d Paulypfanne *f*
e tacho *m* tipo Pauly
n Paulypan *f*

1733 payment for cane
f règlement *m* des cannes
d Zahlung *f* des Zuckerrohrs; Bezahlung *f*
des Zuckerrohrs

e pago *m* de las cañas
n betaling *f* van het suikerriet

1734 pea aphid; green pea louse
(sc; pest)
f puceron *m* vert du pois; puceron *m* du
pois; puceron *m* vert du haricot
d grüne Erbsenblattlaus *f*; Erbsenlaus *f*
e pulgón *m* verde del guisante; pulgón *m*
verde de la arveja
n erwtebladluis *f*
l Acyrtosiphon pisum

**1735 peach-potato aphid; green-peach aphid
(US); greenfly; spinach aphid**
(sb; sc; pest)
f puceron *m* vert du pêcher; puceron *m*
vert; puceron *m* gris du pêcher
d Pfirsichblattlaus *f*; grüne Pfirsichblattlaus *f*; grüne Pfirsichlaus *f*
e pulgón *m* verde del melocotonero; pulgón
m verde del duraznero; pulgón *m* del
melocotonero; pulgón *m* gris del
melocotonero
n groene perzikluis *f*; perzik(blad)luis *f*
l Myzodes persicae; Myzus persicae

1736 pea leaf miner *(US)*
(sb; pest)
d Zichorienminierfliege *f*
l Liriomyza langei; Liriomyza bryoniae

1737 pearly underwing
(sb; pest)
f rubiconde *f*
d Grassteppen-Bodeneule *f*; rötlichbraune
Erdeule *f*; rotgraue Eule *f*
l Peridroma saucia *(adult)*

1738 pectin
f pectine *f*
d Pektin *n*
e pectina *f*
n pectine *f*

1739 pellet
f pellet *m*
d Pellet *n*; Formling *m*; Preßling *m*
e pellet *m*
n pellet *m*

1740 pellet *v*; coat *v (US)*
f enrober
d pillieren; einhüllen
n inhullen; omhullen

1741 pelleted seed; coated seed *(US)*
(agr)
f semence *f* enrobée; graine *f* enrobée

  d pilliertes Saatgut n; Pillensaat n
  e semilla f píldora; semilla f en píldoras;
    semilla f pildorada
  n pillenzaad n; ingehuld zaad n; omgehuld
    zaad n

1742 pelletizing
  f pelletisation f
  d Pelletieren n
  e pelletización f
  n pelletiseren n

1743 pellet station
  f station f de pelletisation
  d Pelletierstation f
  e estación f de pellet
  n pelletiseringsstation n

1744 Pennsylvania smartweed *(US)*
    (sb; weed)
  l Polygonum pensylvanicum

  * **pennycress** → 1003

  * **penny grass** → 1003

1745 pentosan
  f pentosane m
  d Pentosan n
  e pentosano m
  n pentosan n

  * **pepperweed whitetop** → 1252

1746 peptide
  f peptide m
  d Peptid n
  e péptido m
  n peptide n

1747 percolation of juice
  f percolation f du jus
  d Saftperkolation f
  e percolación f del jugo; percolación f del
    guarapo
  n sappercolatie f

  * **perennial ragweed** *(US)* → 2664

  * **perennial sowthistle** → 653

  * **perforated St. John's wort** → 585

1748 perlite
    (A filter aid)
  f perlite f
  d Perlit m
  e perlita f
  n perliet n

1749 permanent invitation to tender
    (EEC)
  f adjudication f permanente
  d Dauerausschreibung f
  e adjudicación f permanente
  n permanente inschrijving f

1750 persicaria; redshank; lady's thumb *(US)*;
    smartweed *(US)*; **spotted lady's-thumb**
    (sb; weed)
  f renouée f persicaire; pied m rouge;
    persicaire f douce
  d Flohknöterich m; Flohkraut n
  e duraznillo m común; pimentilla f;
    persicaria f manchada
  n perzikkruid n
  l Polygonum persicaria

  * **pH-controller** → 1753

1751 pH-meter
  f pH-mètre m
  d pH-Meßgerät n; pH-Messer m
  e medidor m de pH
  n pH-meter m

1752 pH-recorder
  f enregistreur m de pH
  d pH-Diagrammschreiber m; pH-Wert-
    Schreiber m
  e registrador m de pH
  n pH-schrijver m

1753 pH-regulator; pH-controller
  f régulateur m de pH
  d pH-Regler m
  e regulador m de pH
  n pH-regelaar m

1754 phenolphthalein
    (pH control)
  f phénolphtaléine f
  d Phenolphthalein n
  e fenolftaleína f
  n fenolftaleïne n

1755 phenolphthalein paper
    (pH control)
  f papier m de phénolphtaléine
  d Phenolphthaleinpapier n
  e papel m de fenolftaleïna
  n fenolftaleïnepapier n

1756 phenol red
    (pH control)
  f phénol m rouge
  d Phenolrot n
  e rojo m de fenol
  n fenolrood n

* **phoma root rot** *(US)* → **680**

1757 phosphatation
(The defecation process involving the
formation of flocculated calcium
phosphate in the impure liquor)
f phosphatation *f*
d Phosphatation *f*
e fosfatación *f*
n fosfatatie *f*

1758 phosphate
f phosphate *m*
d Phosphat *n*
e fosfato *m*
n fosfaat *n*

* **phosphate deficiency** → **1760**

1759 phosphoric acid
f acide *m* phosphorique
d Phosphorsäure *f*
e ácido *m* fosfórico
n fosforzuur *n*

**1760 phosphorous deficiency; phosphate
deficiency**
(sb; disease)
f carence *f* en phosphore
d Phosphormangel *m*
e deficiencia *f* de fósforo; carencia *f* de
fósforo
n fosforgebrek *n*

1761 photometer
f photomètre *m*
d Fotometer *n*; Photometer *n*
e fotómetro *m*
n fotometer *m*

* **phytophthora root rot** *(US)* → **2462**

1762 phytophthora seed piece rot
(sc; fungous disease)
f pourriture *f* des boutures (due à
Phytophthora megasperma)
d Phytophthora-Stecklingsfäule *f*
e pudrición *f* phytophthora de las estacas
(semilla)
n Phytophthora-stekrot *n*

1763 picking table
(A roller table or broad belt conveyor
designed to spread out the flow of beets
for the manual removal of foreign
objects)
f tamis *m* à secousses
d Schüttelsieb *n*; Ausklaubetisch *m*

e criba *f* de sacudidas
n schudzeef *f*

* **picnometer** → **1871**

1764 picramic acid
f acide *m* picramique
d Pikraminsäure *f*
e ácido *m* picrámico
n picraminezuur *n*

1765 picric acid
f acide *m* picrique
d Pikrinsäure *f*
e ácido *m* pícrico
n picrinezuur *n*

* **pied de cuite** → **1105**

* **piemarker** → **2612**

* **pigeongrass** → **1195**

* **pigeongrass** *(US)* → **2730**

* **pigmy mangold beetle** → **1872**

1766 pineapple disease
(sc; fungous disease)
f maladie *f* de l'ananas
d Ananaskrankheit *f*; Schwarzfäule *f*;
Stecklingsfäule *f*
e enfermedad *f* del corazón negro de la
caña; enfermedad *f* de la piña;
guacatillo *m*
n ananasziekte *f*; zwartrot *n*

* **pin gras** → **586**

1767 pin nematodes
(sc; pest)
l Cacopaurus spp.; Paratylenchus spp.

1768 piston pump; plunger pump
f pompe *f* à piston
d Kolbenpumpe *f*
e bomba *f* de émbolo; bomba *f* de pistón
n zuigerpomp *f*

* **pit** *v* → **910**

1769 pitch of cane knives
f pas *m* du coupe-cannes
d Teilung *f* der Rohrmesser
e paso *m* de las cuchillas cañeras
n steek *m* van de rietmessen

* **pithworms** → **2710**

* pitting → 909

* place of the mill in the set → 1685

* place of the mill in the tandem → 1685

1770 **plantation white (sugar)**
 f blanc m de plantation
 d Plantationsweißzucker m
 e azúcar m blanco de plantación
 n plantagewitsuiker m

1771 **plant breeder for sugar beet**
 f sélectionneur m de betteraves
 d Rübenzüchter m
 e genetista m de remolachas
 n bietenselecteerder m

1772 **plate** *(of filter press)*
 f plateau m
 d Platte f
 e placa f
 n plaat f

1773 **plate-and-frame filter press**
 f filtre-presse m à plaques et à cadres
 d Rahmenfilterpresse f
 e filtro-prensa m de placas y marcos
 n raamfilterpers f

1774 **plate filter**
 f filtre m à plaques
 d Plattenfilter n
 e filtro m de placas
 n plaatfilter n

1775 **plate heat exchanger**
 f échangeur m de chaleur à plaques
 d Plattenwärmeaustauscher m; Platten-
   wärmeübertrager m
 e cambiador m térmico por placas
 n platenwarmteuitwisselaar m

1776 **plate-type air-heater**
 f réchauffair m à plaques; réchauffeur m
   d'air à plaques
 d Plattenlufterhitzer m
 e calentador m de aire de placas
 n platenluchtverhitter m

1777 **pleospora leaf spot**
 (sb; disease)
 f taches fpl de phoma (sur les feuilles)
 d Phoma-Blattflecken mpl
 e foma f de las hojas; phoma f de las
   hojas
 n Phoma-bladvlekkenziekte f

1778 **plumeless thistle; welted thistle;
   acanthus bristle thistle**
 (sb; weed)
 f chardon m à feuilles d'acanthe; chardon
   m acanthoïde
 d Wegdistel f
 n veeldoornige distel m
 l Carduus acanthoides

* **plunger pump** → 1768

1779 **pokkah boeng**
 (sc; fungous disease)
 f "pokkah boeng" *(déformation du
   bourgeon terminal)*
 d Stengelspitzenfäule f; "Pokkahboeng";
   Stengel- und Spitzenfäule f
 e "pokkah boeng"
 n "pokkah boeng"

* **pol** → 1782

1780 **polarimeter**
 f polarimètre m
 d Polarimeter n
 e polarímetro m
 n polarimeter m

1781 **polariscope**
 f polariscope m
 d Polariskop n
 e polariscopio m
 n polariscoop m

1782 **polarization; pol**
 (The amount of rotation given by a
 substance to the plane of plane-polarized
 light under standard conditions
 expressed as equivalent percent sucrose)
 f polarisation f; pol m
 d Polarisation f
 e polarización f; pol m
 n polarisatie f

1783 **pol content**
 f polarisation f; richesse f
 d Polarisation f
 e polarización f
 n polarisatie f

1784 **polished seed**
 (agr)
 f semence f polie; graine f polie
 d poliertes Saatgut n
 e semilla f pulida; semilla f frotada
 n gepolijst zaad n

* **Pol percent spindle Brix** → 78

1785 **polyploid seed**
(agr)
f semence f polyploïde; graine f polyploïde
d polyploides Saatgut n
e semilla f poliploide
n polyploïde zaad n

* **poor man's weatherglass** → 2058

* **poor molasses** → 1198

* **poor running** *(US)* → 1198

1786 **porous filter cake**
f tourteau m poreux
d poröser Filterkuchen m
e torta f porosa; cachaza f porosa
n poreuze filterkoek m

* **position of the mill in the tandem**
→ 1685

1787 **postdefecation**
f postdéfécation f
d Nachdefäkation f; Nachkalkung f; Nach-
scheidung f
e postdefecación f
n nadefecatie f; nakalking f

* **potash** → 1789

1788 **potassium**
f potassium m
d Kalium n; Kali n
e potasio m
n kalium n; potassium n

1789 **potassium carbonate; potash**
f carbonate m de potassium; potasse f
d Kaliumkarbonat n; Pottasche f
e carbonato m de potasa
n kaliumcarbonaat n; potas f

1790 **potassium chloride**
f chlorure m de potassium
d Kaliumchlorid n
e cloruro m de potasio
n kaliumchloride n

1791 **potassium deficiency**
(sb; disease)
f carence f en potassium
d Kalimangel m
e deficiencia f de potasio; carencia f de
potasio
n kaliumgebrek n

1792 **potassium hydroxide**
f hydroxyde m de potassium

d Kaliumhydroxid n
e hidróxido m de potasio; hidróxido m
potásico
n kaliumhydroxyde n

1793 **potassium oxalate**
f oxalate m de potassium
d Kaliumoxalat n
e oxalato m de potasio
n kaliumoxalaat n

1794 **potato capsid**
(sb; pest)
f punaise f norvégienne; punaise f de la
pomme de terre; capside m; calocoris f
de la pomme de terre
d Kartoffelwanze f; gemeine Schmuck-
wanze f; zweipunktige Wiesenwanze f
e chinche f de la patata
l Calocoris norvegicus

* **potato stem borer** *(US)* → 2000

1795 **povertyweed** *(US)*
(sb; weed)
l Iva axillaris

1796 **powdered sugar; pulverized sugar**
(Finely divided sugar produced by
pulverizing granulated sugar, usually
containing an anti-caking additive)
f sucre m en poudre; sucre m poudre
d Puderzucker m; Staubzucker m
e azúcar m en polvo; azúcar m pulverizado
n poedersuiker m

1797 **powdery mildew**
(sb; disease)
f blanc m; oïdium m
d echter Mehltau m
e oídio m
n echte meeldauw m

1798 **powdery scab**
(sb; disease)
f gale f pustuleuse
d Pustelschorf m
e sarna f pustulosa; roña f pustulosa
n schurft f

1799 **prairie grain wireworm** *(US)*
(sb; pest)
l Ctenicera destructor

1800 **prairie pepperweed**
(sb; weed)
f passerage m à fleurs denses
d dichtblütige Kresse f

l Lepidium densiflorum; Lepidium
apetalum

1801 precarbona(ta)tion
f précarbonatation *f*
d Vorcarbonatation *f*
e precarbonatación *f*
n voorcarbonatatie *f*

1802 precipitate *(noun)*
f précipité *m*
d Niederschlag *m*; Präzipitat *n*
e precipitado *m*
n precipitaat *n*; neerslag *m*

1803 precipitate *v*
f précipiter
d fällen; präzipitieren
e precipitar
n neerslaan; precipiteren

1804 precipitation
f précipitation *f*
d Fällen *n*; Ausscheidung *f*
e precipitación *f*
n precipitatie *f*; neerslaan *n*

**1805 precision drill; precision seed drill;
spacing drill; seed spacing drill**
(agr)
f semoir *m* de précision; semoir *m*
monograine
d Einzelkornsägerät *n*; Einzelkorndrill-
maschine *f*
e sembradora *f* de precisión; sembradora *f*
monograno
n precisiezaaimachine *f*;
precisiezaaiapparaat *n*

1806 precision drilling; spaced seeding
(agr)
f semis *m* de précision; semis *m*
monograine
d Einzelkornaussaat *f*; Einzelkornsaat *f*;
Präzisionssaat *f*
e siembra *f* de precisión; siembra *f*
monograno
n precisiezaai *m*

* **precision seed drill** → **1805**

1807 precoat
(A layer of solid particles, calcium
carbonate, filter aid, etc., deposited on a
filter septum to establish initial clarity of
the filtrate)
f précouche *f*
d Anschwemmschicht *f*; Precoatschicht *f*

e precapa *f*
n precoat *m*; grondlaag *f*; stortlaag *f*

1808 precoat filter
f filtre *m* à précouche
d Anschwemmfilter *n*; Precoatfilter *n*
e filtro *m* con precapa
n precoatfilter *n*; grondlaagfilter *n*;
stortlaagfilter *n*

1809 predefecate *v*; **prelime** *v*
f prédéféquer; préchauler
d vorkalken; vorscheiden
e predefecar; preencalar; prealcalinizar;
prealcalizar
n voorkalken; voorscheiden

**1810 predefecated juice; predefecation juice;
prelimed juice; preliming juice**
f jus *m* prédéféqué; jus *m* préchaulé
d Vorkalkungssaft *m*; vorgekalkter Saft *m*;
vorgeschiedener Saft *m*
e jugo *m* predefecado; jugo *m*
preencalado; jugo *m* prealcalizado;
guarapo *m* predefecado; guarapo *m*
preencalado; guarapo *m* prealcalizado
n voorgekalkt sap *n*

* **predefecation** → **1816**

1811 pre-evaporator
f pré-évaporateur *m*
d Vorverdampfer *m*
e preevaporador *m*
n voorverdamper *m*

1812 preferential quantity
(EEC)
f quantité *f* préférentielle
d Präferenzmenge *f*
e cantidad *f* preferencial
n preferentiële hoeveelheid *f*

1813 preferential sugar
(EEC)
f sucre *m* préférentiel
d Präferenzzucker *m*
e azúcar *m* preferencial
n preferentiële suiker *m*

1814 preheater
f préchauffeur *m*
d Vorwärmer *m*
e precalentador *m*
n voorverwarmer *m*; voorwarmer *m*

1815 preheating zone *(of lime kiln)*
f zone *f* de préchauffage
d Vorwärmzone *f*

e zona *f* de precalentamiento
n voorwarmingszone *f*;
   voorwarmingsruimte *f*

* **prelime** *v* → 1809

* **prelimed juice** → 1810

1816 preliming; predefecation
(Treatment of the raw juice with
relatively small amounts of lime, e.g.,
0.2.-0.3% CaO, over a period of about 20
minutes, to precipitate proteinaceous
matter in a relatively stable form, such
that its repeptization is minimal in
subsequent main liming and
carbonatation, or defeco-carbonatation)
f préchaulage *m*
d Vorkalkung *f*; Vorscheidung *f*
e preencalado *m*; prealcalización *f*;
   prealcalinización *f*
n voorkalking *f*; voorscheiding *f*

* **preliming juice** → 1810

1817 premium
(EEC)
f bonification *f*
d Zuschlag *m*
e bonificación *f*
n bonificatie *f*

* **press cake** → 1016

1818 pressed cubes plant
f installation *f* pour morceaux comprimés
d Preßwürfelanlage *f*
e instalación *f* de cuadradillos prensados
n suikertabletteninstallatie *f*

**1819 pressed juice; press juice; expressed
juice**
(The cellular juice obtained from beets
by pressing)
f jus *m* de pression; jus *m* exprimé
d Preßsaft *m*; ausgepreßter Saft *m*
e jugo *m* de presión; guarapo *m* de
   presión; jugo *m* estrujado; guarapo *m*
   estrujado
n perssap *n*; uitgeperst sap *n*

1820 pressed pulp; press pulp
(The pulp leaving the pulp presses)
f pulpes *fpl* pressées; pulpes *fpl*
   surpressées
d Preßschnitzel *npl*; Preßlinge *mpl*; Preß-
   pülpe *f*
e pulpa *f* prensada; cosetas *fpl* prensadas
n geperste pulp *f*; perspulp *f*

* **press juice** → 1819

* **press** *v* **out** → 947

* **press plate** → 1645

1821 press station
f station *f* des presses
d Pressenstation *f*
e estación *f* de las prensas
n persstation *n*

1822 pressure drum filter
f filtre *m* à tambour sous pression
d Druck-Drehfilter *n*
e filtro-tambor *m* de presión
n druktrommelfilter *n*

1823 pressure filter
f filtre *m* sous pression
d Überdruckfilter *n*; Druckfilter *n*
e filtro *m* de presión
n drukfilter *n*

1824 pressure ga(u)ge
f manomètre *m*
d Druckmesser *m*; Manometer *n*
e manómetro *m*
n manometer *m*

1825 pressure regulator
f régulateur *m* de pression; détendeur-
   régulateur *m*
d Druckregler *m*
e regulador *m* de presión
n drukregelaar *m*

1826 press water
(The liquid effluent from the pulp
presses)
f eaux *fpl* de presses
d Preßwasser *n*
e agua *f* de las prensas
n perswater *n*

1827 press water arc screen
f tamis *m* courbe pour eaux de presses
d Preßwasser-Bogensieb *m*
e colador *m* parabólico para las aguas de
   prensas
n boogzeef *f* voor perswater

1828 press water heater
f réchauffeur *m* pour eaux de presses
d Preßwasser-Wärmer *m*
e calentador *m* de aguas de prensas
n perswaterverwarmer *m*

1829 press water pump
f pompe f à eaux de presses
d Preßwasser-Pumpe f
e bomba f de las aguas de prensas
n perswaterpomp f

1830 press water tank
f caisse f à eaux de presses
d Preßwasser-Kasten m
e depósito m de las aguas de prensas
n perswaterkist f

* **prickly grass** → 1722

1831 prickly lettuce; compass lettuce
(sb; weed)
f laitue f scariola; laitue f serriole; laitue f
sauvage; scarole f; escarole f
d Stachellattich m; wilder Lattich m; Zaun-
lattich m
e escarola f
n kompassla f; wilde latuw f; wilde sla f
l Lactuca scariola; Lactuca serriola

* **prickly saltwort** → 2025

1832 prickly sowthistle *(US)*
(sb; weed)
f laiteron m des champs; laiteron m
épineux
d dornige Gänsedistel f; rauhe Gänsedistel f
e cerraja f; lechecino m
n brosse melkdistel m; ruwe melkdistel m
l Sonchus asper (L.) Hill.

1833 primary air *(to boiler)*
f air m primaire
d Primärluft f; Erstluft f
e aire m primario
n primaire lucht f

1834 primary clarifier; primary subsider
f clarificateur m primaire
d Primärdekanteur m
e clarificador m primario
n primaire decanteur m; primair
decanteertoestel n

1835 primary juice
(All the juice expressed undiluted. In
most mills this is the combined crusher
juice and first-mill juice)
f jus m primaire
d Saft m von Vorbrecher und erster Mühle
e guarapo m primario; jugo m primario
n voorperssap n

* **primary subsider** → 1834

1836 principal invitation to tender
(EEC)
f adjudication f principale
d Hauptausschreibung f
e adjudicación f principal
n hoofdinschrijving f

* **processed cane** → 2537

**1837 processed seed; rubbed and graded seed;
technical monogerm seed**
(agr)
f semence f monogerme technique; graine
f monogerme technique; graine f
monogerme artificielle; graine f façonnée
d technisch einkeimiges Saatgut n;
technisch monogermes Saatgut n;
Präzisionssaatgut n; aufbereitetes Saatgut n
e semilla f monogermen mecánica; semilla
f monogermen técnica
n technisch eenkiemig zaad n; mechanisch
eenkiemig zaad n

1838 processing margin
(EEC)
f marge f de fabrication
d Fabrikationsspanne f
e margen m de fabricación
n fabricagemarge f

1839 production levy
(EEC)
f cotisation f à la production
d Produktionsabgabe f
e cotización f a la producción
n produktiebijdrage f

1840 progressive preliming
f préchaulage m progressif
d progressive Vorkalkung f; progressive
Vorscheidung m
e preencalado m adelanto; prealcalización
f adelanta; prealcalinización f adelanta
n trapsgewijze voorkalking f

1841 progressive preliming of Brieghel-Müller
f préchaulage m selon Brieghel-Müller
d verbesserte progressive Vorkalkung f;
Gegenstromvorkalkung f; Vorkalkung f
nach Brieghel-Müller
n voorkalking f volgens Brieghel-Müller

1842 proofcock
f robinet-sonde m; robinet m d'épreuve
d Probehahn m
e grifo m de prueba; llave m de prueba
n monsterkraan f

1843 proofstick
f sonde f d'épreuve; sonde f à tiroir
d Probestock m; Probestecher m
e sonda f de prueba
n proefstok m; proefsonde f

1844 proportional milk-of-lime distributor
f doseur m proportionnel de lait de chaux
d Kalkmilch-Proportionalverteiler m
e dosificador m proporcional de la lechada de cal
n proportionele kalkmelkverdeler m

* **prostrate knotweed** → **1409**

1845 prostrate pigweed *(US)*; **prostrate amaranth**
(sb; weed)
f amarante f blette
d Gemüsefuchsschwanz m; weißrandiger Fuchsschwanz m
e bledo m
n nerfamarant f
l Amaranthus blitoides

1846 protein
f protéine f
d Protein n; Eiweiß n; Eiweißstoff m
e proteína f
n eiwit n; eiwitstof f; proteïne f

* **pseudo scald** *(Australia)* → **480**

1847 pulp collector; pulp catcher; pulp separator
(A dewatering screen, usually with mechanical wipers or carriers, over which beet pulp passes, usually from a diffuser to pulp press or wet silo)
f épulpeur m
d Pülpefänger m; Pülpeabscheider m
e cogedor m de pulpas
n pulpvanger m; pulpverwijderaar m

1848 pulp cooling plant
f refroidisseur m de pulpes séchées
d Trockenschnitzelkühler m
e refrigerador m de pulpa seca
n koelinstallatie f voor droge pulp

1849 pulp drier
f sécheur m à pulpes; sécheur m de pulpes
d Schnitzeltrockner m
e secador m de pulpa
n pulpdroger m

1850 pulp drier with built-in cross-shaped baffles

f sécheur m à pulpes avec chicanes intérieures en forme de croix
d Schnitzeltrockner m mit Kreuzeinbau; Trockentrommel f mit Kreuzeinbau
e secador m de pulpa con construcción interior de chicanas en cruz
n pulpdroger m met ingebouwde kruisvormige keerplaten

1851 pulp drier with built-in quadrant-shaped baffles
f sécheur m à pulpes avec chicanes intérieures en forme de quadrant
d Schnitzeltrockner m mit Quadranten-einbau; Trockentrommel f mit Quadranteneinbau
e secador m de pulpa con construcción interior de chicanas en cuadrante
n pulpdroger m met ingebouwde kwadrantvormige keerplaten

1852 pulp drying station
f station f de séchage des pulpes
d Schnitzeltrocknungsanlage f
e estación f de secado de pulpa; planta f secadora de pulpa
n pulpdrooginstallatie f

1853 pulp drying with waste flue gas; waste flue gas drying; drying with chimney gas
f séchage m des pulpes par gaz de combustion des générateurs de vapeur
d Schnitzelabgastrocknung f
e secado m de pulpa con los gases de las chimeneas
n pulpdroging f met afvalgassen

1854 pulp elevator
f élévateur m de pulpes
d Schnitzelelevator m
e elevador m de pulpa
n pulpelevator m; pulpophaler m

1855 pulp of scalding process; Steffen sugar pulp; Steffen pulp
f pulpe f du procédé Steffen
d Brühschnitzel npl
e pulpa f de procedimiento de escaldado; pulpa f Steffen
n Steffen-pulp f

1856 pulp pelleting press
f presse f à pellets
d Trockenschnitzelpresse f
e prensa f para pulpa seca
n droge-pulppers f

1857 pulp pit
f fosse f à pulpes

   d Schnitzelkanal *m*
   e foso *m* de pulpas
   n pulpgoot *f*

1858 pulp press
   f presse *f* à pulpes
   d Schnitzelpresse *f*; Pülpepresse *f*
   e prensa *f* de pulpas; prensa *f* para pulpa
   n pulppers *f*

1859 pulp pressing station
   f station *f* de pressage des pulpes
   d Schnitzelpreßstation *f*
   e estación *f* de prensado de pulpa
   n pulppersstation *n*

1860 pulp press juice
   f jus *m* de pulpes pressées
   d Schnitzelpreßsaft *m*
   e jugo *m* de pulpa prensada; guarapo *m* de
      pulpa prensada
   n pulpperssap *n*

   * **pulp separator** → **1847**

1861 pulp silo
   f silo *m* à pulpes
   d Schnitzelsilo *m*
   e silo *m* de pulpa
   n pulpsilo *m*

1862 pulverized lime; ground lime
   f chaux *f* en poudre; chaux *f* pulvérisée;
      chaux *f* pulvérulente
   d Puderkalk *m*; Kalkstaub *m*; Staub-
      kalk *m*
   e cal *f* pulverizada; cal *f* pulverulenta
   n poederkalk *m*

   * **pulverized sugar** → **1796**

1863 pumping tank
   f bac *m* de pompage
   d Abpumpgefäß *n*
   e depósito *m* de bombeo
   n pomptank *m*

1864 puncture vine *(US)*; **punctureweed** *(US)*
   (sb; weed)
   f croix *f* de Malte; tribule *m* terrestre;
      tribulus *m*; herse *f*
   d Bürzeldorn *m*; Erdbürzeldorn *m*
   e abrojo *m* terrestre
   n aardangel *f*
   l Tribulus terrestris

   * **pure diffusion** → **809**

   * **purge** *v* → **460**

   * **purging** → **454**

1865 purified juice
   f jus *m* épuré
   d gereinigter Saft *m*
   e jugo *m* purificado; guarapo *m* purificado
   n gezuiverd sap *n*

1866 purity
   (100 x sucrose/dry substance)
   f pureté *f*
   d Reinheit *f*
   e pureza *f*
   n reinheid *f*

1867 purity drop; purity fall
   f chute *f* de pureté
   d Reinheitsabfall *m*
   e caída *f* de pureza
   n reinheidsverschil *n*

   * **purple dead nettle** → **1908**

   * **purple nut sedge** → **1686**

   * **purple spot** → **1911**

1868 purslane; common purslane; purslain;
      pursley
   (sb; weed)
   f pourpier *m* potager; pourpier *m*;
      pourpier *m* des potagers
   d Portulak *m*; Burzelkraut *n*; Kohl-
      portulak *m*; gelber Portulak *m*; gemeiner
      Portulak *m*
   e verdolaga *f* (común)
   n postelein *m*; porselein *m*
   l Portulaca oleracea

   * **pusher-type centrifugal** → **1870**

1869 push rake
   (agr)
   f râteau *m* récolteur
   d Sammelrechen *m*
   e rastrillo *m* amontonador
   n verzamelhark *f*

1870 push-type centrifugal; pusher-type
      centrifugal
   f centrifugeuse *f* pousseuse; centrifuge *m*
      pousseur
   d Schubzentrifuge *f*
   e centrífuga *f* de impulsión; centrífuga *f*
      de tipo de impulsión

1871 pycnometer; picnometer; pyknometer
   f pycnomètre *m*
   d Pyknometer *n*

   e picnómetro *m*
   n pycnometer *m*

1872 pygmy mangold beetle; pigmy mangold beetle; beet beetle
   (sb; pest)
   f atomaire *m* linéaire; atomaire *m* (de la betterave); atomaria *m* de la betterave
   d Moosknopfkäfer *m*; schmaler Geheimfresser *m*; Runkelrübenkäferchen *n*
   e atomaria *f*; escarabajo *m* (de la remolacha)
   n bietekevertje *n*; bietenkevertje *n*
   l Atomaria linearis

   * **pyknometer** → **1871**

Q

* **quack grass** *(US)* → **664**

1873 quadruple effect; quadruple effect evaporator
f quadruple effet *m*
d Vierkörper-Verdampfapparat *m*; Vierfach-verdampfer *m*
e cuádruple efecto *m*
n verdampapparaat *n* met vier verdamplichamen; verdampapparaat *n* à quadruple effet

1874 quadruple effect evaporation
f évaporation *f* à quadruple effet
d vierstufige Verdampfung *f*; Vierfach-verdampfung *f*; Vierkörper-verdampfung *f*
e evaporación *f* de cuádruple efecto
n viervoudige verdamping *f*; viertrapsverdamping *f*

* **quadruple effect evaporator** → **1873**

* **Queen Anne's lace** → **2693**

1875 Quentin run-off
f égout *m* Quentin
d Quentin-Ablauf *m*
e licor *m* del equipo Quentin
n Quentin-stroop *f*

* **quick grass** *(US)* → **664**

1876 quick lime; unslaked lime; burnt lime; quicklime
f chaux *f* vive
d ungelöschter Kalk *m*; Ätzkalk *m*; Brannt-kalk *m*; gebrannter Kalk *m*
e cal *f* viva
n ongebluste kalk *m*

1877 quintuple effect; quintuple effect evaporator
f quintuple effet *m*
d Fünfkörper-Verdampfapparat *m*; Fünffachverdampfer *m*
e quíntuple efecto *m*; quintuplo efecto *m*
n verdampapparaat *n* met vijf verdamplich .men; verdampapparaat *n* à quintuple effet

1878 quintuple effect evaporation
f évaporation *f* à quintuple effet
d fünfstufige Verdampfung *f*; Fünffach-verdampfung *f*; Fünfkörper-verdampfung *f*

e evaporación *f* de quíntuple efecto; evaporación *f* de quintuplo efecto
n vijfvoudige verdamping *f*; vijftrapsverdamping *f*

* **quintuple effect evaporator** → **1877**

* **quitch grass** *(US)* → **664**

1879 quota
(EEC)
f quota *m*
d Quote *f*
e cuota *f*
n quota *f*; kwota *f*

1880 quota system
(EEC)
f régime *m* des quotas; système *m* des quotas
d Quotenregelung *f*; Quotensystem *n*
e régimen *m* de cuotas
n quotaregeling *f*

1881 quotation
(eco)
f cotation *f*; cours *m*
d Notierung *f*
e cotización *f*
n notering *f*

1882 Q.13 disease
(sc; virus disease)
f "Q.13 disease"
d Q.13-Krankheit *f*
e enfermedad *f* de la Q.13
n Q.13-ziekte *f*

* **Q.28 disease** → **1890**

* **Q.28 trouble** → **1890**

R

1883 raffinose; melitriose
f raffinose *m*; mélitose *m*; mélitriose *m*;
  gossypose *m*
d Raffinose *f*; Melitose *f*; Melitriose *f*
e rafinosa *f*
n raffinose *f*; melitose *f*; melitriose *f*

* **ragweed** *(US)* → 581

1884 ragwort; tansy ragwort; common ragwort;
St. James' wort; ragwort groundsel
  (sb; weed)
f séneçon *m* de Jacob; fleur *f* de Jacob;
  séneçon *m* Jacobée; herbe *f* de Saint-
  Jacques; herbe *f* dorée; herbe *f* des
  charpentiers
d Jakobskreuzkraut *n*; Jakobskraut *n*;
  Jakobsgreiskraut *n*; Wiesenkreuzkraut *n*
e hierba *f* jacobina; hierba *f* de Santiago
n jakobskruiskruid *n*; jakobskruid *n*
l Senecio jacobaea

1885 rake
f râteau *m*
d Rechen *m*
e rastrillo *m*
n hark *f*

1886 rake conveyor; rake type conveyor
f transporteur *m* à râteaux; conducteur *m*
  à râteaux
d Rechenförderer *m*
e transportador *m* de rastrillos
n harktransporteur *m*; harkcarrier *m*

1887 ramularia leaf spot
  (sb; disease)
f ramularia *m*; ramulariose *f*
d Ramularia-Blattfleckenkrankheit *f*
e ramularia *f*; ramulariosis *f*
n Ramularia-bladvlekkenziekte *f*

1888 rapidly cooling crystallizer; rapid cooling
crystallizer
f refroidisseur *m* rapide
d Schnellkühlkristallisator *m*; Schnellkühl-
  maische *f*
e cristalizador *m* por refrigeración rápida
n snelkoeler *m*

1889 rasping station; slicing station
f râperie *f*
d Saftfabrik *f*; Saftstation *f*
n rasperij *f*

1890 ratoon stunting disease; Q.28 trouble;
Q.28 disease; R.S.D.
  (sc; virus disease)
f rabougrissement *m* des pousses; R.S.D.
d "Ratoon Stunting Disease"; RSD;
  Stecklingsstauche *f*; Ratoonkrankheit *f*
e enfermedad *f* del raquitismo de la caña
n "ratoon stunting disease"; RSD

1891 raw juice
  (The sugar-bearing juice obtained from
  the beets in the diffuser. The term
  "diffusion juice" is not sufficiently more
  descriptive to offset its greater length)
f jus *m* brut
d Rohsaft *m*
e jugo *m* crudo; guarapo *m* crudo
n ruwsap *n*

1892 raw juice pump
f pompe *f* à jus brut
d Rohsaftpumpe *f*
e bomba *f* de jugo crudo; bomba *f* de
  guarapo crudo
n ruwsappomp *f*

1893 raw sugar
  (1. A partially refined sugar; 2. In a beet-
  sugar factory, the crystalline product of
  the raw boiling)
f sucre *m* brut; sucre *m* roux
d Rohzucker *m*
e azúcar *m* crudo; azúcar *m* mascabado
n ruwe suiker *m*; ruwsuiker *m*

1894 raw sugar factory
f sucrerie *f* travaillant en roux; sucrerie *f*
  travaillant en brut
d Rohzuckerfabrik *f*
e fábrica *f* de azúcar crudo; fábrica *f* de
  azúcar mascabado
n ruwsuikerfabriek *f*

1895 raw sugar value; raw value
  (eco)
f valeur *f* en sucre brut
d Rohzuckerwert *m*
e valor *m* crudo
n ruwsuikerwaarde *f*

* **raw sugar washing** → 25

* **raw syrup** → 30

* **raw washings** → 1198

1896 raw water pomp
f pompe *f* à eau brute
d Rohwasserpumpe *f*

e bomba *f* de agua natural
n ruwwaterpomp *f*

* **RDS → 1943**

1897 reagent
  f réactif *m*
  d Reagens *n*
  e reactivo *m*
  n reagens *n*; reageermiddel *n*

* **receiving laboratory → 2466**

* **recirculation of syrups → 103**

1898 recooler
  f réfrigérant *m* de retour
  d Rückkühler *m*
  e refrigerador *m* de retorno
  n terugkoeler *m*

1899 recoverable sucrose
  f saccharose *m* récupérable
  d gewinnbare Saccharose *f*
  e sacarosa *f* recuperable
  n winbare saccharose *f*

1900 recoverable sugar; available sugar
  f sucre *m* récupérable
  d gewinnbarer Zucker *m*
  e azúcar *m* recuperable; azúcar *m*
     disponible
  n winbare suiker *m*

1901 recoverable white sugar
  f sucre *m* blanc récupérable
  d gewinnbarer Weißzucker *m*
  e azúcar *m* blanco recuperable
  n winbare witte suiker *m*; winbare
     witsuiker *m*

1902 recrystallization
  f recristallisation *f*
  d Umkristallisation *f*; Rekristallisation *f*
  e recristalización *f*
  n herkristallisatie *f*

1903 recrystallize *v*
  f recristalliser
  d umkristallisieren; wieder aus-
     kristallisieren
  e recristalizar
  n herkristalliseren; omkristalliseren

1904 recycling
  f recyclage *m*
  d Rückführung *f*; Rücknahme *f*
  e recirculación *f*; retorno *m*
  n terugbrenging *f*

1905 recycling pump
  f pompe *f* de recyclage
  d Rücknahmepumpe *f*
  e bomba *f* de recirculación
  n recirculatiepomp *f*

* **red archangel → 1908**

1906 red-backed cutworm
     (sb; pest)
  l Euxoa ochrogaster

* **red beet → 226**

1907 red campion; red bird's eye
     (sb; weed)
  f compagnon *m* rouge
  d rote Lichtnelke *f*; rote Nachtnelke *f*
  e colleja *f* roja
  n dagkoekoeksbloem *f*
  l Melandrium rubrum (Weig.) Garcke;
     Melandrium diurnum; Silene dioica (L.)
     Clairv.; Lychnis dioica

* **red chickweed → 2058**

* **red coast blite** *(US)* **→ 1909**

**1908 red dead nettle; red archangel; purple
     dead nettle**
     (sb; weed)
  f lamier *m* pourpré; ortie *f* pourpre
  d rote Taubnessel *f*; Ackertaubnessel *f*;
     kleine Taubnessel *f*
  e ortiga *f* muerta purpúrea; ortiga *f* de
     flores purpúreas; lamio *m* púrpura
  n paarse dovenetel *f*; makke brandnetel *f*
  l Lamium purpureum

1909 red goosefoot; red coast blite *(US)*;
     French spinach *(US)*
     (sb; weed)
  f chénopode *m* rouge
  d roter Gänsefuß *m*
  n rode ganzevoet *m*
  l Chenopodium rubrum

1910 red hemp nettle
     (sb; weed)
  f galéope *m* commun; galéopsis *m*
     ladanum; ortie *f* rouge
  d breitblättriger Hohlzahn *m*; Ackerhohl-
     zahn *m*
  e cáñamo M silvestre
  n breedbladige raai *m*; raai *m*; smalbladige
     hennepnetel *f*
  l Galeopsis ladanum

1911 red leaf spot; purple spot
    (sc; fungous disease)
  f tache *f* rouge de la feuille
  d rote Blattfleckenkrankheit *f*
  e mancha *f* roja de la hoja; mancha *f*
      púrpura de la hoja
  n rode-vlekkenziekte *f* van het blad

1912 red-legged grasshopper *(US)*; **red-legged**
    locust *(US)*
    (sb; pest)
  l Melanoplus femur-rubrum

  * **red mustard** → 282

1913 red orache *(US)*; **red scale** *(US)*;
    tumbling orache
    (sb; weed)
  d Rosenmelde *f*
  l Atriplex rosea

  * **red pimpernel** → 2058

  * **red poppy** → 652

1914 redroot pigweed *(US)*; **rough pigweed**
    (US); **common amaranth; redroot**
    amaranth; beetroot
    (sb; weed)
  f amarante *f* réfléchie; amarante *f*
      recombée; discipline *f* de religieuse; fleur
      f de la jalousie; fleur *f* de l'amour
  d zurückgekrümmter Fuchsschwanz *m*;
      rauhhaariger Fuchsschwanz *m*; Acker-
      fuchsschwanz *m*
  e bledo *m*; amaranto *m*
  n papegaaikruid *n*
  l Amaranthus retroflexus

1915 red rot
    (sc; fungous disease)
  f morve *f* rouge; pourriture *f* rouge
  d Stammrotfäule *f*; Rot-Rotz *m*; Rotfäule *f*
  e muermo *m* rojo; podredumbre *f* roja
  n rood snot *n*

1916 red rot of the leaf sheath
    (sc; fungous disease)
  f pourriture *f* rouge de la gaine (foliaire)
  d Rotfäule *f* der Blattscheide; Fußfäule *f*;
      Sämlingsfäule *f*
  e pudrición *f* roja de la vaina
  n roodrot *n* van de bladschede

  * **red scale** *(US)* → 1913

  * **redshank** → 1750

1917 red-shouldered plant bug *(US)*
    (sb; pest)
  l Thyanta pallido-virens

  * **red sorrel** *(US)* → 2130

1918 red spider (mite)
    (sb; pest)
  f tétranyque *m* tisserand commun
  d Spinnmilbe *f*
  e ácaro *m*
  n spintmijt *f*
  l Tetranychus althaeae

1919 red spider (mite)
    (sb; pest)
  f tétranyque *m* tisserand; araignée *f* rouge
  d Spinnmilbe *f*; rote Spinne *f*
  e araña *f* roja; arañuela *f* roja; arañita *f*
      roja
  n spintmijt *f*
  l Tetranychus sp.

1920 red spider (mite)
    (sb; pest)
  f tétranyque *m*; araignée *f* rouge;
      tétranyque *m* tisserand; tétranyque *m* à
      deux points
  d Spinnmilbe *f*; gemeine Spinnmilbe *f*;
      gemeine Bohnenspinnmilbe *f*
  e ácaro *m* rojo; arañuela *f* roja
  l Tetranychus telarius; Tetranychus urticae

1921 red spot of the leaf sheath
    (sc; fungous disease)
  f tache *f* rouge de la gaine (foliaire)
  d rote Scheidenfleckigkeit *f*
  e mancha *f* roja de la vaina
  n rode-vlekkenziekte *f* van de bladschede

1922 red sprangletop *(US)*
    (sb; weed)
  l Leptochloa filiformis

1923 red stripe
    (sc; bacterial disease)
  f stries *fpl* rouges; maladie *f* des stries
      rouges
  d Rotstreifigkeit *f*
  e raya *f* roja
  n rode-strepenziekte *f*

1924 reduced extraction
  f extraction *f* réduite
  d reduzierte Extraktion *f*; reduzierte
      Mühlenextraktion *f*
  e extracción *f* reducida
  n gereduceerde extractie *f*

1925 reducing substance; RS
(In the sugar industry usually refers to
matter, chiefly reducing saccharides,
which is able to reduce alkaline-copper
reagents such as Fehling's solution)
f réducteur *m*
d reduzierende Substanz *f*
e sustancia *f* reductora
n reducerende stof *f*

* **reducing substance/ash ratio** → 1928

1926 reducing substance content
(The result of an analytical method for
reducing saccharides, usually expressed
as equivalent percent invert sugar)
f teneur *f* en réducteurs
d Gehalt *m* an reduzierenden Substanzen
e contenido *m* de sustancias reductoras
n gehalte *n* aan reducerende stoffen

1927 reducing sugars; RS
f sucres *mpl* réducteurs
d reduzierende Saccharide *npl*;
reduzierende Zuckerstoffe *mpl*
e azúcares *mpl* reductores
n reducerende suikers *mpl*

**1928 reducing sugars/ash ratio; RS/ash ratio;
invert/ash ratio; reducing substance/ash
ratio**
(The percentage ratio between reducing
sugars and ash)
f quotient *m* réducteurs/cendres
d reduzierende Saccharide/Asche-Ver-
hältnis *n*
e razón *m* de azúcares reductores a ceniza;
razón *m* de invertidos a ceniza
n reducerende stoffen/as-verhouding *f*

**1929 reducing sugars ratio; RS ratio; invert
ratio; glucose ratio; glucose quotient**
(The use of "reducing sugars" in such
expressions is cumbersome, and many
prefer "invert". This is identical to the
old term "glucose ratio")
f coefficient *m* glucosique; quotient *m*
glucosique
d Glucosequotient *m*; Glukosequotient *m*
e razón *m* de azúcares reductores; razón
m de invertidos; cociente *m* de glucosa
n glucosecoëfficiënt *m*; glucosefactor *m*;
glucosequotiënt *n*

1930 reed; common reed; common reed grass
(sb; weed)
f roseau *m* commun; roseau *m* à balais;
phragmite *m*; phragmite *m* commun
d Schilfrohr *n*; gemeines Schilfrohr *n*;

Teichrohr *n*
e carrizo *m* común
n riet *n*
l Phragmites communis

1931 reference price
(EEC)
f prix *m* de référence
d Referenzpreis *m*; Bezugspreis *m*
e precio *m* de referencia
n referentieprijs *m*

1932 refilter *v*
f refiltrer
d nachfiltrieren
e refiltrar
n nafilteren; nafiltreren

1933 refine *v*
f raffiner
d raffinieren
e refinar
n raffineren

1934 refined sugar
f sucre *m* raffiné
d Raffinadezucker *m*; Raffinade *f*;
raffinierter Zucker *m*; Zuckerraffinade *f*
e azúcar *m* refinado
n geraffineerde suiker *m*; raffinade *f*

1935 refiner
f raffineur *m*
d Raffineur *m*
e refinador *m*
n raffinadeur *m*

1936 refinery
f raffinerie *f*
d Raffinerie *f*
e refinería *f*
n raffinaderij *f*; raffineerderij *f*

1937 refinery massecuite
f masse *f* cuite de raffinage
d Raffineriefüllmasse *f*
e masa *f* cocida de refinería
n raffinaderijvulmassa *f*; raffinaderij-
masse cuite *f*

1938 refinery molasses
f mélasse *f* de raffinage
d Raffineriemelasse *f*
e melaza *f* de refinería
n raffinaderijmelasse *f*

1939 refining
f raffinage *m*
d Raffination *f*; Raffinieren *n*

e refinación f; refinado m; refinamiento m
n raffinage f; raffineren n

1940 refining margin
(EEC)
f marge f de raffinage
d Raffinationsspanne f
e margen m de refinación; margen m de refinado
n raffinagemarge f

1941 refining qualities *(of sugar)*
f raffinabilité f
d Raffinierbarkeit f
e refinabilidad f
n raffineerbaarheid f

* **refractive dry solids** → 1943

1942 refractometer
f réfractomètre m
d Refraktometer n
e refractómetro m
n refractometer m

1943 refractometer solids; refractometer Brix; refractive dry solids; RDS; refractometric Brix
(Percentage by weight of solids as determined by the refractometer, either by direct sugar scale or by reference to tables of refractive indices and percentage sucrose)
f Brix m réfractométrique
d refraktometrisch bestimmter Feststoffgehalt m
e sólidos mpl al refractómetro; Brix m al refractómetro; Brix m refractométrico
n door refractometrie bepaald vastestofgehalte n

1944 refractometric dry substance
f matières fpl sèches réfractométriques
d refraktometrischer Trockengehalt m; refraktometrisch bestimmter Trockengehalt m; Trockengehalt m (refraktometrisch)
e sólidos mpl refractométricos
n door refractometrie bepaald drogestofgehalte n

1945 refractory cane
f canne f réfractaire
d schwer zu verarbeitendes Rohr n
e caña f refractaria
n moeilijk te verwerken riet n

1946 refractory juice
f jus m réfractaire; jus m difficile à filtrer

d schlecht filtrierbarer Saft m; Saft m ungünstiger Zusammensetzung; schwer verarbeitbarer Saft m; schwer zu reinigender Saft m
e jugo m difícil; jugo m refractario
n slecht filtreerbaar sap n

1947 refunding storage costs
(EEC)
f remboursement m des frais de stockage
d Erstattung f der Lagerhaltungskosten
e reembolso m de los gastos de almacenamiento
n vergoeding f van de opslagkosten

1948 refund on export
(EEC)
f restitution f à l'exportation
d Ausfuhrerstattung f
e restitución f a la exportación
n uitvoerrestitutie f

1949 refund on production
(EEC)
f restitution f à la production
d Produktionserstattung f; Erzeugungserstattung f
e restitución f a la producción
n restitutie f bij de produktie

1950 regenerate v **the resin**
f régénérer la résine
d das Harz regenerieren
e regenerar la resina
n het hars regenereren

1951 regeneration kiln
f four m à régénération
d Regenerierungsofen m; Wiederbelebungsofen m
e horno m regenerador
n regeneratieoven m

1952 regulation
(EEC)
f règlement m
d Verordnung f; VO f
e reglamento m
n verordening f; VO f

1953 re-heating
f réchauffage m
d Wiedererwärmen n
e recalentamiento m
n herverwarming f

* **reintroduction of syrups** → 103

1954 relative humidity
f humidité f relative
d relative Feuchte f; relative Feuchtig-
keit f
e humedad f relativa
n relatieve vochtigheid f

1955 relime v
f rechauler
d wieder kalken
e reencalar; realcalinizar
n opnieuw kalken; herkalken

1956 remelt v
(To redissolve crystallized sugar for the
purpose of recrystallizing it)
f refondre
d umschmelzen; einschmelzen; wieder
auflösen
e refundir
n omsmelten

1957 render v **insoluble**
f insolubiliser
d unlöslich machen
e insolubilizar
n onoplosbaar maken

1958 representative conversation rate
(EEC)
f taux m de conversion représentatif
d repräsentative Umrechnungskurs m
e tipo m de cambio representativo
n representatieve wisselkoers m

1959 reserve quantity
(EEC)
f masse f de manoeuvre
d Manövriermasse f
e margen m de maniobra
n manoeuvreermassa f

**1960 residence time; retention time; detention
time**
f temps m de rétention; durée f de
rétention; durée f de séjour; temps m de
séjour
d Aufenthaltszeit f; Aufenthaltsdauer f;
Verweilzeit f
e tiempo m de permanencia; tiempo m de
residencia
n verwijltijd f

1961 residual alkalinity
f alcalinité f résiduelle
d restliche Alkalität f; Restalkalität f
e alcalinidad f residual
n restalkaliteit f

1962 residual extraction
f extraction f résiduelle
d Restextraktion f
e extracción f residual
n restextractie f

1963 residual juice
(The juice left in the bagasse: bagasse
minus fibre)
f jus m résiduel
d Restsaft m
e guarapo m residual; jugo m residual
n restsap n

1964 residual moisture
f humidité f résiduelle
d Restfeuchtigkeit f
e humedad f residual
n restvochtigheid f

1965 residual molasses
f mélasse f résiduelle
d Restmelasse f
e melaza f residual; miel f residuaria
n restmelasse f

1966 residual nonsugar
f non-sucre m résiduel
d Restnichtzucker m
e no-azúcar m residual
n restnietsuiker m

1967 residual turbidity
f turbidité f résiduelle
d Resttrübung f; restliche Trübung f
e turbidez f residual
n resttroebelheid f

1968 resin
f résine f
d Harz n
e resina f
n hars nm

1969 respiration loss
f pertes fpl par respiration
d Atmungsverluste mpl
e pérdidas fpl por respiración
n ademingsverliezen npl

1970 retention
(Proportion of suspended matter
extracted by a filter, expressed as a
percentage of the suspended matter
contained in the mud arriving at the
filter)
f rétention f
e retención f
n weerhouding f

* retention → 301

* retention time → 1960

1971 **return of juice**
f retour *m* du jus; recyclage *m* du jus
d Rückführung *f* des Saftes
e retorno *m* del jugo; retorno *m* del guarapo
n terugbrenging *f* van het sap

* **reutilization of diffusion waste water** → 816

1972 **revolving cane knives**
f coupe-cannes *m* rotatif
d drehende Rohrmesser *npl*
e cuchillas *fpl* cañeras giratorias
n draaiende rietmessen *npl*; roterende rietmessen *npl*

1973 **revolving scrapers**
f lames *fpl* raclantes; bras *mpl* racleurs
d Krählwerk *n*
e raspadores *mpl* giratorios
n roterende schrapers *mpl*

1974 **Reynolds number**
f nombre *m* de Reynolds
d Reynoldssche Zahl *f*; Reynolds-Zahl *f*
e número *m* de Reynolds
n getal *n* van Reynolds

1975 **rhinoceros beetles; hercules beetles; black beetles** *(US)*; **hard-black beetles**
(sc; pest)
f dynastidés *mpl*
d Nashornkäfer *mpl*; Riesenkäfer *mpl*
l Dynastinae

1976 **Rhizoctonia root rot**
(sb; disease)
f rhizoctone *m* brun
d Rhizoctonia-Fäule *f*
e podredumbre *f* parda
n Rhizoctonia-ziekte *f*

* **ribbed melilot** → 2728

1977 **ribbon calandria; ring element; circular ribbon; ring-shaped hollow heating body**
f faisceau *m* annulaire; faisceau *m* à plaques chauffantes verticales à double paroi disposées concentriquement
d Ringheizkammer *f*
e calandria *f* de tipo de anillo; calandria *f* de anillos concéntricos
n ringvormig verwarmingslichaam *n*

* **ribbon pan** → 1987

* **ribgrass** → 1979

* **rib-roof knife** → 1982

1978 **ribs of a ridge knife; splitters of a V-corrugated knife** *(US)*
f cloisons *fpl* d'un couteau faîtière
d Rippen *fpl* eines Dachrippenmessers
e pantalla *f* de cuchilla de cobija
n ribben *fpl* van een ribbemes

1979 **ribwort; ribgrass; buckhorn plantain** *(US)*; **ribwort plantain; narrow-leaved plantain**
(sb; weed)
f plantain *m* lancéolé
d Spitzwegerich *m*
e llantén *m* menor; llantén *m* lanceolado
n smalle weegbree *f*
l Plantago lanceolata

* **rich molasses** → 2638

* **rich syrup** → 2638

1980 **riddling; sieving; screening**
(The separation of crystals into the appropriate size ranges)
f tamisage *m*
d Absiebung *f*; Siebung *f*
e tamizado *m*
n zeven *n*

1981 **ridge cossettes; grooved cossettes; cossettes with V-shaped cross sections; ridged slices**
f cossettes *fpl* faîtières
d Dachrippenschnitzel *npl*; Rillenschnitzel *npl*; Rinnenschnitzel *npl*; Dachrippen-messerschnitzel *npl*
e cosetas *fpl* de cobija
n dakgootsnijdsels *npl*; dakribbesnijdsels *npl*

1982 **ridge knife; rib-roof knife; splitter knife; V-corrugated knife**
f couteau *m* faîtière; couteau *m* fraisé faîtière
d Dachrippenmesser *n*
e cuchilla *f* de cobija; cuchillo *m* para cosetas en forma de tejas
n ribbemes *n*

1983 **rind disease; sour rot**
(sc; fungous disease)
f maladie *f* de l'écorce
d Rindenkrankheit *f*

e enfermedad *f* de la corteza
n bastziekte *f*; schorsziekte *f*

* **ring element** → 1977

1984 **ring mosaic**
(sc; virus disease)
f mosaïque *f* annulaire
d Ringmosaikkrankheit *f*; Ringmosaik *f*
e mosaico *m* anular
n ringmozaïek *f*

1985 **ring nematode**
(sc; pest)
l Criconemoides spp.

* **ring-shaped hollow heating body** → 1977

1986 **ring spot**
(sc; fungous disease)
f maladie *f* des taches rondes
d Ringfleckenkrankheit *f*
e mancha *f* de anillo
n ringvlekkenziekte *f*

1987 **ring-type calandria pan; ring-type pan; ribbon pan**
(A vacuum pan with a spiral ribbon-form or plate-type heating element)
f appareil *m* à cuire à faisceau annulaire; appareil *m* à cuire à faisceau en spirale; appareil *m* à cuire à faisceau à plaques chauffantes concentriques
d Kochapparat *m* mit Ringheizkammer; Verdampfungskristallisator *m* mit Ringheizkammer
e tacho *m* al vacío de calandria de tipo de anillo
n kookpan *f* met ringvormig verwarmingslichaam

* **ripple grass** → 1193

* **ripple-seed plantain** → 1193

* **rising film evaporator** → 513

1988 **rock catcher; stone catcher; stone separator**
(Apparatus that removes stones and other dense matter from the beet flume, usually by differential flotation)
f épierreur *m*
d Steinabscheider *m*; Steineabscheider *m*; Steinfänger *m*; Steinefänger *m*
e separador *m* de piedras; colector *m* de piedras; despedrador *m*; eliminador *m* de piedras
n steenvanger *m*; stenenvanger *m*

* **rock lime** → 1520

1989 **Rocky Mountain grasshopper** *(US)*; **Rocky Mountain locust** *(US)*
(sb; pest)
l Melanoplus spretus

1990 **rod-chain-type elevator**
(agr)
f élévateur *m* à barrettes
d Stabelevator *m*
e elevador *m* de listones
n staafelevator *m*

1991 **roller conveyor**
f transporteur *m* à rouleaux
d Rollenbahn *f*; Rollenförderer *m*
e transportador *m* de rodillos
n rollenbaan *f*

1992 **roller feeler; roller finder** *(US)*
(agr)
f tâteur *m* à roues
d Tastrad *n*
e ruedas *fpl* palpadoras
n wieltaster *m*

1993 **roller table screen**
f grille *f* à rouleaux
d Rollenrost *m*
e emparrillado *m* de rodillos
n rollenrooster *m*

1994 **root**
(anat)
f racine *f*
d Wurzel *f*
e raíz *f*
n wortel *m*

* **root disease** → 1999

1995 **root galls**
(sb; disease)
f galles *fpl*
d Gallen *fpl* an den Wurzeln der Rübe
e protuberancias *fpl* de las raíces secundarias; agallas *fpl* de las raíces secundarias
n wortelknobbels *mpl*

1996 **root-knot nematode; root-knot eelworm**
(sb; pest)
f nématode *m* des racines; anguillule *f* des racines
d Wurzelgallenälchen *n*
e meloidoginido *m*; nemátodo *m* del género Meloidogyne marioni

n wortelknobbelaaltje *n*
l Meloidogyne marioni

1997 rootlet
f radicelle *f*
d Haarwurzel *f*
e raicilla *f*
n haarwortel *m*

1998 root louse *(US)*
(sb; pest)
d Rübenwurzellaus *f*
l Pemphigus betae

1999 root rot; root disease
(sc; fungous disease)
f pourriture *f* des racines; pourriture *f* de
la racine
d Wurzelfäule *f*; Dongkellkrankheit *f*
e pudrición *f* de la raíz; podredumbre *f* de
la raíz
n wortelrot *n*

* **root rot** → 280

* **root thinner** → 246

**2000 rosy rustic moth; orange ear; potato
stem borer** *(US)*
(sb; pest)
f noctuelle *f* des artichauts; drap *m* d'or;
hérissée *f*
d Uferhochstauden-Markeule *f*; gemeine
Markeule *f*; Kartoffeltriebbohrer *m*;
Kletteneule *f*
n aardappelstengelboorder *m*
l Gortyna ochracea; Gortyna flavago

2001 rotary abutment pump
f pompe *f* à rotors cylindriques
d Kreiskolbenpumpe *f*
e bomba *f* impelente rotatoria

2002 rotary disc
(agr)
f disque *m* rotatif
d Köpfscheibe *f*
e disco *m* giratorio
n draaiende schijf *f*

2003 rotary drum drier
f sécheur *m* à tambour rotatif
d Drehtrommeltrockner *m*
e secador *m* horizontal giratorio
n roterende trommeldroger *m*

2004 rotary feeder *(of mill)*
f alimentateur *m* rotatif
d drehender Zuführer *m*; rotierender

Zuführer *m*
e alimentador *m* rotativo
n draaiende toevoerinrichting *f*; roterende
toevoerinrichting *f*

**2005 rotary lifting share; lifting wheel digger;
digger wheel** *(US)***; lifting wheel**
(agr)
f soc *m* rotatif; roue *f* arracheuse
d rotierendes Rodeschar *n*; Roderad *n*
e rueda *f* arrancadora
n draaiende rooischaar *f*

* **rotary lime slaker** → 1479

2006 rotary pick-up
(agr)
f ramasseur *m* rotatif
d Kreiselrübensammler *m*
e recogedor *m* giratorio
n roterende opraper *m*

2007 rotary piston compressor
f compresseur *m* à piston rotatif
d Drehkolbenkompressor *m*; Drehkolben-
gebläse *n*
e soplador *m* de émbolo giratorio
n roterende zuigercompressor *m*

2008 rotary piston pump
f pompe *f* à pistons rotatifs
d Drehkolbenpumpe *f*
e bomba *f* rotativa de pistón
n roterende zuigerpomp *f*

2009 rotary pump
f pompe *f* rotative
d Rotationspumpe *f*
e bomba *f* rotativa
n roterende pomp *f*

2010 rotary sieve; gyrating sieve
f tamis *m* rotatif
d Trommelsieb *n*; rotierendes Sieb *n*
e tamiz *m* rotativo; tamiz *m* giratorio
n roterende zeef *f*; trommelzeef *f*

2011 rotary sulphur furnace
f four *m* à soufre rotatif
d Rotationsschwefelofen *m*
e horno *m* de azufre rotativo
n draaiende zwaveloven *m*

2012 rotary top feeler
(agr)
f tâteur *m* circulaire
d Radtaster *m*
e palpador *m* circular
n wieltaster *m*

2013 rotary-tower crane
(A cane crane)
_f_ grue _f_ pivotante à tour
_d_ Turmdrehkran _m_
_e_ grúa _f_ de brazo giratorio
_n_ draaitorenkraan _f_

2014 rotary vacuum filter
_f_ filtre _m_ rotatif sous vide; filtre _m_ à
tambour rotatif sous vide
_d_ Vakuumdrehfilter _n_
_e_ filtro _m_ rotativo al vacío; filtro _m_
rotatorio al vacío
_n_ roterend vacuümfilter _n_

* **rotating beet feeder** → 191

2015 rotating brush
_f_ brosse _f_ rotative
_d_ Drehbürste _f_
_e_ cepillo _m_ rotatorio
_n_ draaiende borstel _m_

2016 rotating knife _(of cane mill)_
_f_ couteau _m_ rotatif
_d_ rotierendes Messer _n_
_e_ cuchilla _f_ rotativa
_n_ draaiend mes _n_; roterend mes _n_

2017 rotatory power _(of sugar)_
_f_ pouvoir _m_ rotatoire
_d_ Drehungsvermögen _n_
_e_ poder _m_ rotatorio
_n_ draaiingsvermogen _n_

2018 rotten beet
_f_ betterave _f_ pourrie
_d_ angefaulte Rübe _f_; faule Rübe _f_
_e_ remolacha _f_ podrida
_n_ verrotte biet _f_

* **rough cocklebur** → 1454

* **rough cock's-foot** → 533

* **rough pigweed** _(US)_ → 1914

2019 rough sugar; sugar hail; coarse
_f_ sucre _m_ perlé
_d_ Hagelzucker _m_
_n_ suikerkorrels _mpl_; suikerhagel _m_

* **round rush** → 2199

* **RS** → 1925, 1927

* **RS/ash ratio** → 1928

* **R.S.D.** → 1890

* **RS ratio** → 1929

* **rubbed and graded seed** → 1837

2020 rule
(EEC)
_f_ directive _f_
_d_ Richtlinie _f_
_e_ directiva _f_
_n_ richtlijn _f_

* **runch** → 2697

* **runner finder** _(US)_ → 2163

2021 run-off
_f_ égout _m_
_d_ Ablauf _m_
_e_ melaza _f_ de purga; derrame _m_
_n_ stroop _f_

2022 run-off dilution
_f_ dilution _f_ des égouts
_d_ Ablaufverdünnung _f_
_e_ dilución _f_ de las mieles
_n_ verdunnen _n_ van de stropen

2023 run-off tank
_f_ bac _m_ de purge
_d_ Ablaufbehälter _m_
_e_ depósito _m_ de derrame
_n_ aflooptank _m_

* **run** _v_ **to seed** → 310

* **run** _v_ **up the strike** → 351

* **rush nut** _(US)_ → 1682

* **Russian cactus** _(US)_ → 2025

2024 Russian knapweed _(US)_
(sb; weed)
_l_ Centaurea repens; Centaurea picris

2025 Russian thistle; Russian tumbleweed
_(US)_**; Russian cactus** _(US)_**; prickly**
saltwort; glasswort; kelpwort; common
Russian thistle
(sb; weed)
_f_ soude _f_; soude _f_ kali
_d_ Salzkraut _n_; Kali-Salzkraut _n_; Soda-
kraut _n_; russisches Salzkraut _n_
_e_ barrilla _f_ pinchosa; barrilla _f_ fina
_n_ loogkruid _n_
_l_ Salsola kali

2026 rust
(sc; fungous disease)

f rouille *f*
d Rost *m*; Zuckerrohrrost *m*
e roya *f*
n roest *m*

S

* saccharase → 1348

2027 **saccharate**
(A precipitate formed in the Steffen
process, consisting of lime-sucrose, lime,
and some nonsucrose from the molasses)
f saccharate *m*; sucrate *m*
d Saccharat *n*
e sacarato *m*
n saccharaat *n*

2028 **saccharate cake**
(Filter cake from the washing filters in
the Steffen process, consisting of
saccharate, some adhering filtrate, and
moisture)
f gâteau *m* de saccharate
d Saccharatkuchen *m*
e torta *f* de sacarato
n saccharaatkoek *m*

2029 **saccharate milk**
(A slurry of saccharate cake in
sweetwater used for preliming, main
liming, or defeco-carbonatation in Steffen
factories)
f lait *m* de saccharate
d Saccharatmilch *f*
e lechada *f* de sacarato
n saccharaatmelk *f*

2030 **saccharate of lime; lime sucrate**
f saccharate *m* de chaux; sucrate *m* de
chaux
d Kalksaccharat *n*
e sacarato *m* de cal
n kalksaccharaat *n*

* saccharate process → 2279

2031 **saccharate thickener**
(Apparatus for separating the hot
saccharate from the bulk of its mother
liquor by sedimentation)
f décanteur *m* à saccharate
d Saccharatdekanteur *m*
e precipitador *m* de sacarato
n saccharaatdecanteerbak *m*

2032 **saccharify** *v*
f saccharifier
d verzuckern; in Zucker verwandeln
e sacarificar
n versuikeren

2033 **saccharimeter**
(A polariscope, with a scale reading
percentage of sucrose)
f saccharimètre *m*
d Saccharimeter *n*
e sacarímetro *m*
n saccharimeter *m*

2034 **saccharimetry**
f saccharimétrie *f*
d Saccharimetrie *f*
e sacarimetría *f*
n saccharimetrie *f*

2035 **saccharin(e)**
f saccharine *f*
d Saccharin *n*
e sacarina *f*
n saccharine *fn*

2036 **saccharometer**
(A hydrometer which is used to
determine the concentration of sugar in a
solution)
f saccharomètre *m*
d Saccharometer *n*
e sacarómetro *m*
n saccharometer *m*

2037 **sack piler**
f empileur *m* de sacs
d Sackstapler *m*
e apilador *m* de sacos; apilador *m* de
bolsas; apiladora *f* de sacos; apiladora *f*
de bolsas
n zakkenstapelaar *m*

* **safety factor** *(Australia) (old)* → 821

2038 **safety factor of sugar**
(% moisture/(100 - Pol))
f facteur *m* de sécurité du scure
d Sicherheitsfaktor *m* des Zuckers; SF *m*
des Zuckers
e factor *m* de seguridad del azúcar
n veiligheidsfactor *m*

* **Saint-John's-wort** → 585

* **salsify** → 2611

* **saltbush** → 1696

2039 **salt grass** *(US)*
(sb; weed)
l Distichlis stricta

2040 salt-marsh caterpillar *(US)*
(sb; pest)
l Estigmene acrea

2041 sample catcher
f échantillonneur *m*
d Probenehmer *m*
e sacamuestras *m*; tomapruebas *m*
n monstertrekker *m*

2042 sandbur(r) *(US)*
(sb; weed)
f bardanette *f*
l Cenchrus pauciflorus

2043 sand catcher
(Apparatus which separates sand and
earth from a liquid stream, such as raw
juice, press water, or flume water)
f séparateur *m* de sable; dessableur *m*
d Sandabscheider *m*
e desarenador *m*
n zandvanger *m*; zandafscheider *m*

2044 sandweevil
(sb; pest)
f charançon *m*
d grauer Kugelrüßler *m*; grauer Erdbeer-
rüßler *m*; Sandgraurüßler *m*
l Philopedon plagiatum

2045 sandwich gauze *(in centrifugal)*
f toile *f* sandwich
d Sandwichgewebe *n*
e malla *f* sandwich
n sandwichgaas *n*

2046 saturated juice
f jus *m* saturé
d gesättigter Saft *m*
e guarapo *m* saturado; jugo *m* saturado
n gesatureerd sap *n*; verzadigd sap *n*

2047 saturated solution
f solution *f* saturée
d gesättigte Lösung *f*
e solución *f* saturada
n verzadigde oplossing *f*

2048 saturated steam
f vapeur *f* saturée
d Sattdampf *m*
e vapor *m* saturado
n verzadigde stoom *m*

2049 saturation
f saturation *f*
d Sättigung *f*

e saturación *f*
n verzadiging *f*

* **saturation** → 1290

* **saturation coefficient** → 2204

2050 saturation point
f point *m* de saturation
d Sättigungspunkt *m*
e punto *m* de saturación
n verzadigingspunt *n*

2051 saturation temperature
f température *f* de saturation
d Sättigungstemperatur *f*
e temperatura *f* de saturación
n verzadigingstemperatuur *f*

* **saturation water** → 1293

* **save-all** → 438

2052 savoy disease
(sb; disease)
f "Beet Savoy Disease"
d Savoy-Krankheit *f*
e "Beet Savoy Disease"
n "Beet Savoy Disease"

2053 say stink bug; say's stinkbug
(sb; pest)
l Chlorochroa sayi

2054 scalding
(Preheating of the cossettes before
diffusion by mixing them with hot raw
juice)
f échaudage *m*
d Brühung *f*
e escaldamiento *m*; escaldado *m*
n broeien *n*

* **scale removal** → 781

* **scales** → 1303

* **scalings** → 1303

2055 scalper
(agr)
f scalpeuse *f*
d Rübenkopfskalpierer *m*
e mecanismo *m* de deshojado; mecanismo
m descoronador-desmenuzador
n scalpeermachine *f*

2056 scalper holder
(agr)

f porte-scalper *m*
d Skalpiermesserträger *m*
e portacuchilla *m*
n scalpeermeshouder *m*

2057 scalper knife
(agr)
f scalper *m*
d Skalpiermesser *n*
e cuchilla *f* descoronadora
n scalpeermes *n*

**2058 scarlet pimpernel; poor man's
weatherglass; red pimpernel; red
chickweed; cure all**
(sb; weed)
f mouron *m* rouge; mouron *m* des
champs; anagallide *f* des champs
d Ackergauchheil *m*; roter Gauchheil *m*
e murajes *mpl*
n rood guichelheil *n*; rode basterdmuur *f*;
rode bastaardmuur *f*; guichelheil *n*;
guichelkruid *n*; gewoon guichelheil *n*
l Anagallis arvensis

* **scented mayweed** → **2694**

**2059 scentless mayweed; scentless camomile;
corn mayweed; scentless chamomile;
horsy daisy**
(sb; weed)
f matricaire *f* inodore; camomille *f*
romaine
d geruchlose Kamille *f*; duftlose Kamille *f*;
falsche Kamille *f*
e manzanilla *f* inodora; magarza *f* inodora
n reukloze kamille *f*; reukeloze kamille *f*
l Matricaria inodora; Chrysanthemum
inodorum

**2060 sclerophthora disease; sclerospora
disease**
(sc; fungous disease)
d falscher Mehltau *m*
e enfermedad *f* esclerophora; enfermedad
f de la esclerophora
n valse meeldauw *m*

* **sclerotial rot** *(US)* → **2211**

2061 scraper; comb
f raclette *f*; peigne *m*
d Abstreifer *m*
e raspador *m*; peine *m*
n schraper *m*

2062 scraper *(in clarifier)*
f lame *f* raclante
d Krählwerk *n*

e lámina *f* raspadora
n schraper *m*

2063 scraper conveyor; scraper-type carrier
f transporteur *m* racleur; transporteur *m*
à raclettes
d Kratzertransporteur *m*
e transportador *m* de rastrillo; conductor
m de rascadores
n schrapertransporteur *m*

2064 screen
f tamis *m*
d Sieb *n*
e tamiz *m*
n zeef *f*

2065 screen area
f surface *f* criblante; surface *f* tamisante
d Siebfläche *f*
e superficie *f* del tamiz; superficie *f*
cribante
n zeefoppervlak *n*

2066 screen box *(of cush cush elevator)*
f châssis *m* du tamis
d Siebkasten *m*
e caja *f* del colador
n zeefkist *f*

2067 screener
f cribleur *m*
d Siebanlage *f*
e criba *f* mecánica
n zeefapparaat *n*

* **screening** → **1980**

2068 screening fabrics
f tissu *m* de tamisage
d Siebgewebe *n*
e tejido *m* para tamices
n zeefgaas *n*

2069 screening machine *(for sugar)*
f tamiseuse *f*
d Siebmaschine *f*
e cribadora *f*
n zeefmachine *f*

2070 screw conveyor
f transporteur *m* à hélice; vis *f*
transporteuse; transporteur *m* à vis
d Förderschnecke *f*; Schneckenförderer *m*;
Transportschnecke *f*
e tornillo *m* transportador; tornillo *m* sin
fin; conductor *m* de gusano
n transportschroef *f*;
schroeftransporteur *m*

2071 **screw press**
   f presse *f* à hélice
   d Schneckenpresse *f*
   e prensa *f* helicoidal
   n schroefpers *f*

2072 **screw pump**
   f pompe *f* à vis
   d Schraubenspindelpumpe *f*
   e bomba *f* de husillo
   n schroefpomp *f*

2073 **scum**
   f écume *f*
   d Schaum *m*
   e espuma *f*
   n schuim *n*

2074 **scum gutter**
   f gouttière *f* d'écumes
   d Schaumrinne *f*
   e canal *m* para espumas
   n schuimgoot *f*

2075 **scum washing**
   f lavage *m* des écumes
   d Schlammabsüßung *f*
   e lavado *m* de las espumas
   n schuimafzoeting *f*

   * **scutch grass** *(US)* → 664

   * **sea beet** → 2692

2076 **seal tank; atmospheric tank; hot well**
   (The seal on the bottom of a barometric
   leg pipe)
   f bac *m* du pied du condenseur
   d Fallwasserkasten *m*
   e pozo *m* caliente; sello *m*
   n valwaterbak *m*

2077 **Searby shredder; hammer mill of Searby
   type; Searby type pulverizer**
   (A swing hammer pulverizer)
   f broyeur *m* Searby; broyeur *m* à
   marteaux type Searby; shredder *m* type
   Searby
   d Searby-Shredder *m*
   e desfibradora *f* tipo Searby; desfibradora
   f de martillos tipo Searby
   n Searby-shredder *m*; rafelmachine *f* van
   Searby

2078 **seaside heliotrope** *(US)***; Chinese pursley**
   (sb; weed)
   l Heliotropium curassavicum

   * **season** → 374

2079 **secondary air** *(to boiler)*
   f air *m* secondaire
   d Sekundärluft *f*; Zweitluft *f*
   e aire *m* secundario
   n secundaire lucht *f*

2080 **secondary clarifier; secondary subsider**
   f clarificateur *m* secondaire
   d Sekundärdekanteur *m*
   e clarificador *m* secundario
   n secundaire decanteur *m*; secundair
   decanteertoestel *n*

   * **secondary grain** → 966

2081 **secondary juice**
   (The diluted juice which joins the
   primary juice to form mixed juice)
   f jus *m* secondaire
   d Saft *m* der zweiten und dritten Mühlen
   e guarapo *m* secundario; jugo *m*
   secundario
   n sap *n* van de tweede en derde molen

2082 **secondary liming**
   f chaulage *m* secondaire
   d sekundäre Kalkung *f*
   e encalado *m* secundario; alcalización *f*
   secundaria; alcalinización *f* secundaria
   n secundaire kalking *f*

   * **secondary subsider** → 2080

2083 **second carbona(ta)tion; second carb.**
   f deuxième carbonatation *f*; seconde
   carbonatation *f*
   d zweite Carbonatation *f*; zweite
   Saturation *f*
   e segunda carbonatación *f*
   n tweede carbonatatie *f*

2084 **second carbona(ta)tion juice**
   f jus *m* de deuxième carbonatation
   d zweiter Carbonatationssaft *m*
   e jugo *m* de segunda carbonatación;
   guarapo *m* de segunda carbonatación
   n tweede-carbonatatiesap *n*

   * **second massecuite** → 293

2085 **second mill bagasse**
   f bagasse *f* du deuxième moulin
   d Bagasse *f* der zweiten Mühle
   e bagazo *m* del segundo molino
   n bagasse *m* van de tweede molen; ampas
   m van de tweede molen

   * **second molasses** → 294

* **second sugar** → 343

2086 sedimentation
f sédimentation f
d Absetzung f
e sedimentación f
n neerslaan n; afzetting f

2087 seed *(noun)*
(Magma or fine grained massecuite used as a footing for boiling a massecuite)
f semence f
d Saatgut n
e semilla f; siembra f
n entgoed n

2088 seed v
f ensemencer
d impfen
e semillar
n enten

* **seed ball** *(US)* → 523

2089 seed bearer; seed plant
(agr)
f porte-graines m
d Samenträger m; Samenpflanze f
e planta-semilla f; portagranos m
n zaaddrager m

2090 seed bed
(agr)
f lit m de germination
d Saatbett n
e cama f de siembra
n zaaibed n; zaadbed n; kiembed n

2091 seed bed preparation
(agr)
f préparation f du lit de germination
d Saatbettbereitung f; Saatbett-vorbereitung f
e preparación f de la cama de siembra
n gereedmaken n van het zaaibed; voorbereiding f van het zaaibed

* **seed beet** → 196

2092 seed breeder
(agr)
f sélectionneur m de semences
d Saatzüchter m
e seleccionador m de estirpes
n zaadselecteerder m

2093 seed breeding station
(agr)
f institut m de sélection des semences;
centre m de sélection des semences
d Saatzuchtanstalt f
e criadero m de semillas; semillero m;
centro m de selección de semillas;
empresa f de selección de estirpes
n zaadveredelingscentrum n

2094 seed crystal
f centre m de cristallisation
d Kristallisationskeim m; Kristallkeim m;
Keimkristall m; Impfkristall m; Saat-kristall m; Kornfußkristall m
e centro m de cristalización
n kristallisatiekern f

2095 seed density
(agr)
f densité f de semis
d Aussaatdichte f
e densidad f de siembra
n zaaidichtheid f

2096 seed drier
(agr)
f séchoir m à graines
d Saatgutdarre f; Saatguttrocknungs-anlage f
e desecador m de semillas
n zaaddroger m

2097 seed grain
f grain m d'ensemencement
d Impfkorn n
e grano m semilla
n entgrein n

2098 seed grower; seed multiplier; seed producer
(agr)
f multiplicateur m de graines;
multiplicateur m de semences;
producteur m de semences
d Saatgutvermehrer m
e productor m de semillas
n zaadvermeerderaar m

2099 seeding
(Supplying crystal nuclei to the graining charge, after the latter has been concentrated to the metastable zone of supersaturation, in numbers equal to the desired crystal population of the finished massecuite)
f ensemencement m; amorçage m
d Saatimpfen n; Saatimpfung f
e semillamiento m; siembra f
n enten n

2100 seedling
(agr)
f plantule f
d Keimling m; Keimpflanze f
e plántula f
n kiemplant f

* **seed multiplier** → 2098

* **seed plant** → 2089

* **seed producer** → 2098

* **seed slurry** → 2183

* **seed spacing drill** → 1805

2101 seed sugar
f amorce f; semence-poudre f cristalline *(à introduire dans l'appareil à cuire)*
d Impfzucker m; Zucker m als Kornfuß; Zucker m als Kristallfuß
e polvo m semilla
n entsuiker m

* **seed volume** → 1107

2102 segmented seed
(agr)
f semence f segmentée; graine f segmentée
d segmentiertes Saatgut n
e semilla f segmentada
n gesegmenteerd zaad n

2103 segmenting machine
(agr)
f segmenteuse f
d Segmentiermaschine f
n segmenteermachine f

2104 self-aligning share
(agr)
f soc m autoguidé
d selbstführendes Rodeschar n
e reja f autolineadora
n zich zelf richtende rooischaar f

2105 self-discharging centrifugal
f centrifugeuse f à vidange automatique; centrifugeuse f à vidange naturelle
d selbstentleerende Zentrifuge f; Zentrifuge f mit automatischer Entleerung
e centrífuga f de descarga automática; centrífuga f autodescargable
n zelflossende centrifuge f

* **self-evaporation** → 2228

* **self-setting mill** → 610

2106 self-sharpening *(of knives)*
f auto-affûtage m
d Selbstschärfen n
e autoafilado m
n zelfscherpen n; zelfslijpen n

2107 self-sharpening knives
f couteaux mpl à auto-affûtage
d selbstschärfende Messer npl
e cuchillas fpl autoafilantes
n zelfslijpende messen npl; zelfscherpende messen npl

2108 sembur
(sc; virus disease)
f "sembur"
d Semburkrankheit f
e "sembur"
n sembur-ziekte f

2109 semi-automatic centrifugal
f centrifugeuse f semi-automatique
d halbautomatische Zentrifuge f
e centrífuga f semiautomática
n halfautomatische centrifuge f

* **semolina sugar** → 531

2110 sensible heat
f chaleur f propre; chaleur f sensible
d fühlbare Wärme f
e calor m sensible
n voelbare warmte f

* **sensing device** → 986

2111 sensing tracks
(agr)
f chenille f de palpage
d Raupentaster m; Kettentaster m; Laufbandtaster m
e cadena f palpadora
n kettingtaster m

* **separation of green and wash** *(US)* → 2454

* **separation of wash and run-off** → 2454

2112 sereh
(sc; virus disease)
f "sereh"; maladie f du sereh
d Serehkrankheit f
e "sereh"
n sereh-ziekte f

2113 setaceous Hebrew character
(sb; pest)
f C *m* noir; C-noir *m*
d Frischkräuterrasen-Bodeneule f;
   schwarzes C *n*; schwarze C-Erdeule f
e c-negra f
l Amathes c-nigrum; Diarsia c-nigrum;
   Agrotis c-nigrum

2114 set of knives
f jeu *m* de couteaux
d Messersatz *m*
e juego *m* de cuchillas
n stel *n* messen

2115 settling; subsidation
f décantation f
d Absetzen *n*; Klären *n*
e decantación f
n bezinking f

**2116 settling basin; settling pond; mud
settling pond**
(A basin or pond for the elimination by
sedimentation, and also the storage of
suspended solids from flume effluent or
main waste)
f bassin *m* de décantation; bassin *m* de
   sédimentation
d Absetzteich *m*; Klärbecken *n*; Absetz-
   becken *n*; Schlammteich *m*
e balsa f de decantación; pila f de
   sedimentación
n bezinkvijver *m*

2117 settling compartment *(of subsider)*
f compartiment *m* de décantation
d Absetzkammer f
e compartimiento *m* de asentamiento
n bezinkruimte f; bezinkcompartiment *n*

* **settling pond** → **2116**

2118 settling tank
f bac *m* de décantation
d Absetzgefäß *n*; Absetzbehälter *m*
e tanque *m* de decantación
n bezinkbak *m*; decanteerbak *m*

**2119 sewage; waste water; mud water; dirty
water; effluent**
f eaux *fpl* résiduaires
d Abwasser *n*; Schmutzwasser *n*
e aguas *fpl* residuales; aguas *fpl* sucias;
   aguas *fpl* servidas
n afvalwater *n*; vuilwater *n*

**2120 sextuple effect; sextuple effect
evaporator**

f sextuple effet *m*
d Sechskörper-Verdampfapparat *m*;
   Sechsfachverdampfer *m*
e séxtuplo efecto *m*
n verdampapparaat *n* met zes
   verdamplichamen; verdampapparaat *n* à
   sextuple effet

2121 sextuple effect evaporation
f évaporation f à sextuple effet
d sechsstufige Verdampfung f
e evaporación f de séxtuplo efecto
n zesvoudige verdamping f;
   zestrapsverdamping f

* **sextuple effect evaporator** → **2120**

2122 shaft lime kiln; vertical lime kiln
f four *m* à chaux à cuve; four *m* à chaux
   droit
d Kalkschachtofen *m*
e horno *m* de cuba de cal
n kalkschachtoven *m*

* **shaker conveyor** → **1189**

2123 shaker screen
f tamis *m* à secousses; tamis *m* secoueur
d Rüttelsieb *n*; Schüttelsieb *n*
e tamiz *m* de sacudidas; tamiz *m*
   vibratorio; criba f de sacudidas; criba f
   oscilante
n schudzeef f

* **shamrock** → **2714**

2124 sharpen *v*
f affûter
d schärfen
e afilar
n scherpen; slijpen

2125 sharpening
f affûtage *m*
d Schärfen *n*
e afilado *m*
n scherpen *n*; slijpen *n*

2126 sharpening machine
f affûteuse f
d Schärfmaschine f
e afiladora f
n scherpmachine f; slijpmachine f

2127 sheath nematodes
(sc; pest)
l Hemicycliophora spp.

2128 sheath rot
(sc; fungous disease)
f pourriture f de la gaine (foliaire)
d Cytospora-Fäule f der Blattscheide
e pudrición f cytospora de la vaina
n cytospora-rot n van de bladschede

2129 sheath rot; stem canker
(sc; fungous disease)
f pourriture f de la gaine (foliaire)
d Blattscheidenfäule f
e pudrición f de la vaina
n rot n van de bladschede

2130 sheep's sorrel; red sorrel *(US)*; **sheep sorrel; field sorrel**
(sb; weed)
f petite oseille f sauvage; petite oseille f; vinette f; oseille f des brebis
d kleiner Ampfer m; kleiner Sauerampfer m; Feldampfer m
e acederilla f; acedera f menor; vinagrillo m
n schapezuring f
l Rumex acetosella

2131 shepherd's needle; lady's-comb; Venus' comb
(sb; weed)
f scandix m peigne de Vénus; aiguille f de berger; aiguille-de-berger f; peigne m de Vénus
d Venuskamm m; Nadelkerbel n; echter Venuskamm m
e aguja f de pastor; peine m de Venus
n naaldekervel m
l Scandix pecten-veneris

2132 shepherd's purse
(sb; weed)
f bourse-à-pasteur f; bourse-à-berger f; capselle f bourse-à-pasteur
d Hirtentäschelkraut n; gemeines Hirtentäschelkraut n
e bolsa f de pastor; pan m y quesillo; zurrón m de pastor
n herderstasje n
l Capsella bursa-pastoris

2133 shocking *(seeding)*
f choc m
d Schock m
e choque m
n enten n

2134 short counter-current condenser
f condenseur m court à contre-courants
d kurzer Gegenstromkondensator m
e condensador m corto de contracorriente
n korte tegenstroomcondensor m

2135 short-tube vertical evaporator
f évaporateur m à grimpage à tubes verticaux courts
d Verdampfapparat m mit senkrechten kurzen Heizrohren
e evaporador m de película ascendente de tubo corto
n verdamper m met verticale korte verwarmingsbuizen; verdamptoestel n met verticale korte verwarmingsbuizen

2136 showy milkweed *(US)*
(sb; weed)
l Asclepias speciosa

2137 shred v *(cane)*
f couper en petits morceaux; déchiqueter; mettre en lambeaux
d zerfasern; zerkleinern
e cortar en pedazos pequeños; desfibrar
n uiteenrafelen

2138 side-elevator; transverse transporter
(agr)
f transporteur m transversal
d Querförderband n
e transportador m transversal; banda f transportadora transversal
n dwarselevator m

2139 sieve grating
f grille f de tamisage
d Siebrost m
e emparrillado m de criba
n zeefrooster mn

* sieving → 1980

2140 sift v
f tamiser
d sichten
e cribar; tamizar
n zeven

2141 sight glass; window
f lunette f; regard m
d Beobachtungsfenster n
e mirilla f
n kijkglas n

* silage → 909

2142 silage additive *(for pressed pulp)*
f additif m d'ensilage
d Silierzusatz m

e aditivo *m* para el silaje
n sileeradditief *n*

2143 silicate
f silicate *m*
d Silikat *n*
e silicato *m*
n silicaat *n*

2144 silky bent grass; silky agrostis; wind bent grass; wind grass; corn grass
(sb; weed)
f agrostis *f* épi du vent; plumette *f*; jouet *m* du vent; épi *m* du vent; agrostide *f* des champs
d gemeiner Windhalm *m*; Ackerschmiele *f*; Ackerwindhalm *m*
n windhalm *m*
l Apera spica-venti; Agrostis spica-venti

2145 silo
(A pit, vat, or circular structure of wood, concrete, etc., for packing away fodder to convert it into silage by fermentation. Frequently applied colloquially to storage bins for granulated sugar)
f silo *m*
d Silo *m*
e silo *m*
n silo *m*

2146 silversheath knotweed *(US)*
(sb; weed)
l Polygonum argyrocoleon

* **silver Y moth** → 1156

2147 simple beet harvester
(agr)
f machine *f* simple
d einfache Erntemaschine *f*
e máquina *f* simple
n eenvoudige bietenoogstmachine *f*

2148 simple clarification
f clarification *f* simple
d einfache Klärung *f*; einstufige Klärung *f*
e clarificación *f* simple
n eenvoudige klaring *f*

2149 simple defecation
f défécation *f* simple
d einfache Defäkation *f*
e defecación *f* simple
n eenvoudige defecatie *f*

2150 simple imbibition
f imbibition *f* simple
d einfache Imbibition *f*

e imbibición *f* simple
n eenvoudige imbibitie *f*

2151 simple washing *(of filter cake)*
f lavage *m* simple
d einfaches Auswaschen *n*
e lavado *m* simple
n eenvoudig uitwassen *n*

2152 single *v*; **thin** *v*
(agr)
f démarier; éclaircir; dégarnir; distancer
d vereinzeln; verziehen; ausdünnen; verhacken; auslichten
e aclarear; ralear; arralar; entresacar
n uitdunnen; dunnen; opeenzetten; op afstand zetten

2153 single carbona(ta)tion
f carbonatation *f* simple
d einfache Carbonatation *f*; einmalige Carbonatation *f*; einstufige Carbonatation *f*
e carbonatación *f* simple
n enkele carbonatatie *f*

2154 single crusher
f simple défibreur *m*
d einfacher Crusher *m*; einfacher Vorverbrecher *m*
e desmenuzadora *f* simple
n enkele crusher *m*; enkele kneuzer *m*

2155 single effect; single effect evaporator
f simple effet *m*
d Einfachverdampfer *m*; Einkörperverdampfer *m*
e evaporador *m* de un solo cuerpo *m*; simple efecto *m*
n verdampapparaat *n* met één verdamplichaam; verdampapparaat *n* à simple effet

2156 single effect evaporation
f évaporation *f* à simple effet
d einstufige Verdampfung *f*; Einfachverdampfung *f*; Einkörperverdampfung *f*
e evaporación *f* de simple efecto
n eenvoudige verdamping *f*; eentrapsverdamping *f*

* **single effect evaporator** → 2155

2157 single feeler; single finder *(US)*
(agr)
f tâteur *m* simple
d einfacher Taster *m*
e palpador *m* sencillo
n enkele taster *m*; eenvoudige taster *m*

* **singleness** *(US)* → 1636

2158 single preliming; single predefecation
f préchaulage m simple
d einstufige Vorkalkung f; einstufige Vorscheidung f
e preencalado m simple; prealcalización f
simple; prealcalinización f simple
n eenvoudige voorkalking f; eenvoudige
voorscheiding f

2159 single purging
f simple turbinage m
e purgado m simple; simple purga f
n enkelvoudig centrifugeren n

2160 single simple imbibition
f imbibition f simple unique
d einmalige einfache Imbibition f
e imbibición f simple única
n enkele eenvoudige imbibitie f

2161 singling; thinning
(agr)
f démariage m; éclaircissage m;
distançage m; dégarnissage m
d Vereinzelung f; Vereinzeln n;
Verziehen n; Ausdünnen n;
Verhacken n; Auslichten n
e aclareo m; raleo m; entresaca f;
entresaque m
n uitdunnen n; dunnen n; opeenzetten n;
op afstand zetten n (Belg.)

2162 singling hoe; beet hoe
(agr)
f rasette f à betteraves
d Rübenkrehl m
e escardillo m para remolachas
n kleine bietenhark f

2163 skid feeler; runner finder *(US)*
(agr)
f tâteur m à patin
d Kufentaster m
e palpador m de patín
n sleetaster m

* **skimmed juice** → 758

2164 skimming
f démoussage m
d Entschäumen n
e desespumación f
n ontschuimen n; afschuimen n

* **skimming agent** → 72

* **skimming medium** → 72

2165 skimming station
f installation f de démoussage
d Entschäumungsanlage f
e instalación f desespumadora
n ontschuiminstallatie f;
afschuiminstallatie f

2166 skimming tank
f ballon m de démoussage
d Entschäumungsgefäß n
e aparato m cortador de espuma
n ontschuimbak m; afschuimbak m

* **skip** → 2300

* **skipjack** → 2705

* **skipping** → 2300

2167 slab of sugar
f lingot m de sucre
d Plattenzucker m
e plancha f de azúcar; tabla f de azúcar;
losa f de azúcar
n plaat f van suiker

* **slacked lime** → 2169

* **slack lime** → 2169

* **slack season** → 1688

2168 slake *v*; **slack** *v (lime)*
f éteindre
d löschen
e apagar
n blussen

**2169 slaked lime; slack lime; hydrated lime;
slacked lime**
f chaux f éteinte; chaux f hydratée
d gelöschter Kalk m; Löschkalk m
e cal f apagada; cal f hidratada
n gebluste kalk m; zuivere gebluste kalk m

**2170 slat conveyor; apron conveyor; slat
carrier; apron carrier**
f conducteur m à tablier; conducteur m à
persiennes
d Plattenbandtransporteur m
e conductor m con cinta de persianas
n lattentransporteur m

2171 sledge-type feeler; sledge-type finder
(US)
(agr)
f tâteur m fixe
d feststehender Taster m

e palpador *m* fijo
n vaste taster *m*

2172 sleeve bearing sugar cane mill
  f moulin *m* à cannes à paliers glisseurs
  d Gleitlager-Zuckerrohrmühle *f*
  e molino *m* de caña de azúcar con
     cojinetes de fricción
  n rietsuikermolen *m* met glijlagers

2173 sleeve feeler; sleeve finder *(US)*
     (agr)
  f tâteur *m* glissant
  d Schleiftaster *m*
  e palpador *m* deslizante

2174 sleeve topper
     (agr)
  f traîneau *m* décolleteur
  d Rübenköpfschlitten *m*
  e descoronadora *f* con palpador de patín
  n bietenkopslee *f*

  *   **slender foxtail** → 279

2175 sliced beets
  f betteraves *fpl* découpées (en cossettes)
  d geschnittene Rüben *fpl*
  e remolachas *fpl* cortadas en cosetas
  n gesneden bieten *fpl*

  *   **slices conveyor** → 656

  *   **slices elevator** → 657

2176 slicing blade
     (agr)
  f lame *f* tranchante
  d Häckselmesser *n*
  e cuchilla *f* troceadora
  n hakselmes *n*

2177 slicing capacity
  f capacité *f* de découpage; débit *m* de
     découpage
  d Schneidleistung *f*
  e capacidad *f* de cortar
  n snijcapaciteit *f*

  *   **slicing station** → 1889

  *   **slim amaranth** → 2191

2178 slop utilization plant
  f installation *f* d'utilisation des vinasses
  d Schlempenverwertungsanlage *f*
  e planta *f* para la utilización de melaza
     residual; planta *f* para la utilización de

     vinaza de melaza
  n vinasseverwerkingsinstallatie *f*

2179 slotted gauze *(in centrifugal)*
  f toile *f* perforée
  d Schlitzgewebe *n*
  e tela *f* perforada
  n geperforeerd gaas *n*

2180 slotted screen
  f tamis *m* à fentes
  d Spaltsieb *n*
  e tamizador *m* de rendijas
  n spleetzeef *f*

2181 sludge
     (The settled mud from the clarifier,
     before filtration and washing)
  f boue *f*
  d Schlamm *m*; Abschlamm *m*
  e lodo *m*
  n slijk *n*; slib *n*

2182 sluicing device
  f dispositif *m* de rinçage
  d Spülvorrichtung *f*
  e dispositivo *m* de lavado
  n doorspoelapparaat *n*

2183 slurry of ground sugar *(for seeding)*;
     seed slurry
  f pâte *f* de sucre moulu; "slurry"
  d "Slurry"
  e pasta *f* de azúcar molido; "slurry"
  n "slurry"

  *   **small bindweed** → 997

  *   **small bugloss** → 349

  *   **smaller burdock** → 1454

**2184 small flowered cranesbill; small
     cranesbill; small geranium**
     (sb; weed)
  f géranium *m* à tiges grêles; géranium *m*
     à tige grêle; bec-de-grue *m* mauvin;
     géranium *m* nain
  d kleiner Storchschnabel *m*; kleinblütiger
     Storchschnabel *m*
  n kleine ooievaarsbek *m*
  l Geranium pusillum

2185 small flowered melilot; sour clover *(US)*;
     bitter clover *(US)*; **small melilot; annual
     yellow clover**
     (sb; weed)
  f petit mélilot *m*; mélilot *m* à petite fleur
  d kleinblütiger Steinklee *m*

e meliloto *m* de flor pequeña
n kleinbloemige honingklaver *f*
l Melilotus indicus (L.) All.

* **smallflower galinsoga** → 1154

* **small geranium** → 2184

2186 small mallow
(sb; weed)
f petite mauve *f*
d kleinblütige Malve *f*
e malva *f* de flor pequeña
n rond kaasjeskruid *n*; rondbladig
kaasjeskruid *n*
l Malva pusilla Sm.; Malva borealis
Wallmann

* **small melilot** → 2185

2187 small nettle; burning nettle *(US)*; annual nettle; dog nettle
(sb; weed)
f petite ortie *f*; ortie *f* brûlante; ortie *f* romaine
d kleine Brennessel *f*; kleine Nessel *f*
e ortiga *f* romana; ortiga *f* menor
n kleine brandnetel *f*
l Urtica urens

2188 small-seeded false flax *(US)*; hairy gold-of-pleasure; little-pod false flax
(sb; weed)
f caméline *f* à petits fruits
d kleinfrüchtiger Leindotter *m*; kleinfrüchtiger Dotter *m*; Wilddotter *m*
e bolinas *fpl* de mieses
n kleinzadige huttentut *f*
l Camelina microcarpa

* **small tortoise beetle** → 2306

* **smartweed *(US)*** → 1750

2189 smooth groundcherry *(US)*
(sb; weed)
l Physalis subglabrata

2190 smooth hawk's-beard
(sb; weed)
f crépis *m* verdâtre
d grüner Pippau *m*; dünnästiger Pippau *m*; kleinköpfiger Pippau *m*
n klein streepzaad *n*; groen streepzaad *n*
l Crepis capillaris; Crepis virens

2191 smooth pigweed *(US)*; green amaranth; slim amaranth
(sb; weed)

f amarante *f* hybride; amarante *f* paniculée
d Bastard-Fuchsschwanz *m*; grünähriger Fuchsschwanz *m*
e bledo *m*
n basterdamarant *f*
l Amaranthus hybridus

* **smooth sowthistle** → 583

2192 smut
(sc; fungous disease)
f charbon *m*
d "Smut"; Zuckerrohrbrand *m*
e carbón *m*
n suikerrietbrand *m*

* **snapping beetle *(US)*** → 959

* **snout beetles *(US)*** → 2661

* **soapwort** → 672

2193 soda
f soude *f*
d Natriumkarbonat *n*; Soda *fn*
e sosa *f*
n natriumcarbonaat *n*; soda *mf*

2194 sodium deficiency
(sb; disease)
f carence *f* en sodium
d Natriummangel *m*
e deficiencia *f* de sodio; carencia *f* de sodio
n natriumgebrek *n*

2195 soft crystal
f cristal *m* mou
d weicher Kristall *m*
e cristal *m* suave
n zacht kristal *n*; week kristal *n*

2196 soften *v*
f adoucir
d enthärten
e endulzar; ablandar
n ontharden

2197 softening; dehardening
f adoucissement *m*
d Enthärtung *f*
e endulzamiento *m*; ablandamiento *m*
n ontharding *f*

2198 soft grain
f grain *m* tendre
d weiches Korn *n*

e grano *m* dulce; grano *m* blando
n zacht grein *n*

2199 soft rush; round rush; common rush
(sb; weed)
f jonc *m* commun; jonc *m* épars; jonc *m* glauque; jonc *m* des jardiniers
d Flatterbinse *f*
e junco *m* de esteras
n pitrus *m*
l Juncus effusus

2200 solid-liquid-extraction
f extraction *f* solide-liquide
d Fest-Flüssig-Extraktion *f*
e extracción *f* sólida-líquida
n vast-vloeistof-extractie *f*

2201 solids
f solides *mpl*
d Feststoff *m*
e sólidos *mpl*
n vaste stof *f*

* **solids by drying** → 2526

2202 solids content
f teneur *f* en solides
d Feststoffanteil *m*; Feststoffgehalt *m*
e contenido *m* de sólidos
n vaste-stofgehalte *n*

2203 solubility
f solubilité *f*
d Löslichkeit *f*
e solubilidad *f*
n oplosbaarheid *f*

2204 solubility coefficient; saturation coefficient
(Ratio of the quantity of sucrose soluble in a given weight of water in an impure solution, at a certain temperature, to the quantity soluble in pure water at the same temperature)
f coefficient *m* de solubilité; coefficient *m* de saturation
d Löslichkeitszahl *f*
e coeficiente *m* de solubilidad; coeficiente *m* de saturación
n oplosbaarheidscoëfficiënt *m*

2205 soluble
f soluble
d löslich
e soluble
n oplosbaar

2206 soluble lime salts
f sels *mpl* de chaux solubles
d lösliche Kalksalze *npl*
e sales *fpl* de cal solubles
n oplosbare kalkzouten *npl*

2207 soluble sugar
f sucre *m* soluble
d löslicher Zucker *m*
e azúcar *m* soluble
n oplosbare suiker *m*

2208 solvent
f solvant *m*
d Lösungsmittel *n*
e solvente *m*
n oplosmiddel *n*

2209 sooty mold
(sc; fungous disease)
f fumagine *f*
d Rußtau *m*
e fumagina *f*

2210 sound beet
f betterave *f* saine
d gesunde Rübe *f*
e remolacha *f* sana
n gezonde biet *f*

* **sour clover** *(US)* → 2185

* **sour dock** → 582

* **sour rot** → 1983

2211 Southern blight *(US)*; **Southern sclerotium root rot** *(US)*; **sclerotial rot** *(US)*
(sb; disease)
f maladie *f* sclérotique des régions irriguées; sclerotium *m*
d Sklerotienkrankheit *f*
e mal *m* del esclerocio; esclerotina *f*
n sclerotiënrot *n*

* **Southern garden leafhopper** *(US)* → 1196

2212 Southern sandbur(r) *(US)*
(sb; weed)
f bardanette *f*
l Cenchrus echinatus

* **Southern sclerotium root rot** *(US)* → 2211

2213 sowing in situ; drilling to a stand
(agr)
f semis *m* en place

d Endstandsaat *f*
e siembra *f* de asiento; sembrado *m* directo; siembra *f* definitiva
n eindafstandzaai *m*; bezaaiing *f* op eindafstand; uitzaai *m* op eindafstand

2214 sow thistle
(sb; weed)
f laiteron *m*
d Gänsedistel *f*; Saudistel *f*
e cerraja *f*
n melkdistel *m*
l Sonchus sp.

* **spaced seeding** → 1806

* **spacing drill** → 1805

* **spear-leaved fat hen saltbush** → 1223

* **spear-leaved orache** → 1223

2215 spear thistle; bull thistle *(US)*; fuller's thistle
(sb; weed)
f chardon *m* lancéolé; cirse *m* lancéolé
d gemeine Kratzdistel *f*; Wegdistel *f*; gewöhnliche Kratzdistel *f*; lanzen-blättrige Kratzdistel *f*; Lanzetkratz-distel *f*
e cardo *m* común; cardo *m* de toro; cardo *m* negro; cardo *m* lanceolado
n speerdistel *m*
l Cirsium lanceolatum; Cirsium vulgare

2216 specific heat
f chaleur *f* spécifique
d spezifische Wärme *f*
e calor *m* específico
n soortelijke warmte *f*

2217 specific volume
f volume *m* spécifique
d spezifisches Volumen *n*
e volumen *m* específico
n soortelijk volume *n*

2218 spectrophotometer
f spectrophotomètre *m*
d Spektralfotometer *n*; Spektrofotometer *n*
e espectrofotómetro *m*
n spectrofotometer *m*

2219 speed of rotation
f vitesse *f* de rotation
d Drehzahlgeschwindigkeit *f*; Drehzahl *f*
e velocidad *f* de rotación
n omwentelingssnelheid *f*; toerental *n*

2220 speed of settling
f vitesse *f* de décantation
d Sedimentationsgeschwindigkeit *f*; Absetz-geschwindigkeit *f*
e velocidad *f* de decantación
n bezinksnelheid *f*

2221 speedwell
(sb; weed)
f véronique *f*
d Ehrenpreis *m*
e verónica *f*
n ereprijs *m*
l Veronica sp.

2222 spike
(sc; virus disease)
d Spike-Krankheit *f*
e enfermedad *f* de la banderilla
n spike-ziekte *f*

* **spin** *v* → 460

* **spinach aphid** → 1735

2223 spinach carrion beetle
(sb; pest)
l Silpha bituberosa

2224 spinach flea beetle *(US)*
(sb; pest)
l Disonycha xanthomelas

* **spinach leaf miner *(US)*** → 1545

2225 spiral chute
f goulotte *f* hélicoïdale
d Wendelrutsche *f*
e lanzadero *m* helicoidal
n schroefgoot *f*

2226 spiral nematodes
(sc; pest)
d Spiralälchen *npl*
l Helicotylenchus spp.

* **spittlebugs** → 1134

* **spittle insects** → 1134

* **splitter knife** → 1982

* **splitters of a V-corrugated knife *(US)*** → 1978

2227 spontaneous crystal formation
f formation *f* spontanée des cristaux
d spontane Keimbildung *f*; spontane Kristallbildung *f*

e formación f espontánea de cristales
n spontane kristalvorming f

**2228 spontaneous evaporation; self-
        evaporation**
        (Extraction of condensate)
f auto-évaporation f; évaporation f
spontanée
d Selbstverdampfung f
e autoevaporación f; evaporación f
espontánea
n zelfverdamping f

2229 spot market
        (eco)
f marché m spot; marché m en disponible
d Lokomarkt m; Spotmarkt m
e mercado m al contado; mercado m de
actuales
n spotmarkt f; locomarkt f

2230 spot price
        (eco)
f prix m spot
d Lokopreis m; Spotkurs m
e precio m al contado
n spotprijs m; locoprijs m

2231 spotted blister beetle *(US)*
        (sb; pest)
l Epicauta maculata

2232 spotted cutworm *(US) (larva)*
        (sb; pest)
f ver m gris tacheté
d Larve f der Frischkräuterrasen-Bodeneule; Larve f des schwarzen C; Larve f
der schwarzen C-Erdeule
e larva f de la c-negra
l Amathes c-nigrum; Diarsia c-nigrum;
Agrotis c-nigrum

* **spotted lady's-thumb** → 1750

**2233 spotted millipede; spotted snake
        millipede**
        (sb; pest)
f blaniule m moucheté; blaniule m
tacheté; iule m à taches rouges
d getüpfelter Tausendfüßler m; getüpfelter
Tausendfuß m; gemeiner Tüpfeltausend-
fuß m; getüpfelter Schnurfüßler m
e cardador m manchado
l Blaniulus guttulatus

* **sprangled beet** → 968

2234 spray
f pulvérisateur m

d Sprühdüse f
e aspersor m
n verstuiver m

* **spray nozzles** → 1635

2235 spray pipe
f rampe f de clairçage
d Sprührohr n
e aspersor m
n sproeibuis f

**2236 spreading orache; common orache;
        creeping fat hen; fat hen**
        (sb; weed)
f arroche f étalée; arroche f des champs
d Rutenmelde f; gemeine Melde f; aus-
gebreitete Melde f; spreizende Melde f
e armuelle m silvestre
n uitstaande melde f
l Atriplex patula

2237 springtail
        (sb; pest)
f collembole m; collembole m ordinaire
d Kugelspringer m; Springschwanz m
e colémbolo m
n springstaart m
l Sminthurus sp.

**2238 springtail; collembolous insect;
        collembolan**
        (sc; pest)
f collembole m
d Springschwanz m
e colémbolo m
n springstaart m
l Collembola

2239 spring wild oat; wild oat
        (sb; weed)
f folle avoine f; avéneron m
d Flughafer m; Windhafer m; Wildhafer m
e avena f loca; avena f fatua; cugula f;
ballueca f; avena f silvestre común
n wilde haver m; oot f
l Avena fatua

* **spun sugar** → 710

2240 squirrel-cage motor
        (An a-c machine of the induction type,
        consisting essentially of the stator and
        rotor. The rotor assembly resembles a
        squirrel cage treadmill, and has no
        electrical connections to a power source)
f moteur m à cage d'écureuil
d Kurzschlußläufermotor m; Kurzschluß-
motor m

e motor *m* de jaula de ardilla
n kooiankermotor *m*

* **squirreltail** → 1116

* **squirreltail barley** → 1116

* **squirreltail grass** → 1116

2241 standard liquor
(The principal feed for the white boiling. In a conventional three-boiling scheme it consists of thick juice plus remelt syrup. In a conventional four-boiling scheme it consists of remelt syrup containing only water as the solvent)
f liqueur *f* standard
d Standardkläre *f*
e licor *m* standard
n standaardstroop *f*

2242 standard quality
(EEC)
f qualité *f* type
d Standardqualität *f*
e calidad *f* tipo
n standaardkwaliteit *f*

2243 standard rate of storage costs; standard storage costs
(EEC)
f montant *m* forfaitaire des frais de stockage
d Lagerkostenpauschale *f*
e montante *m* global de los gastos de almacenamiento
n forfaitair bedrag *n* voor de opslagkosten

2244 standard solution
f solution *f* normale
d Standardlösung *f*; Titerlösung *f*; Normal-lösung *f*
e solución *f* normal
n standaardoplossing *f*; normaaloplossing *f*

* **standard storage costs** → 2243

2245 stand-by pump
f pompe *f* de secours
d Hilfspumpe *f*
e bomba *f* de refacción
n hulppomp *f*

2246 starch
f amidon *m*
d Stärke *f*
e almidón *m*
n zetmeel *n*

2247 starch sugar; crude glucose
f glucose *m* massé; sucre *m* d'amidon
d Stärkezucker *m*
e azúcar *m* de almidón
n glucose *f* massé; vaste glucose *f*

2248 starch syrup; confectioner's glucose
f sirop *m* de glucose
d Stärkesirup *m*; Stärkezuckersirup *m*; Kapillarsirup *m*
e jarabe *m* de glucosa
n glucosestroop *f*; confiseursstroop *f*

* **star grass** → 268

* **starry cerastium** → 1001

2249 stationary dry unloading plant
(beet)
f point *m* fixe à sec
d stationäre Trockenentladeanlage *f*
e instalación *f* fija de descarga en seco
n stationaire droge lossing *f*

2250 steam accumulator
f accumulateur *m* de vapeur
d Dampfspeicher *m*
e acumulador *m* de vapor
n stoomaccumulator *m*

2251 steam balance
f bilan *m* de vapeur
d Dampfbilanz *f*
e balance *m* de vapor
n stoombalans *f*

* **steam belt** *(US)* → 2253

2252 steam boiler; steam generator
f chaudière *f* à vapeur
d Dampfkessel *m*
e caldera *f* de vapor
n stoomketel *m*

2253 steam chamber; steam chest *(US)*; **steam belt** *(US)*
f chambre *f* de vapeur
d Dampfraum *m*
e cuarto *m* de vapor
n stoomkamer *f*; stoomruimte *f*

2254 steam cock
f robinet *m* de vapeur
d Dampfventil *n*
e llave *m* de vapor
n stoomkraan *f*

2255 steam coil
f serpentin *m* à vapeur

d Dampfheizschlange f; Dampfschlange f
e serpentín m de vapor
n stoomverwarmingsslang f;
  stoomserpentijn m

2256 steam conduct; steam main
f conduite f de vapeur
d Dampfleitung f
e cañería f de vapor
n stoomleiding f

2257 steam consumption
f consommation f de vapeur
d Dampfverbrauch m
e consumo m de vapor
n stoomverbruik n

2258 steam engine
f machine f à vapeur
d Dampfmaschine f
e máquina f de vapor
n stoommachine f

2259 steam exhaust side
f côté m d'échappement de la vapeur
d Dampfauslaßseite f
e lado m de escape del vapor
n stoomuitlaatzijde f

* **steam generator** → **2252**

2260 steam heating
f chauffage m à la vapeur
d Dampfheizung f
e calefacción f de vapor; calefacción f a
  vapor
n stoomverwarming f

2261 steam hose
f tuyau m à vapeur
d Dampfschlauch m
e tubería f para vapor
n stoomslang f

2262 steam imbibition
f imbibition f à la vapeur
d Dampfimbibition f
e imbibición f de vapor
n stoomimbibitie f

2263 steam injection
f injection f de vapeur
d Dampfinjektion f
e inyección f de vapor
n stoominjectie f

2264 steam injector; steam jet injector
f injecteur m de vapeur
d Dampfstrahlinjektor m

e inyector m de vapor
n stoominjecteur m

2265 steam jacket
f chemise f à vapeur; double enveloppe f
  de vapeur; manteau m chauffant à
  vapeur
d Dampfmantel m
e camisa f de vapor
n stoommantel m

2266 steam jet cleaning machine
f installation f de nettoyage à jet de
  vapeur
d Dampfstrahl-Reinigungsgerät n
e máquina f de limpieza a chorro de vapor
n stoomstraalreinigingsinstallatie f

* **steam jet injector** → **2264**

* **steam main** → **2256**

2267 steam v out
f traiter au jet de vapeur; jeter de la
  vapeur; rincer à la vapeur
d ausdämpfen
e limpiezar con vapor; evaporar
n uitstomen

2268 steam pressure
f pression f de vapeur
d Dampfdruck m
e presión f de vapor
n stoomdruk m

2269 steam tight
f étanche à la vapeur; imperméable à la
  vapeur
d dampfdicht
e impermeable al vapor
n stoomdicht

2270 steam trap
(Extraction of condensate)
f purgeur m de vapeur; séparateur m
  d'eau de condensation; purgeur m d'eau
  de condensation
d Kondensatableiter m; Kondenstopf m;
  Wassersammler m
e separador m de agua condensada;
  trampa f de vapor; purgador m de vapor;
  colector m de condensado
n condenspot m

2271 steam turbine
f turbine f à vapeur
d Dampfturbine f
e turbina f de vapor
n stoomturbine f

2272 steam washing
(Centrifuging)
f clairçage *m* à la vapeur
d Waschen *n* mit Dampf; Decken *n* mit
Dampf; Dampfdecke *f*
e lavado *m* con vapor; lavado *m* a vapor
n dekken *n* met stoom; afdekken *n* met
stoom

2273 steckling
(agr)
f planchon *m*
d Steckling *m*
e remolacha *f* joven
n stekbietje *n*; stekling *m*

2274 Steffen factory
(A factory that incorporates a version of
the Steffen process)
f sucraterie *f*
d Fabrik *f* mit Melasseentzuckerung nach
dem Steffen-Verfahren
e fábrica *f* de azúcar con instalación de
sacarato
n fabriek *f* met melasseontsuikering
volgens het Steffen-proces

2275 Steffen filtrate
(The liquid effluent from the Steffen
process, consisting of the decantate
(*overflow*) from the saccharate thickener
plus the hot filtrate, if any. If the hot-
saccharate stage is not used, the cold
filtrate becomes the Steffen filtrate)
f filtrat *m* du décanteur à saccharate
d Filtrat *n* der Saccharatfällung
e filtrado *m* del precipitador de sacarato
n Steffen-filtraat *n*

2276 Steffen house
(The section of the factory that
encompasses the Steffen process)
f atelier *m* de sucratage
d Melasseentzuckerungsstation *f*
e estación *f* de sacarato
n melasseontsuikeringsstation *n*

2277 Steffen loss
(The sucrose leaving the process in the
Steffen filtrate; usually expressed as a
percentage of the sucrose in the molasses
worked, of the weight of beets sliced, or
of the total sucrose entering the factory)
f pertes *fpl* de sucraterie
d Steffenverluste *mpl*
e pérdidas *fpl* de la instalación de sacarato
n Steffen-verliezen *npl*

2278 Steffen molasses
f mélasse *f* de sucraterie
d Steffen-Melasse *f*; Restmelasse *f* nach
dem Steffen-Verfahren; Ablauge *f* nach
dem Steffen-Verfahren; Abfallauge *f*
nach dem Steffen-Verfahren
e melaza *f* de instalación de sacarato;
melaza *f* de fábrica de sacaratos
n Steffen-melasse *f*; restmelasse *f* na het
Steffen-proces

**2279 Steffen process; Steffen separation
process; saccharate process; calcium
saccharate process**
(A process for the extraction of sucrose
from molasses by combination with CaO
and Ca(OH)$_2$. This term should be
retained as a memorial to the inventors,
Steffen *pere et fils*, in preference to the
terms "saccharate process" and "calcium
saccharate process". "Steffens" and
"Steffens process" are incorrect)
f procédé *m* Steffen; désucrage *m* de la
mélasse selon le procédé Steffen;
dessucrage *m* de la mélasse selon le
procédé Steffen
d Steffensches Ausscheidungsverfahren *n*;
Steffen-Verfahren *n* zur Melasseentzuckerung; Kalksaccharatverfahren *n*
e procedimiento *m* de Steffen
n Steffen-proces *n*

* **Steffen pulp** → 1855

* **Steffen sugar pulp** → 1855

2280 Steffen waste water
f eaux *fpl* résiduaires de sucraterie
d Steffen-Abwasser *n*
e aguas *fpl* residuales de fábrica de
sacaratos por el procedimiento de
Steffen; aguas *fpl* residuales de la
instalación de sacarato
n Steffen-afvalwater *n*

* **stem-and-bulb eelworm** → 352

* **stem canker** → 2129

* **stem eelworm** → 352

* **stem nematode** → 352

2281 stem rot
(sb; disease)
f taches *fpl* de phoma (sur la tige)
d Phoma-Stengelflecken *mpl*
e foma *f* del tallo; phoma *f* del tallo
n Phoma-stengelvlekken *fpl*

* **stem rot** → 1148

* **stem rot disease** → 1148

2282 step-wise boiling
 f cuisson f par étages
 d Stufenkochung f
 e cocimiento m por etapas
 n trapsgewijs koken n

2283 stinging nettle; greater nettle; common nettle; big-sting nettle
 (sb; weed)
 f grande ortie f; ortie f dioïque
 d große Brennessel f; große Nessel f
 e ortiga f mayor; ortiga f grande
 n grote brandnetel f
 l Urtica dioica

2284 sting nematode
 (sc; pest)
 l Belonolaimus gracilis

2285 stink grass *(US)*; **stinking lovegrass**
 (sb; weed)
 f éragrostis f à gros épi; éragrostide f à gros epi
 d großähriges Liebesgras n; bewimpertes Liebesgras n
 n groot liefdegras n
 l Eragrostis megastachya; Eragrostis cilianensis

2286 stinking mayweed; dog fennel; dog chamomile; dog camomile; mayweed chamomile; mayweed camomile
 (sb; weed)
 f camomille f puante
 d stinkende Hundskamille f; Stinkhunds-kamille f
 e manzanilla f hedionda; manzanilla f fétida; magarzuela f
 n stinkende kamille f; stinkbloem f; paddebloem f; koedille f
 l Anthemis cotula

* **stinkweed** → 1003

2287 stirrer; agitator
 f agitateur m
 d Rührwerk n
 e agitador m
 n roerwerk n

2288 stirring; agitation
 f agitation f
 d Rühren n
 e agitación f
 n roeren n

2289 stirring arm
 f bras m agitateur
 d Rührarm m
 e brazo m agitador
 n roerarm m

2290 stirring crystallizer
 f malaxeur-agitateur m
 d Rührmaische f
 e cristalizador-enfriador m
 n koeltrog m met roerwerk; roermalaxeur m

* **St. James' wort** → 1884

* **stone catcher** → 1988

* **stoneseed** → 649

* **stone separator** → 1988

* **storability** *(US)* → 2294

2291 storage contract
 (EEC)
 f contrat m de stockage
 d Lagervertrag m
 e contrato m de almacenamiento
 n opslagovereenkomst f

2292 storage levy
 (EEC)
 f cotisation f des frais de stockage; cotisation f de stockage
 d Lagerkostenabgabe f
 e cotización f de almacenamiento
 n bijdrage f tot de opslagkosten

* **storage rot** → 495

* **storage rot** *(US)* → 680

* **storage tank** → 2432

2293 storage yard
 f aire f de stockage
 d Stapelplatz m
 e área f de almacenamiento
 n opslagplaats f

2294 storing properties; storability *(US)*
 f stockabilité f; aptitude f au stockage
 d Lagerfähigkeit f
 e conservabilidad f
 n bewaarkwaliteit f bij het opslaan; houdbaarheid f

2295 strained juice
 f jus m tamisé

d gesiebter Saft *m*
e jugo *m* colado; guarapo *m* colado
n gezeefd sap *n*

2296 strainer
f tamiseur *m*
d Sieb *n*
e colador *m*
n zeef *f*

2297 strangles
(Deformation of roots; sb; disease)
f étranglement *m* de la racine
d Umfallkrankheit *f* der Rübe
e estrangulación *f* de la raíz;
estrangulamiento *m* de la raíz
n insnoeren *n* en omvallen van de biet

* **strangle weed → 843**

* **straw catcher → 2532**

2298 streak disease
(sc; virus disease)
f "streak"; maladie *f* des stries
d chlorotische Streifenkrankheit *f*
e rayado *m* de la hoja
n strepenziekte *f*; chlorotische strepenziekte *f*

2299 striate mosaic
(sc; virus disease)
f "striate mosaic"
d Strichelkrankheit *f*
e mosaico *m* estriado
n "striate mosaic"

2300 strike; skip; pan boiling; skipping
(The finished batch of massecuite when
discharged from the pan prior to
centrifuging)
f cuite *f*
d Sud *m*
e templa *f*
n kooksel *n*

2301 string proof; string test
(In pan boiling, a condition of the mother
liquor, determined by placing a quantity
between the thumb and first finger, and
pulling the fingers apart rapidly. A
thread will form if the desired
consistency has been reached)
f preuve *f* au filet
d Fadenprobe *f*
e prueba *f* al hilo
n draadproef *f*

**2302 string-proof boiling; boiling to string-
proof**
f cuisson *f* au filet; cuisson *f* par preuve
au filet
d Kochen *n* auf Faden; Fadenkochen *n*
e cocimiento *m* al hilo
n koken *n* op draad; koken *n* op filet

* **string test → 2301**

2303 striped blister beetle *(US)*
(sb; pest)
l Epicauta vittata

2304 striped flea beetle *(US)*
(sb; pest)
l Phyllotreta striolata

2305 striped mealybug *(US)*
(sc; pest)
d weiße Lamtorolaus *f*
l Ferrisiana virgata

**2306 striped tortoise beetle; small tortoise
beetle**
(sb; pest)
f casside *f* noble; casside *f* de la betterave
d goldstreifiger Schildkäfer *m*; kleiner
glanzstreifiger Schildkäfer *m*; edler
Schildkäfer *m*
e casida *f* noble
n gestreepte schildpadtor *f*
l Cassida nobilis

2307 sublimation
f sublimation *f*
d Sublimation *f*
e sublimación *f*
n sublimatie *f*

2308 sublimation chamber *(of sulphur
furnace)*
f sublimateur *m*
d Sublimationskammer *f*
e sublimador *m*
n sublimatiekamer *f*

* **subsidation → 2115**

* **subsider → 2484**

2309 subsider juice
f jus *m* décanté; jus *m* de décantation
d Klarsaft *m*
e guarapo *m* de decantación; jugo *m* de
decantación
n klaarsap *n*; helder indiksap *n*

2310 subterranean springtail
(sb; pest)
f collembole *m* souterrain
d unterirdisch lebender Springschwanz *m*
e colémbolo *m* subterráneo
n ondergronds levende springstaart *m*
l Onychiurus campatus

* **sucrase** → 1348

2311 sucro-carbonate of lime
(Carbonatation)
f sucro-carbonate *m* de chaux
d Sukrose-Kalziumkarbonat *n*
e sucrocarbonato *m* de cal
n kalk-suikercarbonaat *n*

2312 sucrochemistry
f sucrochimie *f*
d Zuckerchemie *f*; Sucrochemie *f*
e sucroquímica *f*
n suikerchemie *f*; sucrochemie *f*

2313 sucrose
(The disaccharide α-D-glucopyranosyl-β-
D-fructofuranoside)
f saccharose *m*
d Saccharose *f*
e sacarosa *f*
n saccharose *f*

2314 sucrose content
f teneur *f* en saccharose
d Saccharosegehalt *m*
e contenido *m* de sacarosa
n saccharosegehalte *n*

2315 sucrose crystal
f cristal *m* de saccharose
d Saccharosekristall *m*
e cristal *m* de sacarosa
n saccharosekristal *n*

2316 sucrose destruction; breakdown of
sucrose
f décomposition *f* du saccharose;
dégradation *f* du saccharose
d Saccharoseabbau *m*; Saccharose-
zersetzung *f*
e degradación *f* de sacarosa;
decomposición *f* de sacarosa
n suikeraftrek *m*

2317 sucrose extraction
(Sucrose in mixed juice percent sucrose
in cane)
f extraction *f* du saccharose
d Saccharoseextraktion *f*

e extracción *f* de sacarosa
n saccharose-extractie *f*

* **sucrose/fibre content** → 1595

2318 sucrose inversion
f inversion *f* du saccharose
d Saccharose-Inversion *f*
e inversión *f* de la sacarosa
n saccharose-inversie *f*

2319 sucrose loss
f perte *f* de saccharose
d Saccharoseverlust *m*
e pérdida *f* de sacarosa
n saccharoseverlies *n*

2320 suction filter
f filtre *m* à aspiration; filtre *m* aspirateur
d Saugfilter *n*
e filtro *m* por succión
n zuigfilter *n*

2321 suction pipe *(for vacuum)*
f tube *m* d'aspiration
d Saugrohr *n*; Saugleitung *f*
e tubo *m* de succión
n zuigpijp *f*; zuigleiding *f*

2322 sugar
f sucre *m*
d Zucker *m*
e azúcar *m*
n suiker *m*

2323 sugar balance *(of a factory)*
f bilan *m* sucre
d Zuckerbilanz *f*
e balance *m* de azúcar
n suikerbalans *f*

2324 sugar balance *(of a country)*
(eco)
f bilan *m* de sucre
d Zuckerbilanz *f*
e balance *m* azucarero
n suikerbalans *f*

2325 sugar-bearing pulp
f pulpe *f* sucrée
d zuckerhaltige Pülpe *f*
e pulpa *f* saturada de azúcar
n suikerhoudende pulp *f*

2326 sugar beet
f betterave *f* sucrière; betterave *f* à sucre
d Zuckerrübe *f*
e remolacha *f* azucarera
n suikerbiet *f*

l Beta vulgaris ssp. vulgaris var. altissima
(Doell); Beta vulgaris altissima; Beta
vulgaris saccharifera

**2327 sugar beet area to be harvested; sugar
beet acreage to be harvested**
f surface *f* betteravière; surface *f*
ensemencée en betteraves sucrières
d Zuckerrübenerntefläche *f*; Zuckerrüben-
anbaufläche *f*; Rübenfläche *f*
e superficie *f* para remolacha azucarera a
ser cosechada
n bietenareaal *n*; bietenoppervlakte *f*

* **sugar beet armyworm** *(US)* → 176

2328 sugar beet crop
f récolte *f* betteravière; récolte *f* des
betteraves sucrières
d Zuckerrübenernte *f*
e cosecha *f* de las remolachas azucareras
n suikerbietenoogst *m*

2329 sugar beet crown borer
(sb; pest)
l Hulstia undulatella

* **sugar beet cultivation** → 2331

2330 sugar beet disease
f maladie *f* de la betterave sucrière
d Zuckerrübenkrankheit *f*
e enfermedad *f* de la remolacha azucarera
n suikerbietziekte *f*

* **sugar beet eelworm** → 188

* **sugar beet factory** → 240

**2331 sugar beet growing; sugar beet
cultivation**
f culture *f* betteravière; culture *f* de la
betterave sucrière
d Zuckerrübenbau *m*; Zuckerrüben-
anbau *m*
e cultivo *m* remolachero; cultivo *m* de
remolacha azucarera
n suikerbietenteelt *f*; suikerbietencultuur *f*

2332 sugar beet molasses
f mélasse *f* de betterave sucrière
d Zuckerrübenmelasse *f*
e melaza *f* de remolacha azucarera
n suikerbietmelasse *f*

* **sugar beet nematode** *(US)* → 188

2333 sugar beet pest
f ennemi *m* de la betterave sucrière;

ravageur *m* de la betterave sucrière
d Schädling *m* der Zuckerrübe
e plaga *f* de la remolacha azucarera;
parásito *m* de la remolacha azucarera
n plaag *f* van de suikerbiet

2334 sugar beet root-aphid *(US)*
(sb; pest)
l Pemphigus betae; Pemphigus
populivenae

2335 sugar beet root maggot *(US)*
(sb; pest)
f tétranyque *m*; araignée *f* rouge
d Spinnmilbe *f*
l Tetanops myopaeformis

2336 sugar beet seed
f graine *f* de betterave sucrière
d Zuckerrübensamen *m*
e semilla *f* de remolacha azucarera
n suikerbietenzaad *n*

* **sugar beet weevil** → 228

2337 sugar beet wireworm *(US)*
(sb; pest)
l Limonius californicus

* **sugar boiler** → 1723

2338 sugar boiling
(The transformation of syrup into a
mixture of crystals and mother liquor by
simultaneous evaporation and
crystallization)
f cuite *f*
d Zuckerkochen *n*
e cocimiento *m*; cocción *f*
n koken *n*; suikerkoken *n*

* **sugar cake** → 1433

2339 sugar cane
f canne *f* à sucre
d Zuckerrohr *n*
e caña *f* de azúcar
n suikerriet *n*
l Saccharum officinarum

**2340 sugar cane area to be harvested; sugar
cane acreage to be harvested**
f surface *f* ensemencée en cannes à sucre
d Zuckerrohrerntefläche *f*; Zuckerrohr-
anbaufläche *f*; Rohrfläche *f*
e área *f* de caña de azúcar a ser cosechada
n rietareaal *n*; rietoppervlak *n*

2341 sugar cane beetle
(sc; pest)
l Euetheola rugiceps

2342 sugar cane borer
(sc; pest)
f borer *m* de la tige; chenille *f* mineuse de
la canne à sucre; ver *m* rongeur de la
tige
d Zuckerrohrstengelbohrer *m*
e gusano *m* minador de la caña de azúcar;
taladrador *m* de la caña
n stengelboorder *m*
l Diatraea saccharalis

2343 sugar cane disease
f maladie *f* de la canne à sucre
d Zuckerrohrkrankheit *f*
e enfermedad *f* de la caña de azúcar
n suikerrietziekte *f*

2344 sugar cane leafhopper
(sc; pest)
l Perkinsiella saccharicida

2345 sugar cane molasses
f mélasse *f* de canne à sucre
d Zuckerrohrmelasse *f*
e melaza *f* de caña de azúcar
n suikerrietmelasse *f*

2346 sugar cane mosaic; yellow-stripe disease
(sc; virus disease)
f mosaïque *f*
d Mosaik *n*; Mosaikkrankheit *f*; Gelb-
streifenkrankheit *f*
e mosaico *m*
n mozaïek *f*

2347 sugar cane pest
f ennemi *m* de la canne à sucre; ravageur
m de la canne à sucre
d Schädling *m* des Zuckerrohrs
e plaga *f* de la caña de azúcar; parásito *m*
de la caña de azúcar
n plaag *f* van het suikerriet

2348 sugar centrifugal
f centrifugeuse *f* à sucre; essoreuse *f* à
sucre
d Zuckerzentrifuge *f*
e centrífuga *f* de azúcar; centrífuga *f*
azucarera
n suikercentrifuge *f*

2349 sugar compartment *(of centrifugal)*
f chambre *f* à sucre
d Zuckerraum *m*

e cámara *f* de azúcar
n suikerruimte *f*

2350 sugar concentration transducer
f régulateur *m* de la concentration du
sucre
d Zuckerkonzentrationsregler *m*
e regulador *m* de concentración del azúcar
n regelaar *m* van de suikerconcentratie

* **sugar cone** → 2379

2351 sugar consumption
(eco)
f consommation *f* de sucre
d Zuckerverbrauch *m*
e consumo *m* de azúcar
n suikerverbruik *n*; suikerconsumptie *f*

2352 sugar containing product
f produit *m* sucré; produit *m* contenant du
sucre
d zuckerhaltiges Erzeugnis *n*
e producto *m* azucarado
n suikerhoudend produkt *n*

2353 sugar content *(of beet, cane)*
f teneur *f* en sucre; richesse *f* saccharine
d Zuckergehalt *m*
e contenido *m* de azúcar
n suikergehalte *n*

2354 sugar content of a juice
f richesse *f* d'un jus
d Zuckergehalt *m* eines Saftes
e riqueza *f* de un jugo
n suikergehalte *n* van een sap

2355 sugar conveyor *(below centrifugals)*
f transporteur *m* de sucre
d Zuckerförderer *m*; Zucker-
transporteur *m*
e conductor *m* de azúcar
n suikertransporteur *m*

* **sugar cook** → 1723

2356 sugar cooling plant
f refroidisseur *m* de sucre
d Zuckerkühler *m*
e refrigerador *m* para azúcar
n suikerkoelinrichting *f*

2357 sugar country; sugar producing country
f pays *m* sucrier; pays *m* à sucre; pays *m*
producteur de sucre
d Zuckerland *n*
e país *m* azucarero
n suikerland *n*

2358 sugar crusher
  f broyeur *m* de sucre
  d Zuckerbrecher *m*
  e rompedor *m* de azúcar
  n suikerbreker *m*

2359 sugar crystal
  f cristal *m* de sucre
  d Zuckerkristall *m*
  e cristal *m* de azúcar
  n suikerkristal *n*

2360 sugar cubes
  f morceaux *mpl* de sucre
  d Zuckerwürfel *mpl*
  e cubos *mpl* de azúcar
  n suikerklontjes *npl*

2361 sugar dealer
  f grossiste *m* en sucres
  d Zuckergroßhändler *m*
  e mayorista *m* de azúcares; grosista *m* de azúcares
  n suikergroothandelaar *m*

2362 sugar deliveries
  (eco)
  f livraisons *fpl* de sucre
  d Zuckerablieferungen *fpl*
  e suministro *m* de azúcar
  n suikerleveringen *fpl*

2363 sugar distribution
  (eco)
  f distribution *f* de sucre
  d Zuckerabsatz *m*
  e distribución *f* del azúcar; venta *f* del azúcar
  n suikerafzet *m*

2364 sugar drier
  (Apparatus for drying the damp-sugar product of the white centrifugals)
  f sécheur *m* de sucre
  d Zuckertrockner *m*
  e secador *m* de azúcar
  n suikerdroger *m*

  * sugar drier → 456

2365 sugar-drying room
  f sécherie *f*
  d Trocknungskammer *f*
  e cámara *f* de secado
  n droogkamer *m*

2366 sugar dust
  f poussière *f* de sucre
  d Zuckerstaub *m*

  e polvo *m* de azúcar; polvillo *m* de azúcar
  n suikerstof *f*

2367 sugar economy
  f économie *f* sucrière
  d Zuckerwirtschaft *f*
  e economía *f* azucarera
  n suikereconomie *f*

2368 sugar elevator
  f élévateur *m* de sucre; élévateur *m* à sucre
  d Zuckerelevator *m*; Zuckersenkrecht-förderer *m*
  e elevador *m* de azúcar
  n suikerelevator *m*; suikerophaler *m*

2369 sugar end; sugar house; back-end of the factory
  (The section of the factory that includes the processes for separating the final product from the purified juice by crystallization and recrystallization. Many include the evaporator station(s), especially in factories which make extensive use of vapo(u)rs for heating the vacuum pans)
  f atelier *m* de la cristallisation; chantier *m* de la cristallisation; arrière-usine *f*; atelier *m* du sucre; chantier *m* sucre
  d Zuckerhaus *n*; Kristallisationsstation *f*
  e estación *f* de cristalizadores; estación *f* de tachos; casa *f* de azúcar
  n kristallisatiewerf *f*; achterfabriek *f*

2370 sugar exchange
  (eco)
  f bourse *f* du sucre
  d Zuckerbörse *f*
  e lonja *f* del azúcar
  n suikerbeurs *f*

2371 sugar exporter
  (eco)
  f exportateur *m* de sucre
  d Zuckerexporteur *m*
  e exportador *m* de azúcar
  n suikeruitvoerder *m*; suikerexporteur *m*

2372 sugar exporting country
  f pays *m* exportateur de sucre
  d Zuckerexportland *n*
  e país *m* exportador de azúcar
  n suikerexporterend land *n*; suikerexportland *n*

2373 sugar factory
  (A processing facility which extracts and refines sugar from sugar beets or sugar

cane)
f sucrerie *f*
d Zuckerfabrik *f*; Zfb. *f*
e fábrica *f* de azúcar; azucarera *f*
n suikerfabriek *f*

2374 sugar factory product
   f produit *m* de sucrerie
   d Zuckerfabriksprodukt *n*
   e producto *m* de fábrica de azúcar
   n suikerfabrieksprodukt *n*

2375 sugar grinding plant
   f installation *f* de broyage du sucre
   d Zuckermahlanlage *f*
   e instalación *f* de molienda para azúcar
   n maalinstallatie *f* voor suiker

   * **sugar hail** → **2019**

   * **sugar head** → **2379**

2376 sugar hopper
   f trémie *f* à sucre
   d Zuckertrichter *m*
   e tolva *f* de azúcar
   n suikertrechter *m*

   * **sugar house** → **2369**

2377 sugar importer
   (eco)
   f importateur *m* de sucre
   d Zuckerimporteur *m*
   e importador *m* de azúcar
   n suikerimporteur *m*; suikerinvoerder *m*

2378 sugar importing country
   f pays *m* importateur de sucre
   d Zuckerimportland *n*
   e país *m* importador de azúcar
   n suikerimporterend land *n*;
     suikerimportland *n*

   * **sugar layer** → **1433**

2379 sugar loaf; sugar cone; sugar head
   f pain *m* de sucre; cône *m* de sucre
   d Zuckerhut *m*; Zuckerbrot *n*; Zucker-
     kegel *m*
   e pan *m* de azúcar; pilón *m* de azúcar;
     azucarillo *m*
   n suikerbrood *n*

2380 sugar loss
   f perte *f* en sucre
   d Zuckerverlust *m*
   e pérdida *f* de azúcar
   n suikerverlies *n*

2381 sugar lump crusher
   f broyeur *m* à grugeons
   d Klumpenbrecher *m*
   e triturador *m* de terrones
   n klontjesbreker *m*

2382 sugar manufacturer; sugar producer
   f producteur *m* de sucre; sucrier *m*;
     fabricant *m* de sucre
   d Zuckerhersteller *m*; Zuckerproduzent *m*;
     Zuckerfabrikant *m*
   e azucarero *m*; fabricante *m* de azúcar;
     productor *m* de azúcar
   n suikerfabrikant *m*; suikerproducent *m*

2383 sugar market
   f marché *m* du sucre; marché *m* sucrier
   d Zuckermarkt *m*
   e mercado *m* del azúcar; mercado *m*
     azucarero
   n suikermarkt *f*

2384 sugar price
   f prix *m* du sucre
   d Zuckerpreis *m*
   e precio *m* del azúcar
   n suikerprijs *m*

   * **sugar producer** → **2382**

   * **sugar producing country** → **2357**

2385 sugar production
   f production *f* de sucre; production *f*
     sucrière
   d Zuckererzeugung *f*; Zuckerproduktion *f*
   e producción *f* azucarera; producción *f* de
     azúcar
   n suikerproduktie *f*

2386 sugar recovery
   f rendement *m* en sucre
   d Zuckerausbeute *f*
   e recuperación *f* del azúcar; recobrado *m*
     de azúcar
   n suikerwinning *f*

2387 sugar refinery
   (A processing facility which produces
   refined sugar from raw sugar or syrup)
   f raffinerie *f* de sucre
   d Zuckerraffinerie *f*
   e refinería *f* de azúcar; refinería *f*
     azucarera
   n suikerraffinaderij *f*;
     suikerraffineerderij *f*

2388 sugar rope
   f boudin *m* de sucre

d Zuckerstrang *m*
n suikerstreng *f*

2389 sugar scale
f bascule *f* à sucre
d Zuckerwaage *f*
e balanza *f* de azúcar; báscula *f* de azúcar
n suikerbalans *f*; suikerbascule *f*

2390 sugar sifting plant
f installation *f* de classement du sucre
d Zuckersichtanlage *f*
e instalación *f* para la clasificación del azúcar
n suikerzeefinstallatie *f*

2391 sugar silo
f silo *m* à sucre
d Zuckersilo *m*
e silo *m* de azúcar
n suikersilo *m*

2392 sugar solution
f solution *f* sucrée
d Zuckerlösung *f*; Zuckerlsg. *f*
e solución *f* de azúcar; solución *f* azucarada
n suikeroplossing *f*; suikerhoudende oplossing *f*

2393 sugar storage
f stockage *m* du sucre
d Zuckerlagerung *f*
e almacenaje *m* del azúcar
n opslaan *n* van suiker

2394 sugar store
f magasin *m* à sucre
d Zuckerlager *n*
e almacén *m* de azúcar
n suikerpakhuis *n*

2395 sugar supplies
(eco)
f disponibilité *f* de sucre; approvisionnement *f* en sucre
d Zuckerversorgung *f*
e abastecimiento *m* de azúcar
n suikervoorziening *f*

2396 sugar technologist
f technicien *m* de sucrerie
d Zuckertechniker *m*; Zuckertechnologe *m*
e técnico *m* azucarero; tecnólogo *m* azucarero
n suikertechnicus *m*

2397 sugar technology
f technologie *f* sucrière

d Zuckertechnologie *f*
e tecnología *f* azucarera
n suikertechnologie *f*

2398 sugar trade
f commerce *m* du sucre; commerce *m* des sucres
d Zuckerhandel *m*
e comercio *m* del azúcar; comercio *m* azucarero
n suikerhandel *m*

2399 sugar turbidity
f turbidité *f* du sucre
d Trübung *f* des Zuckers
e turbidez *f* del azúcar
n troebelheid *f* van de suiker

2400 sugar vermicelli
f vermicelles *mpl* au sucre; sucre *m* à saupoudrer
d Zuckerstreusel *mn*
n suikerstrooisel *n*

2401 sugar year
(EEC)
f campagne *f* sucrière
d Zuckerwirtschaftsjahr *n*; ZWJ *n*
e campaña *f* azucarera
n suikercampagne *f*

* **sulfate** *(US)* → **2403**

* **sulfitator** *(US)* → **2408**

* **sulfitor** *(US)* → **2408**

2402 sulphamic acid
(Scale removal)
f acide *m* sulfamique
d Sulfaminsäure *f*
e ácido *m* sulfámico
n sulfaminezuur *n*

2403 sulphate; sulfate *(US)*
f sulfate *m*
d Sulfat *n*
e sulfato *m*
n sulfaat *n*

2404 sulphated ash
f cendres *fpl* comptées en sulfates
d Sulfatasche *f*
e cenizas *fpl* sulfatadas
n sulfaatas *f*

2405 sulphitate *v*; **sulphite** *v*
f sulfiter
d sulfitieren; schwefeln

e sulfitar
n sulfiteren; zwavelen

2406 sulphitated juice
f jus *m* sulfité
d sulfitierter Saft *m*; geschwefelter Saft *m*
e jugo *m* sulfitado; guarapo *m* sulfitado
n gesulfiteerd sap *n*; gezwaveld sap *n*

2407 sulphitation
(The addition of SO_2 to the juice)
f sulfitation *f*
d Sulfitation *f*; Schwefelung *f*
e sulfitación *f*
n sulfitatie *f*; zwaveling *f*

2408 sulphitation apparatus; sulfitor *(US)*; sulfitator *(US)*
f appareil *m* à sulfiter
d Sulfitationsapparat *m*
e aparato *m* de sulfitación; sulfitador *m*
n sulfitatietoestel *n*; sulfiteur *m*

2409 sulphitation column
f colonne *f* de sulfitation; colonne *f* à sulfiter
d Sulfitationsturm *m*
e columna *f* de sulfitación
n sulfitatietoren *m*

2410 sulphitation factory
f sucrerie *f* à sulfitation
d Sulfitationsfabrik *f*
e fábrica *f* con sulfitación
n sulfitatiefabriek *f*; sulfiterende fabriek *f*

* **sulphite** *v* → 2405

2411 sulphur
f soufre *m*
d Schwefel *m*
e azufre *m*
n zwavel *m*

2412 sulphur box
f caisse *f* à sulfiter
d Sulfitationskasten *m*
e caja *f* de sulfitación
n sulfitatiekist *f*

2413 sulphur deficiency
(sb; disease)
f carence *f* en soufre
d Schwefelmangel *m*
e deficiencia *f* de azufre; carencia *f* de azufre
n zwavelgebrek *n*

2414 sulphur dioxide; sulphurous anhydride
f anhydride *m* sulfureux
d Schwefeldioxid *n*
e anhídrido *m* sulfuroso; dióxido *m* de azufre
n zwaveldioxyde *n*

* **sulphur furnace** → 2417

2415 sulphuric acid
f acide *m* sulfurique
d Schwefelsäure *f*
e ácido *m* sulfúrico
n zwavelzuur *n*

2416 sulphuric anhydride
f anhydride *m* sulfurique
d Schwefeltrioxid *n*
e anhídrido *m* sulfúrico
n zwavelzuuranhydride *n*

* **sulphurous anhydride** → 2414

2417 sulphur stove; sulphur kiln; sulphur furnace
f four *m* à soufre
d Schwefelofen *m*
e horno *m* de azufre
n zwaveloven *m*

2418 sulphur tower
f tour *f* de sulfitation
d Schwefelturm *m*
e torre *f* de sulfitación
n zwaveltoren *m*

2419 summer chafer; June bug; June beetle; European June beetle; European June bug
(sb; pest)
f hanneton *m* de la Saint-Jean; hanneton *m* de juin; hanneton *m* solsticial
d Junikäfer *m*; gemeiner Junikäfer *m*; Brachkäfer *m*; Sonn(en)wendkäfer *m*
e escarabajo *m* de San Juan; escarabajo *m* solsticial; rizotrogo *m*
n junikever *m*
l Amphimallon solstitialis

* **summer cypress** → 1411

* **sun euphorbia** → 2421

2420 sunflower; common sunflower
(sb; weed)
f hélianthe *m* annuel; grand soleil *m*; soleil *m*; tournesol *m*
d gemeine Sonnenblume *f*; Sonnenblume *f*; einährige Sonnen-

blume *f*
e girasol *m*; mirasol *m*; tornasol *m*; hierba
  f del sol
n zonnebloem *f*
l Helianthus annuus

2421 sun spurge; wartweed; sun euphorbia
  (sb; weed)
  f euphorbe *f* réveille-matin; herbe *f* aux
  verrues; réveille-matin *m*; euphorbe *f*
  hélioscope; euphorbe *f* des vignes
  d Sonnen-Wolfsmilch *f*; Garten-
  Wolfsmilch *f*
  e lechetrezna *f* común; lecherulla *f*
  n kroontjeskruid *n*
  l Euphorbia helioscopia

2422 superheat *v*
  f surchauffer
  d überhitzen
  e sobrecalentar
  n oververhitten

2423 superheated steam
  f vapeur *f* surchauffée
  d überhitzter Dampf *m*
  e vapor *m* sobrecalentado
  n oververhitte stoom *m*

2424 superheater
  f surchauffeur *m*
  d Überhitzer *m*
  e sobrecalentador *m*
  n oververhitter *m*

2425 superheating
  f surchauffe *f*
  d Überhitzung *f*
  e sobrecalentamiento *m*
  n oververhitting *f*

2426 superior blackstrap molasses
  (Cane molasses containing 23.4% or less
  water and 53.5% or more total sugars)
  f "Superior Blackstrap Molasses"
  d "Superior Blackstrap Molasses"
  e melazas *fpl* superiores
  n "Superior Blackstrap Molasses"

2427 supernatant liquor; supernatant liquid
  f liqueur *f* surnageante; liquide *m*
  surnageant
  d überstehende Flüssigkeit *f*
  e licor *m* flotante; líquido *m* flotante
  n bovenstaande vloeistof *f*; bovendrijvende
  vloeistof *f*

2428 supersaturated solution
  f solution *f* sursaturée

  d übersättigte Lösung *f*
  e solución *f* sobresaturada
  n ooververzadigde oplossing *f*

2429 supersaturation
  (The quotient of the sucrose/water ratio
  in the given solution divided by the
  sucrose/water ratio of a solution that is
  saturated with sucrose under given
  conditions)
  f sursaturation *f*
  d Übersättigung *f*; Übersaturation *f*
  e sobresaturación *f*; supersaturación *f*
  n oververzadiging *f*

2430 supersaturation coefficient
  (Ratio of the weight of sucrose per cent
  water contained in a supersaturated
  solution to the weight of sucrose per cent
  water which would be present in a
  saturated solution having the same
  temperature and the same purity)
  f coefficient *m* de sursaturation
  d Übersättigungszahl *f*
  e coeficiente *m* de sobresaturación
  n oververzadigingscoëfficiënt *m*

2431 supplementary invitation to tender
  (EEC)
  f adjudication *f* complémentaire
  d Ergänzungsausschreibung *f*
  e adjudicación *f* complementaria
  n aanvullende inschrijving *f*

2432 supply tank; storage tank
  f bac *m* d'attente; bac *m* d'alimentation
  d Einziehkasten *m*; Vorratskasten *m*;
  Vorratstank *m*; Einzugskasten *m*
  e tanque *m* de depósito; tanque *m* de
  alimentación
  n trekbak *m*; wachtbak *m*

* **supporting screen** *(in centrifugal basket)*
  → **104**

2433 surface condenser
  f condenseur *m* à surface
  d Oberflächenkondensator *m*
  e condensador *m* a superficie; condensador
  m de superficie
  n oppervlaktecondensor *m*

2434 surface tension
  f tension *f* superficielle
  d Oberflächenspannung *f*
  e tensión *f* superficial
  n oppervlaktespanning *f*

2435 surplus quantity
(EEC)
f quantité f excédentaire
d Überschußmenge f
e cantidad f excedentaria
n overschothoeveelheid f

2436 suspended beet slicer
f coupe-racines m à arbre suspendu
d hängende Schneidmaschine f
e cortadora f suspendida
n hangende bietensnijmolen m

* **suspended calandria** → 1090

* **suspended calandria pan** → 1091

2437 suspended matter
f matières fpl en suspension
d Schwebestoffe mpl; suspendierte Stoffe
   mpl
e sustancias fpl en suspensión
n gesuspendeerde stoffen fpl

* **sutch grass** → 268

2438 swamp milkweed *(US);* **swamp silkweed**
   (US)
   (sb; weed)
f asclépiade f incarnate
d fleischrote Seidenpflanze f
n rode zijdeplant f
l Asclepias incarnata

2439 sweeten v
   (To introduce a sugar-containing liquid to
   a substance, replacing the liquid phase
   previously present, if any)
f sucrer
d süßen
e endulzar; azucarar; dulcificar
n zoet maken

2440 sweetener
f édulcorant m; agent m sucrant
d Süßungsmittel n; Süßstoff m
e endulzante m
n zoetmiddel n; zoetmakend middel n

2441 sweetening off
   (The process of decreasing the
   concentration of liquor passing over the
   adsorbent at the end of the
   decolo(u)rizing cycle)
f dessucrage m; désucrage m
d Absüßen n
e desendulzamiento m
n afzoeten n

2442 sweetening on
   (The process of increasing the
   concentration of liquor passing over the
   adsorbent at the beginning of the
   decolo(u)rizing cycle)
f sucrage m
d Ansüßen n
e endulzamiento m
n zoeten n

2443 sweetening power
f pouvoir m sucrant
d Süßkraft f
n zoetkracht f; zoetend vermogen n

2444 sweeten v **off the filter cake; wash** v **out**
   the filter cake
f laver le tourteau; dessucrer le tourteau;
   désucrer le tourteau
d den Filterkuchen absüßen; den Filterkuchen auswaschen
e lavar la torta
n de filterkoek afzoeten; de filterkoek
   uitwassen

* **sweet plantain** → 1253

2445 sweet water channel
f nochère f à petit jus; canal m des petites
   eaux de lavage; canal m pour l'eau de
   dessucrage
d Absüßwasserkanal m; Waschwasser-
   kanal m
e canaleta f para aguas endulzadas;
   canaleta f para aguas dulces
n afzoetwaterkanaal n

2446 sweet waters
   (The liquid effluent from the filters that
   desweeten the carbonatation-filter feed,
   consisting of the displaced juice and
   some of the wash water)
f petites eaux fpl de lavage; petits jus
   mpl; eau f de dessucrage; eau f de
   désucrage
d Absüßwasser n; Waschwasser n
e aguas fpl endulzadas; aguas fpl para el
   lavado; aguas fpl dulces
n afzoetwater n; afzoetsap n

2447 swine cress; wart cress; swine's cress;
   creeping wart cress
   (sb; weed)
f senebière f; corne f de cerf (commune)
d niederliegender Krähenfuß m; gemeiner
   Krähenfuß m; liegender Krähenfuß m
e mastuerzo m silvestre; cerezo m de
   cerdo
n varkenskers f; grote varkenskers f

l Coronopus procumbens; Coronopus ruellii; Coronopus squamatus; Senebiera coronopus

* **sword-grass moth** *(US)* → **520**

2448 synthetic resin
f résine *f* synthétique
d Kunstharz *n*
e resina *f* sintética
n kunsthars *n*

2449 syrphid fly
(sb)
f syrphe *m*
d Schwebfliege *f*
n zweefvlieg *f*
l Syrphus corollae

2450 syrup
(The general term for sugar solutions of relatively high concentration)
f sirop *m*
d Sirup *m*
e jarabe *m*; sirope *m*
n stroop *f*

2451 syrup; machine syrup
f égout *m* de la turbine
d Zentrifugenablauf *m*; Schleuderablauf *m*
e jarabe *m*; meladura *f*
n afloop *m*; afloopstroop *f*

* **syrup intake** → **1323**

2452 syrup pump
f pompe *f* à sirop
d Siruppumpe *f*
e bomba *f* de meladura
n strooppomp *f*

2453 syrup separating device
f dispositif *m* de séparation des égouts
d Ablauftrennvorrichtung *f*; Sirupablauf-vorrichtung *f*
e dispositivo *m* separador de mieles
n stroopscheidingsinrichting *f*

2454 syrup separation; separation of green and wash *(US)*; separation of wash and run-off
f séparation *f* des égouts
d Ablauftrennung *f*; Siruptrennung *f*
e separación *f* del lavado y de la melaza de purga; separación *f* de jarabe
n stroopscheiding *f*

2455 systematic washing *(of filter cake)*
f lavage *m* méthodique

d systematisches Auswaschen *n*
e lavado *m* sistemático
n systematisch uitwassen *n*

2456 system of minimum stocks (EEC)
f régime *m* du stock minimal
d Mindestbestandsregelung *f*
e régimen *m* de stock mínimo
n regeling *f* voor de minimumvoorraad

T

2457 tail-bar
(Connection between gearing and cane mill)
f tail-bar *m*
d Verbindungsstange *f*
e barra *f* de acoplamiento
n verbindingsstang *f*

2458 tailings separator; beet tail catcher; tail separator; beet tail separator
(An apparatus that recovers beet tails and pieces from the separated beet tops, weeds, and other unwanted matter; usually utilizes the differential flotation or bouncing property of the beet-root matter)
f séparateur *m* de radicelles; récupérateur *m* de radicelles; ramasse-radicelles *m*
d Schwänzeabscheider *m*; Rübenschwänze-abscheider *m*; Rübenschwanzfänger *m*
e colector *m* de nabos de remolachas; separador *m* de colas de remolachas; recuperador *m* de rabillos; cogedor *m* de colas
n bietenstaartjesvanger *m*

2459 tail rot
(sb; disease)
f pourriture *f* des radicelles
d Rübenschwanzfäule *f*
e podredumbre *f* de las raicillas
n staartjesrot

* tail separator → 2458

2460 tall rocket; tumble mustard *(US)*
(sb; weed)
f herbe *f* aux chantres; grand vélar *m*
d ungarische Rauke *f*
e jaramago *m*
n Hongaarse raket *f*
l Sisymbrium altissimum

* tank with stirrer → 41

2461 tansy mustard; flixweed; flixweed tansy mustard; hedge mustard
(sb; weed)
f sagesse *f*; sagesse *f* des chirurgiens; science *f* des chirurgiens; sisymbre *m* sagesse
d Besenrauke *f*; gemeine Besenrauke *f*; Sophienrauke *f*; gemeine Sophienrauke *f*
e sofia *f* de cirujanos
n sofiekruid *n*; fiekruid *n*; vuurkruid *n*
l Descurainia sophia; Sisymbrium sophia

* tansy ragwort → 1884

2462 taproot rot; phytophthora root rot *(US)*
(sb; disease)
f phytophtora *f*
d Phytophtora-Wurzelfäule *f*
e phytophtora *f*
n phytophtora *f*

2463 tare
(Material which must be discarded)
f tare *f*
d Schmutz *m*
e tara *f*
n tarra *f*

* tare → 2621

2464 tare control
f contrôle *m* de la tare
d Schmutzgehaltskontrolle *f*
e control *m* de la tara
n tarracontrole *f*

2465 tarehouse
f salle *f* de tare
d Prozentstube *f*; Erdprozentbestimmungs-haus *n*; Probewäsche *f*; Anlage *f* für qualitative Rübenübernahme
e sala *f* de tara
n tarreerlokaal *n*

2466 tare laboratory; receiving laboratory
(The station at which the beet samples taken at the receiving station are analyzed)
f laboratoire *m* de réception
d Rübenannahmelaboratorium *n*
e laboratorio *m* de recepción
n bietenontvangstlaboratorium *n*

2467 target blotch
(sc; fungous disease)
d Braunstreifigkeit *f*
e mancha *f* concéntrica
n bruine-strepenziekte *f*

2468 target price
(EEC)
f prix *m* indicatif
d Richtpreis *m*
e precio *m* indicativo
n richtprijs *m*

2469 tariff quota
(EEC)
f contingent *m* tarifaire
d Tarifkontingent *n*

e contingente *m* arancelario
n tariefcontingent *n*

2470 tarnished plant bug; common capsid bug
(US)
(sb; pest)
f capside *m*; punaise *f*
d Wiesenwanze *f*
l Lygus rugulipennis; Lygus pratensis

2471 tarnish plant bug *(US)*; **tarnished plant**
bug; European tarnished plant bug
(sb; pest)
f punaise *f*; punaise *f* terne
d gemeine Wiesenwanze *f*; graue Wald-
wanze *f*; trübe Feldwanze *f*
l Lygus lineolaris

2472 technical molasses
f mélasse *f* technique; mélasse *f*
industrielle
d technische Melasse *f*
e melaza *f* técnica
n technische melasse *f*

* **technical monogerm seed** → 1837

2473 technical sugar solution
f solution *f* technique de sucre
d technische Zuckerlösung *f*
e solución *f* técnica de azúcar
n technische suikeroplossing *f*

2474 technological value of sugar beet
f valeur *f* technique de la betterave
sucrière
d technischer Wert *m* der Zuckerrübe
e valor *m* técnico de la remolacha
azucarera
n technologische waarde *f* van de
suikerbiet

2475 tender
(EEC)
f adjudication *f*
d Ausschreibung *f*
e adjudicación *f*
n inschrijving *f*

2476 thermal balance; heat balance
f bilan *m* thermique
d Wärmebilanz *f*
e balance *m* térmico
n warmtebalans *f*

* **thermal stability of the juice** → 2480

2477 thermo-compression
f thermo-compression *f*

d Thermokompression *f*
e termocompresión *f*
n thermocompressie *f*

2478 thermo-compressor
f thermo-compresseur *m*
d Thermokompressor *m*
e termocompresor *m*
n thermocompressor *m*

2479 thermolabile juice
f jus *m* thermolabile
d thermolabiler Saft *m*
e jugo *m* termolábil; guarapo *m* termolábil
n thermolabiel sap *n*

2480 thermostability of the juice; thermal
stability of the juice
f thermostabilité *f* du jus
d Thermostabilität *f* des Saftes
e termoestabilidad *f* del jugo;
termoestabilidad *f* del guarapo
n thermostabiliteit *f* van het sap

2481 thermostable juice
f jus *m* thermostable
d thermostabiler Saft *m*
e jugo *m* termoestable; guarapo *m*
termoestable
n thermostabiel sap *n*

2482 thicken *v*; **concentrate** *v*
f épaissir; concentrer
d eindicken
e concentrar
n indikken

2483 thickened mud
f boues *fpl* épaissies
d Dickschlamm *m*; Schlammkonzentrat *n*;
Schlammsuppe *f*
e cachaza *f* espesa
n ingedikt slib *n*

2484 thickener; subsider; decanter
f épaississeur *m*; décanteur *m*
d Eindicker *m*; Dekanteur *m*; Absetzer *m*
e espesador *m*; decantador *m*
n indikker *m*; decanteur *m*;
decanteertoestel *n*

2485 thickener for waste water
f décanteur *m* pour eaux boueuses
d Eindicker *m* für Abwasser; Absetzer *m*
für Abwasser
e decantador *m* para agua sucia
n indikker *m* voor afvalwater

189

2486 thickening
f épaississement *m*; concentration *f*
d Eindickung *f*
e espesamiento *m*
n indikking *f*

2487 thickening ratio
f degré *m* de concentration; degré *m* d'épaississement
d Eindickungsgrad *m*
e grado *m* de espesamiento; grado *m* de concentración
n indikkingsgraad *m*

2488 thicken *v* up the strike; bring *v* in the skipping; boil *v* off the massecuite; tighten *v* the massecuite
f serrer la masse cuite
d den Sud fertigkochen; den Sud stramm abkochen
e cerrar la masa cocida; apretar la masa codida
n het kooksel afkoken

2489 thick juice
(Concentrated purified juice; the sugar-bearing product of the juice evaporator system)
f jus *m* dense; sirop *m* vierge
d Dicksaft *m*
e jugo *m* denso; guarapo *m* denso; jugo *m* concentrado; jarabe *m* espeso; arrobe *m*
n diksap *n*

2490 thick mud pump
f pompe *f* à boue épaissie
d Dickschlammpumpe *f*
e bomba *f* aspiradora de lodo espeso
n ingedikt-slibpomp *f*

* **thin *v* → 2152**

2491 thin juice
(The purified juice before evaporation)
f jus *m* épuré; jus *m* léger; jus *m* non concentré; premier jus *m*
d Dünnsaft *m*
e jugo *m* clarificado; jugo *m* claro; guarapo *m* clarificado; guarapo *m* claro
n dunsap *n*

2492 thin layer evaporator
f évaporateur *m* à couche mince; évaporateur *m* à ruissellement
d Dünnschichtverdampfer *m*
e evaporador *m* de película fina; evaporador *m* de capa delgada; evaporador *m* pelicular
n dunlaagverdamper *m*

* **thinning → 2161**

2493 thiourea
f thiourée *f*
d Thiocarbamid *n*; Thiokarbamid *n*; Thioharnstoff *m*
e tiourea *f*; tiocarbamida *f*
n thioureum *n*

2494 third countries
(EEC)
f pays *mpl* tiers
d Drittländer *npl*
e países *mpl* terceros; terceros *mpl* países
n derde landen *npl*

* **third massecuite → 526**

* **third molasses → 527**

2495 third saturation
(Refers to sulphitation of the thin juice, although this is by no means carried out to the point of saturation)
f sulfitation *f* du jus léger
d Sulfitation *f* des Klarsaftes
e sulfitación *f* del jugo claro; sulfitación *f* del guarapo claro
n sulfitatie *f* van het klaarsap

* **third sugar → 701**

2496 thistle; bristle thistle
(sb; weed)
f chardon *m*
d Distel *f*
e cardo *m*
n distel *m*
l Carduus sp.

2497 thorn apple; jimson(weed); jimpson(weed); jamestown weed; jimson-weed datura; jimson-weed datura
(sb; weed)
f datura *m*; pomme *f* épineuse; stramoine *f* commune; herbe *f* à la taupe
d Stechapfel *m*; gemeiner Stechapfel *m*; Dornapfel *m*; Fliegenkraut *n*; Knöten-melde *f*
e estramonio *m*; hierba *f* hedionda; higuera *f* laca; manzana *f* espinosa
n doornappel *m*
l Datura stramonium

2498 three-boiling system; three-product boiling scheme; three-massecuite system
f schéma *m* de cristallisation en trois jets; travail *m* en trois jets; procédé *m* des trois masses cuites; procédé *m*

d'épuisement à trois masses cuites;
schéma *m* de cuisson en trois jets
d Dreiprodukte-Kristallisationsschema *n*;
dreistufiges Kristallisationsschema *n*;
Drei-Produkte-Schema *n*; Dreiprodukten-
schema *n*
e esquema *m* de cristalización de tres
productos; esquema *m* de tres productos;
sistema *m* de tres templas; sistema *m* de
tres cocciones; sistema *m* de
cristalización de tres productos; esquema
m de tres templas
n kristallisatieschema *n* met drie
produkten; drieproduktenkookschema *n*;
koken *n* in drie trappen

2499 three-roller crusher; mill-crusher
f défibreur *m* à trois cylindres; moulin-
défibreur *m*
d Drei-Roller-Crusher *m*; Drei-Walzen-
crusher *m*; Drei-Roller-Brecher *m*; Drei-
Walzenbrecher *m*
e desmenuzadora *f* de tres mazas;
desmenuzadora *f* de tres cilindros;
molino *m* desmenuzador
n driecilindercrusher *m*;
driecilinderkneuzer *m*

2500 three-roller mill
f moulin *m* à trois cylindres
d Drei-Roller-Mühle *f*; Drei-Walzen-
Mühle *f*
e molino *m* de tres cilindros
n driecilindermolen *m*

2501 three-spotted flea beetle *(US)*
(sb; pest)
l Disonycha triangularis

2502 threshold price
(EEC)
f prix *m* de seuil
d Schwellenpreis *m*
e precio *m* de umbral
n drempelprijs *m*

2503 thrips
(sb; pest)
f thrips *m*
d Blasenfuß *m*; Thrips *m*
e trips *m*
n thrips *m*
l Thrips spp.

2504 thymol; thyme camphor
f thymol *m*
d Thymol *n*; Thymiankampfer *m*
e timol *m*
n thymol *n*

**2505 tightening the strike; tightening the
massecuite**
(The final phase of batch sugar boiling in
which the syrup feed is stopped and
evaporation continued to further exhaust
the mother liquor of sucrose and to
adjust the massecuite to the desired final
density)
f serrage *m* de la masse cuite; serrage *m*
de la cuite
d Eindicken *n* der Kochmasse;
Abkochen *n*
e cerrado *m* de la masa cocida
n indikken *n* van de masse cuite; indikken
n van de vulmassa

* **tighten** *v* **the massecuite** → 2488

2506 tilt *v*
f basculer
d kippen
e volcar
n kippen; kantelen

2507 tilting platform; tipping platform
f plate-forme *f* de bascule; plate-forme *f* à
bascule
d Kipp-Plattform *f*; Kippbühne *f*
e plataforma *f* de volteo; plataforma *f*
basculante
n kipplatform *n*

* **tipula** → 674

2508 titrate *v*
f titrer
d titrieren
e titular; titrar
n titreren

2509 titrated soap solution
f liqueur *f* titrée de savon
d Titerseifenlösung *f*; Pellet-Seifen-
lösung *f*; Seifenlösung *f* Boutron-Boudet
e solución *f* de jabón para titración
n titerzeepoplossing *f*

2510 titration
f titration *f*; titrage *m*
d Titration *f*
e titulación *f*; titración *f*
n titratie *f*

2511 titration apparatus
f appareil *m* de titrage
d Titriergerät *n*
e aparato *m* de titulación
n titreertoestel *n*

* **toadpipe** → 576

2512 toad rush
(sb; weed)
f jonc m des crapauds
d Krötenbinse f
e junco m de sapo
n greppelrus m; paddegras n; mothaar n;
motgras n
l Juncus bufonius

2513 toluene
(An antiseptic)
f toluène m
d Toluol n; Methylbenzol n
e tolueno m
n tolueen n; methylbenzeen n; toluol n

* **toothed-legged turnip beetle** → 1544

2514 toothed spurge *(US)*
(sb; weed)
l Euphorbia dentata

2515 top v
(agr)
f décolleter
d köpfen
e descoronar; descabezar
n koppen

2516 top-fed shredder
f shredder m à alimentation des cannes
par le haut
d Shredder m für Zuckerrohraufgabe von
oben
e desfibradora f con carga de la caña
desde arriba
n shredder m met riettoevoer van
bovenaan

* **top feeler** → 986

* **topper loader** → 248

2517 topping knife
(agr)
f couteau m décolleteur
d Köpfmesser n
e cuchilla f descoronadora
n kopmes n

2518 topping unit
(agr)
f tâteur-décolleteur m
d Köpftaster m
e conjunto m palpador
n koptaster m

* **top rake** → 2521

2519 top roll(er); upper roll(er) *(of cane mill, crusher)*
f cylindre m supérieur
d Oberroller m; oberer Roller m; obere
Walze f; Oberwalze f
e cilindro m superior; rodillo m mayor;
rodillo m superior; maza f superior
n topcilinder m; toprol f

2520 top roll(er) scraper
f raclette f supérieure
d oberer Schraper m
e raspador m superior
n bovenschraper m

2521 top spinner; beet top rake; top rake
(agr)
f andaineur m de verts; balayeur m de
verts; andaineuse f de verts
d Rübenblattschwader m; Schwadrechen
m für Rübenblatt; Blattrechen m;
Schwadenrechen m
e rastrillo m hilerador de coronas
n zwadenhark f voor bietenloof

2522 top tare
(Beet reception)
f tare f collets
d Abzug m für Rübenblätter und -köpfe
e tara f de hojas y coronas
n tarra f aan bietbladeren en -koppen

2523 total acidity
f acidité f totale
d Gesamtacidität f
e acidez f total
n totale aciditeit f

2524 total nitrogen; total N
f azote m total
d Gesamtstickstoff m
e nitrógeno m total
n totale stikstof f

2525 total nonsugars
f non-sucres mpl totaux
d Gesamtnichtzucker m
e no-azúcares mpl totales
n totale nietsuikerstoffen fpl; totale
nietsuiker m

* **total recovery** → 1704

2526 total solids; solids by drying; dry substance
(The material remaining after drying the
product examined)

f matières *fpl* sèches totales
d Gesamttrockengehalt *m*
e sólidos *mpl* totales; sólidos *mpl* por
desecación; materia *f* seca
n totale droge stof *f*

2527 total sugars
f sucres *mpl* totaux
d Gesamtzucker *m*; Gesamtzuckerstoffe
mpl
e azúcares *mpl* totales
n totale suikerstoffen *fpl*

2528 total-sugars content
(Content of saccharides as defined by the
method(s) of analysis used and the basis
for reporting)
f teneur *f* en sucres totaux
d Gesamtzuckergehalt *m*
e contenido *m* de azúcares totales
n gehalte *n* aan totale suikerstoffen

2529 total tare
(Beet reception)
f tare *f* totale
d Gesamtabzug *m*
e tara *f* total
n totale tarra *f*

* **T. Pur. → 2555**

* **train of mills → 1601**

**2530 tramp iron separator; magnetic
separator**
f séparateur *m* magnétique
d Abscheider *m* für eiserne Fremdkörper
e separador *m* magnético
n magneetafscheider *m*; magnetische
afscheider *m*

2531 transformed product
(EEC)
f produit *m* transformé
d Verarbeitungserzeugnis *n*
e producto *m* transformado
n verwerkt produkt *n*

* **transverse transporter → 2138**

**2532 trash catcher; grass catcher; weed
catcher; straw catcher**
(Apparatus that removes beet tops and
weeds from the beet flume; usually a
mechanism of travelling, self-cleaning
rakes or combs)
f séparateur *m* d'herbes; désherbeur *m*
d Krautfänger *m*; Krautabscheider *m*;
Rübenkrautfänger *m*

e separador *m* de hojas; eliminador *m* de
hojas; despastador *m*; separador *m* de
tallos
n bladvanger *m*

2533 trash plate; turnplate; trash turner
f bagassière *f*
d Bagassebalken *m*; Bagassemesser *n*
e cuchilla *f* central; cuchilla *f*;
tornabagazo *m*; parihuela *f*; bagacera *f*

2534 travelling crane
(A cane crane)
f grue *f* roulante
d Laufkran *m*
e grúa *f* corrediza; grúa *f* rodante
n loopkraan *f*; rolkraan *f*

2535 travelling grate
f grille *f* mobile; grille *f* tournante
d Wanderrost *m*
e emparrillado *m* móvil
n kettingrooster *m*

* **treacle hare's-ear → 1221**

**2536 treacle mustard; wormseed mustard
(US); English wormseed; treacle
erysimum**
(sb; weed)
f fausse giroflée *f*; giroflée *f* sauvage; vélai
m fausse-giroflée; caraflée *f*
d Ackerschöterich *m*; Ackerschoten-
dotter *m*; lackartiger Schotendotter *m*
e erísimo *m*; irión *m*; jaramago *m*
n gewone steenraket *f*; wilde dragon *m*
l Erysimum cheiranthoides

2537 treated cane; processed cane
f canne *f* travaillée
d verarbeitetes Rohr *n*
e caña *f* trabajada
n verwerkt riet *n*

2538 tricalcium phosphate
f phosphate *m* tricalcique
d Trikalziumphosphat *n*
e fosfato *m* tricálcico
n tricalciumfosfaat *n*

2539 tricalcium sucrate
f saccharate *m* tricalcique
d Trikalziumsaccharat *n*
e sacarato *m* tricálcico
n tricalciumsaccharaat *n*

* **trimethylglycocol → 270**

2540 triose
  f triose *m*
  d Triose *f*
  e triosa *f*
  n triose *f*

2541 triple crusher
  f triple défibreur *m*
  d dreifacher Crusher *m*
  e triple desmenuzadora *f*
  n driedubbele crusher *m*

2542 triple effect; triple effect evaporator
  f triple effet *m*
  d Drei-Körper-Verdampfapparat *m*;
    Dreifachverdampfer *m*
  e triple efecto *m*
  n verdampapparaat *n* met drie
    verdamplichamen; verdampapparaat *n* à
    triple effet

2543 triple effect evaporation
  f évaporation *f* à triple effet
  d dreistufige Verdampfung *f*; Dreifach-
    verdampfung *f*; Dreikörper-
    verdampfung *f*
  e evaporación *f* de triple efecto
  n drievoudige verdamping *f*;
    drietrapsverdamping *f*

  * triple effect evaporator → 2542

2544 triple simple imbibition
  f imbibition *f* simple triple
  d dreistufige einfache Imbibition *f*
  e imbibición *f* simple triple
  n driedubbele eenvoudige imbibitie *f*

2545 tripper
  f chariot *m* déverseur; chariot-verseur *m*
  d Abwurfwagen *m*
  e carro *m* de descarga
  n afwerpwagen *m*

2546 trisodium phosphate
  (Alkalizing agent)
  f phosphate *m* trisodique
  d Trinatriumphosphat *n*
  e fosfato *m* trisódico
  n trinatriumfosfaat *n*

2547 trough
  f nochère *f*; auge *f*
  d Trog *m*; Mulde *f*
  e cubeta *f*; artesa *f*
  n trog *m*

2548 trough chain conveyor
  f transporteur *m* en auge à chaînes

  d Trogkettenförderer *m*
  e transportador *m* de cadena con artesa
  n trogkettingtransporteur *m*

2549 troughed belt
  f courroie *f* en auge
  d muldenförmiger Gurt *m*
  e cinta *f* transportadora cóncava
  n trogvormige band *m*

2550 trough formed conveyor belt
  f convoyeur *m* à courroie à section en
    auge; courroie *f* transporteuse à section
    en auge
  d Trogbandförderer *m*; Muldenförder-
    band *n*
  e correa *f* transportadora de sección de
    artesa
  n trogvormige transportband *m*

2551 truck weighbridge
  (Beet unloading)
  f pont *m* à bascule pour camions
  d LKW-Brückenwaage *f*
  e puente-báscula *m* de camión; puente *m*
    pesador para camiones
  n weegbrug *f* voor vrachtauto's

2552 true delivery opening *(between mill
    rollers)*
  f ouverture *f* arrière réelle
  d wahre Austragsöffnung *f*
  e abertura *f* verdadera de salida; abertura
    f trasera verdadera
  n ware afvoeropening *f*

2553 true density
  f densité *f* réelle
  d wahre Dichte *f*
  e densidad *f* verdadera
  n ware densiteit *f*

2554 true diffusion
  f diffusion *f* réelle
  d wahre Diffusion *f*
  e difusión *f* verdadera
  n ware diffusie *f*

2555 true purity; T. Pur.
  (The percentage proportion of true
    sucrose to total dry substance)
  f pureté *f* réelle
  d wahre Reinheit *f*
  e pureza *f* verdadera
  n ware reinheid *f*; ware zuiverheid *f*

2556 true temperature drop
  f chute *f* réelle de température
  d wahres Temperaturgefälle *n*

e caída f real de temperatura
n waar temperatuurverschil n

2557 tube plate
f plaque f tubulaire
d Rohrplatte f; Rohrboden m; Rohrplattenboden m
e placa f tubular; fondo m tubular;
plancha f de tubos
n pijpenplaat f; pijpplaat f; tubenplaat f

2558 tubular air-heater
f réchauffair m tubulaire; réchauffeur m
d'air tubulaire
d Röhrenlufterhitzer m
e calentador m de aire tubular

2559 tubular calandria
f faisceau m tubulaire
d Röhrenheizkammer f
e calandria f tubular; calandria f de tubos
n pijpenstoomkamer f

2560 tubular preheater
f réchauffeur m tubulaire
d Röhrenvorwärmer m
e precalentador m tubular
n pijpenbundelvoorverwarmer m

* **tule mint** *(US)* → **651**

* **tumble grass** *(US)* → **2711**

* **tumble mustard** *(US)* → **2460**

2561 tumble pigweed *(US)*; **white pigweed;
tumbleweed amaranth**
(sb; weed)
f amarante f blanche
d weißer Fuchsschwanz m; weißer
Amarant m
e bledo m blanco
n witte amarant f
l Amaranthus albus

* **tumbleweed** *(US)* → **1411**

* **tumbling orache** → **1913**

2562 turbidity
f turbidité f; trouble m
d Trübung f
e turbidez f; enturbiamiento m
n troebeling f

* **turbine drier** *(US)* → **2564**

2563 turbo-generator
f turbogénérateur m

d Turbogenerator m
e turbogenerador m
n turbogenerator m

**2564 turbo tray drier; Büttner drier; turbine
drier** *(US)*
f turboSécheur m
d Turbinentrockner m
e granulador m de turbina; secador m de
bandeja; secador m de turbinas
n Büttnerdroger m; trommeldroger m

2565 turnip dart; turnip moth; dart moth
(sb; pest)
f noctuelle f des moissons;
moissonneuse f; testacée f
d Wintersaateule f; Raupe f der Wintersaateule; Feldflur-Bodeneule f; Nageleule f
e noctua f de los sembrados; noctua f
común
l Agrotis segetum

2566 turnip flea beetle
(sb; pest)
f petite altise f
d gelbstreifiger Erdfloh m
e pulguilla f
l Phyllotreta nemorum

* **turnip moth** → **2565**

* **turnplate** → **2533**

* **twitch** → **664**

* **twitch grass** → **664**

**2567 two-boiling system; two-massecuite
system; A-C system**
f procédé m des deux masses cuites;
travail m en deux jets; schéma m de
cristallisation en deux jets; schéma m de
cuisson en deux jets
d zweistufiges Kristallisationsschema n;
Zwei-Produkte-Kristallisationsschema n;
Zwei-Produkte-Schema n; Zwei-
produktenschema n
e sistema m de dos templas; sistema m A-
C; sistema m de cristalización de dos
productos; sistema m de dos cocciones;
esquema m de cristalización de dos
productos; esquema m de dos productos;
esquema m de dos templas
n kristallisatieschema n met twee
produkten; tweeproduktenkookschema n;
koken n in twee trappen

2568 two-lined grasshopper *(US)*
   (sb; pest)
   l Melanoplus bivittatus

 * **two-massecuite system** → **2567**

2569 two-roller crusher
   f défibreur *m* à deux cylindres
   d Zwei-Roller-Crusher *m*; Zwei-Walzen-
      crusher *m*; Zwei-Roller-Brecher *m*;
      Zwei-Walzenbrecher *m*
   e desmenuzadora *f* de dos mazas;
      desmenuzadora *f* de dos cilindros
   n tweecilindercrusher *m*;
      tweecilinderkneuzer *m*

U

2570 ultra-centrifuge
(Juice purification)
f ultracentrifugeuse *f*
d Ultrazentrifuge *f*
e ultracentrífuga *f*
n ultracentrifuge *f*

2571 ultrafilter
(Juice purification)
f ultrafiltre *m*
d Ultrafilter *n*
e ultrafiltro *m*
n ultrafilter *n*

2572 ultrafiltration
(Filtration in which particles of colloidal
size are separated from molecular and
ionic substances by drawing the liquid
through a membrane, in which the
capillaries are very small)
f ultrafiltration *f*
d Ultrafiltration *f*
e ultrafiltración *f*
n ultrafiltratie *f*

2573 ultramarine (blue)
f bleu *m* d'outremer
d Ultramarin *n*; Ultramarinblau *n*
e ultramarino *m*; azul *m* ultramarino; azul
   m ultramar
n ultramarijn *n*; ultramarijnblauw *n*

2574 umbrella-type separator
(Entrainment separator)
f séparateur *m* parapluie
d Schirmsaftabscheider *m*; Schirmsaft-
   fänger *m*
e separador *m* tipo sombrilla
n paraplu-sapvanger *m*

**2575 unaccounted losses; unknown losses;
undetermined losses**
(The deficit in the sucrose balance of the
process after the production and
accounted losses are subtracted from the
quantity of sucrose entered)
f pertes *fpl* indéterminées; pertes *fpl*
   inconnues
d unbestimmte Verluste *mpl*; unbekannte
   Verluste *mpl*
e pérdidas *fpl* indeterminadas
n onbepaalde verliezen *npl*; onbekende
   verliezen *npl*

2576 unburnt stone
f incuit *m*

d ungebrannter Stein *m*
e material *m* no calcinado
n ongebrand steen *m*

**2577 unchokable pump; chokeless pump;
dredging pump**
f pompe *f* inengorgeable; pompe *f* à
   draguer
d Freistrompumpe *f*
e bomba *f* inatascable

2578 under-driven beet slicer
f coupe-racines *m* à commande inférieure
d stehende Schneidmaschine *f*
e cortadora *f* con accionamiento inferior
n staande bietensnijmolen *m*

2579 underfeed roll(er) *(of cane carrier)*
f rouleau *m* d'alimentation inférieur
d untere Zuführungswalze *f*; unterer
   Zuführungsroller *m*; untere Speisewalze *f*; Unter-Zuführungsroller *m*
e rodillo *m* alimentador inferior
n onderste toevoercilinder *m*

* **undetermined losses → 2575**

**2580 undetermined water; hygroscopic water;
water of constitution**
(Cane minus fiber minus undiluted juice)
f eau *f* de constitution
d Konstitutionswasser *n*
e agua *f* indeterminada; agua *f* de
   constitución; agua *f* higroscópica
n constitutiewater *n*

2581 undiluted juice; normal juice
f jus *m* non dilué; jus *m* normal
d nichtverdünnter Saft *m*; Normalsaft *m*
e guarapo *m* sin diluir; guarapo *m* normal;
   jugo *m* sin diluir; jugo *m* normal
n onverdund sap *n*; normaal sap *n*

2582 unit of account
(EEC)
f unité *f* de compte; UC *f*
d Rechnungseinheit *f*; RE *f*
e unidad *f* de cuenta; UC *f*
n rekeneenheid *f*; RE *f*

2583 universal roll(er)
f cylindre *m* universel
d Universalroller *m*; Universalwalze *f*
e cilindro *m* universal; rodillo *m* universal;
   maza *f* universal
n universele cilinder *m*

* **unknown losses → 2575**

* **unloader** → **829**

* **unloader** *(in centrifugal)* → **830**

* **unslaked lime** → **1876**

* **upper roll(er)** *(of cane mill, crusher)*
  → **2519**

2584 upper tube plate
  f plaque *f* tubulaire supérieure
  d oberer Rohrboden *m*; oberer Rohrplatten-
    boden *m*
  e placa *f* tubular superior; plancha *f* de
    tubos superior; fondo *m* tubular superior
  n bovenpijpenplaat *f*; bovenpijpplaat *f*;
    boventubenplaat *f*; bovenste pijpen-
    plaat *f*

**2585 upright yellow-sorrel; yellow wood
  sorrel; common yellow oxalis**
    (sb; weed)
  f oxalis *f* droite; oxalis *f* raide
  d steifer Sauerklee *m*
  n stijve klaverzuring *f*
  l Oxalis stricta auct.; Oxalis europaea Jord.

2586 utility blackstrap
    (Cane molasses containing 26.5% or more
    water and 42.5% to 48.4% total sugars)
  f "Utility Blackstrap"
  d "Utility Blackstrap"
  e melazas *fpl* "utility" *(aceptables para
    uso general)*
  n "Utility Blackstrap"

V

2587 vacuum
 f vide *m*
 d Vakuum *n*
 e vacío *m*
 n vacuüm *n*

2588 vacuum breaker
 f robinet *m* à air; robinet *m* casse-vide;
   casse-vide *m*
 d Lufthahn *m*
 e rompedor *m* de vacío; grifo *m* de aire
 n vacuümverbreker *m*; luchtkraan *m*

 * **vacuum drying oven → 2594**

2589 vacuum evaporation
 f évaporation *f* sous vide
 d Vakuumverdampfung *f*
 e evaporación *f* de vacío
 n vacuümverdamping *f*

2590 vacuum filter
 f filtre *m* à vide
 d Vakuumfilter *n*
 e filtro *m* al vacío
 n vacuümfilter *n*

2591 vacuum filtration
 f filtration *f* sous vide
 d Vakuumfiltration *f*
 e filtración *f* al vacío
 n vacuümfiltratie *f*

2592 vacuum ga(u)ge; vacuum meter
 f indicateur *m* de vide; vacuomètre *m*;
   videmètre *m*
 d Vakuummeter *n*; Vakuummeßgerät *n*
 e vacuómetro *m*
 n vacuümmeter *m*

2593 vacuum house
 f salle *f* de vide
 d Vakuumsaal *m*; Vakuumstation *f*
 e casa *f* de vacío
 n vacuümzaal *f*; vacuümstation *n*

 * **vacuum meter → 2592**

2594 vacuum oven; vacuum drying oven
 f étuve *f* à vide
 d Vakuumtrockenschrank *m*
 e secadero *m* al vacío
 n vacuümdroogstoof *f*

2595 vacuum pan; pan
   (The unit of process plant in which sugar

is crystallized under vacuum from supersaturated liquor)
 f appareil *m* à cuire; chaudière *f* à cuire
 d Kochapparat *m*; Verdampfungskristallisator *m*; Eindampf-
   kristallisator *m*; Kristallisations-
   kochapparat *m*; Vakuumkochapparat *m*
 e tacho *m* al vacío; tacho *m* de vacío;
   evapocristalizador *m*
 n kookpan *f*; kookapparaat *n*;
   vacuümpan *f*; vacuümkookpan *f*

 * **vacuum pan control instrument → 709**

**2596 vacuum pan with centre well; vacuum
   pan with central downtake**
 f appareil *m* à cuire à puits central
 d Kochapparat *m* mit Zentralrohr
 e tacho *m* al vacío con tubo central
 n kookpan *f* met centrale buis; kookpan *f*
   met middenbuis

2597 vacuum pan with inner circulation
 f appareil *m* à cuire à circulation
   intérieure
 d Kochapparat *m* mit Innenzirkulation;
   Kochapparat *m* mit Innenzirkulations-
   heizkammer
 e tacho *m* al vacío con circulación interior
 n kookpan *f* met binnencirculatie

2598 vacuum pan with outer circulation
 f appareil *m* à cuire à circulation
   extérieure
 d Kochapparat *m* mit Außenzirkulation
 e tacho *m* al vacío con circulación exterior
 n kookpan *f* met buitencirculatie

2599 vacuum pump
 f pompe *f* à vide
 d Vakuumpumpe *f*
 e bomba *f* de vacío
 n vacuümpomp *f*

2600 vacuum valve
 f vanne *f* de vide
 d Vakuumventil *n*
 e válvula *f* de vacío
 n vacuümventiel *n*

2601 valuation of sugar beets
 f détermination *f* de la valeur des
   betteraves sucrières
 d Bewertung *f* von Zuckerrüben
 e cálculo *m* de valor de las remolachas
   azucareras
 n kwaliteitsbepaling *f* van suikerbieten

2602 valveless milk-of-lime distributor
f distributeur *m* de lait de chaux sans
  soupape
d ventilloser Kalkmilchverteiler *m*
e distribuidor *m* de lechada de cal sin
  válvula
n kalkmelkverdeler *m* zonder klep

2603 vanilla sugar
f sucre *m* vanillé
d Vanillezucker *m*
e azúcar *m* de vainilla
n vanillesuiker *m*

2604 vanillin sugar
f sucre *m* vanilliné
d Vanillinzucker *m*
e azúcar *m* de vainillina
n vanillinesuiker *m*

* **vapo(u)r belt** *(in vessel of multiple effect)*
  → 2609

2605 vapo(u)r bleeding
f prélèvement *m* de vapeur
d Brüdenentnahme *f*; Brüdenabnahme *f*
e toma *f* de vapor

2606 vapo(u)r bubbles
f bulles *fpl* de vapeur
d Dampfblasen *fpl*
e burbujas *fpl* de vapor
n stoombellen *fpl*

* **vapo(u)r chamber** → 2609

2607 vapo(u)r pipe
f tuyau *m* de vapeur
d Dampfrohr *n*
e tubo *m* de vapor
n stoompijp *f*; stoomleiding *f*

2608 vapo(u)r separator
f séparateur *m* de vapeur
d Dampfabscheider *m*
e separador *m* de vapor
n stoomafscheider *m*

**2609 vapo(u)r space; vapo(u)r chamber;
       vapo(u)r belt** *(in vessel of multiple effect)*
f espace-vapeur *m*; chambre *f* à vapeur;
  calandre *f*
d Brüdenraum *m*
e espacio *m* vapor; cámara *f* de vapor;
  banda *f* de vapor
n stoomkamer *f*; stoomruimte *f*

2610 variegated cutworm *(US)*
  (sb; pest)

f noctuelle *f*; ver *m* gris panaché
d Larve *f* der großen Grassteppen-Bodeneule; Larve *f* der rötlichbraunen Erdeule
l Peridroma saucia *(larva)*

* **V-corrugated knife** → 1982

2611 vegetable-oyster salsify *(US)*; **salsify;
       oyster plant; vegetable oyster** *(US)*
  (sb; weed)
f salsifis *m* blanc; salsifis *m* à feuilles de
  poireau
d Haferwurz *f*; Haferwurzel *f*
e salsifí *m* (blanco)
n paarse morgenster *f*; blauwe
  morgenster *f*
l Tragopogon porrifolius

* **velocity of crystallization** → 692

* **velvet grass** → 2737

2612 velvetleaf *(US)*; **Indian mallow** *(US)*;
       **China jute; butter print; chingma
       abutilon; piemarker**
  (sb; weed)
f abutilon *m* ordinaire; jute *m* de Chine;
  fausse guimauve *f*
d Chinajute *f*; chinesischer Hanf *m*;
  chinesische Jute *f*; gelbe Schönmalve *f*
e abutilón *m*; yute *m* de China
l Abutilon theophrasti; Abutilon avicennae

2613 veneer blotch
  (sc; fungous disease)
d Veneer blotch-Krankheit *f*
e mancha *f* alada
n veneer blotch-ziekte *f*

2614 Venice mallow *(US)*; **flower-of-an-hour**
  (sb; weed)
d Stundeneibisch *m*; Stundenblume *f*
n drie-urenbloem *f*
l Hibiscus trionum

2615 ventilation; airing
f ventilation *f*; aération *f*
d Lüftung *f*; Belüftung *f*
e ventilación *f*
n ventilatie *f*

* **Venus' comb** → 2131

* **veronica** → 584

* **vertical diffuser** → 958

**2616 vertical evaporator; vertical tube
       evaporator**

f évaporateur *m* vertical
d Vertikalrohrverdampfer *m*; stehender
  Verdampfapparat *m*; stehender
  Verdampfer *m*
e evaporador *m* vertical
n verticaal verdamplichaam *n*

* **vertical lime kiln** → 2122

2617 vertical tube boiler
f chaudière f tubulaire verticale
d Steilrohrkessel *m*
e caldera f tubular vertical
n steilpijpketel *m*

* **vertical tube evaporator** → 2616

**2618 verticillium wilt; verticillium vascular
disease**
  (sb; disease)
f verticilliose f
d Verticillium-Welke f
e verticiliosis f
n verwelkingsziekte f; geelzucht f

2619 very high pol sugar; VHP sugar
d VHP-Zucker *m*; Zucker *m* hohen
  optischen Drehwertes
e azúcar *m* de tan alta polarización
n suiker *m* van zeer hoge polarisatie

2620 vessel of multiple effect
f caisse f de multiple effet
d Körper *m* einer mehrstufigen Verdampfanlage
e cuerpo *m* de múltiple efecto
n verdamplichaam *n* van een meervoudige
  verdampingsinstallatie

2621 vetch; tare
  (sb; weed)
f vesce f
d Wicke f
e veza f; arveza f
n wikke f
l Vicia sp.

* **VHP sugar** → 2619

**2622 vibrating sieve; vibrating sift *(US)*;
vibrating screen**
f tamis *m* vibrant
d Rüttelsieb *n*; Schüttelsieb *n*
e tamiz *m* vibratorio; colador *m* vibratorio;
  criba f vibratoria
n trilzeef f

2623 vibratory conveyor
f transporteur *m* à mouvements

  oscillatoires
d Schwingförderer *m*
e transportador *m* oscilante
n triltransporteur *m*

**2624 vigorous liming; energetic liming; heavy
liming**
f chaulage *m* massif; chaulage *m*
  énergique; chaulage *m* vigoureux;
  chaulage *m* poussé
d energische Kalkung f; energische
  Scheidung f
e encalado *m* en exceso
n krachtige kalking f

2625 violet root rot
  (sb; disease)
f rhizoctone *m* violet
d Rotfäule f
e mal *m* vinoso
n violetwortelrot *n*

2626 Virginia pepperweed
  (sb; weed)
f passerage *m* de Virginie
d virginische Kresse f
n virginische kruidkers f; Amerikaanse
  kruidkers f
l Lepidium virginicum

2627 virus disease; virosis
f maladie f à virus; maladie f virale;
  virose f
d Viruskrankheit f
e enfermedad f de virus; enfermedad f
  virótica; virosis f; enfermedad f viral
n virusziekte f; door virussen veroorzaakte
  ziekte f

2628 virus yellows
  (sb; disease)
f jaunisse f
d Vergilbungskrankheit f; viröse
  Vergilbung f
e amarillez f virosa
n vergelingsziekte f

2629 viscosity
f viscosité f
d Viskosität f; Zähflüssigkeit f; Zähigkeit f
e viscosidad f
n viscositeit f

2630 vole
  (sb; pest)
f campagnol *m*; mulot *m*; gros
  campagnol *m*; ratte f; grand campagnol
  m (d'Europe)
d Wühlmaus f; große (europäische) Wühl-

    maus *f*
e rata *f* de agua grande
n woelrat *f*
  l Arvicola terrestris

W

2631 waggon tipping machine; waggon unloader
f culbuteur m de wagons
d Waggonkippeinrichtung f; Waggon-
kipper m
e volcador m de vagones
n wagontipinrichting f; wagontipper m

* **wall of sugar** *(in the centrifugal)* → 1433

* **wart cress** → 2447

* **wartweed** → 2421

* **wash** *(noun)* → 2633

2632 wash v
f claircer
d waschen; decken; abdecken
e lavar
n dekken; afdekken

* **wash** v *(US)* → 32

* **washed raw sugar** → 24

* **washed sugar liquor** → 1576

* **washer with water nozzles** → 1365

2633 washing; wash *(noun)*
(Centrifuging)
f clairçage m
d Waschen n; Decken n; Abdecken n
e lavado m
n dekken n; afdekken n

2634 washing by sprays
f lavage m par ruissellement
d Waschen n mit Sprühdüsen; Decken n
mit Sprühdüsen
e lavado m con aspersores
n dekken n met verstuivers

2635 washing filtrate
f filtrat m de lavage
d Waschfiltrat m
e filtrado m diluido del lavado
n wasfiltraat n

* **washings** → 2638

2636 washing spray
f pulvérisateur m de clairçage
d Waschbrause f; Deckdüse f

e aspersor m de lavado
n deksproeier m; dekverstuiver m

* **washing water** → 2639

* **wash molasses** → 2638

2637 wash v **out**
f laver; rincer
d spülen; auswaschen
e lavar; enjugar
n uitwassen; spoelen

* **wash** v **out the filter cake** → 2444

**2638 wash syrup; high syrup; washings;
centrifuge wash; high wash syrup; rich
syrup; light molasses; rich molasses;
wash molasses**
(The filtrate leaving a centrifugal after
the initiation of the wash, consisting of
displaced green syrup plus the solution
of sugar crystals dissolved by the wash)
f égout m riche
d Deckablauf m; Decksirup m;
Waschablauf m
e miel f ligera; jarabe m rico; miel f rica
n blanke stroop f

2639 wash water; washing water
(Beet washing)
f eau f de lavage
d Waschwasser n
e agua f de lavado
n waswater n

2640 wash water
(The water used to wash the mother
syrup from the crystal in the machine
(centrifugal))
f eau f de clairçage
d Deckwasser n; Waschwasser n
e agua f de lavado
n dekwater n

2641 wash water *(for washing the filter cake)*
f petites eaux fpl de lavage
d Waschwasser n
e agua f de lavado; agua f lavadora; agua f
de lavadura
n waswater n

* **waste flue gas drying** → 1853

2642 waste heat
f chaleur f perdue
d Abgaswärme f; Abwärme f;
überschüssige Wärme f
e calor m perdido; calor m de escape;

calor *m* desperdiciado
n verloren warmte *f*

2643 waste pipe
f tuyau *m* de décharge
d Abflußrohr *n*
e caño *m* de desagüe; tubo *m* de desagüe
n afvoerpijp *f*

* **waste water** → **2119**

2644 waste water clarifying plant
f installation *f* d'épuration des eaux
  résiduaires
d Abwasserkläranlage *f*
e instalación *f* clarificadora de aguas
  residuales
n afvalwaterzuiveringsinstallatie *f*

2645 water conditioning
f conditionnement *m* de l'eau
d Wasseraufbereitung *f*
e tratamiento *m* del agua; preparación *f*
  del agua
n waterbereiding *f*

2646 water-cooled counter-current crystallizer
f malaxeur *m* à refroidissement par eau et
  à contre-courant
d wassergekühlte Gegenstrommaische *f*
e cristalizador *m* con agua de enfriamiento
  a contracorriente
n tegenstroommalaxeur *m* met
  waterkoeling

2647 water-cooled crystallizer
f malaxeur *m* à circulation d'eau
d wassergekühlte Maische *f*
e cristalizador *m* con enfriamiento de
  agua; cristalizador *m* enfriado por agua
n malaxeur *m* met waterkoeling

2648 water cooling
f refroidissement *m* par eau
d Wasserkühlung *f*
e refrigeración *f* de agua
n waterkoeling *f*

2649 water-driven centrifugal
f centrifugeuse *f* hydraulique
d Zentrifuge *f* mit Wasserantrieb
e centrífuga *f* con accionamiento
  hidráulico
n centrifuge *f* met wateraandrijving

2650 water grass *(US)*; canary grass
  (sb; weed)
f phalaris *m*; baldingera *f*
d Glanzgras *n*; Kanariengras *n*; Kanarien-

hirse *f*
n kanariegras *n*
l Phalaris sp.

* **water jet** → **1363**

* **water-jet discharge of beets** → **1102**

2651 water meter
f compteur *m* d'eau
d Wasserzähler *m*
e contador *m* de agua
n watermeter *m*

* **water of constitution** → **2580**

2652 water pressure
f pression *f* d'eau
d Wasserdruck *m*
e presión *f* de agua
n waterdruk *m*

2653 water pump
f pompe *f* à eau
d Wasserpumpe *f*
e bomba *f* de agua
n waterpomp *f*

2654 water ring centrifugal pump; water ring pump
f pompe *f* centrifuge à anneau liquide
d Wasserringpumpe *f*
e bomba *f* centrífuga con anillo líquido
n waterringpomp *f*; vloeistofringpomp *f*;
  vloeistofring-centrifugaalpomp *f*

2655 water separator
f séparateur *m* d'eau
d Wasserabscheider *m*
e separador *m* de agua
n waterafscheider *m*

2656 water-tube boiler
f chaudière *f* à tube d'eau; chaudière *f*
  aquatubulaire
d Wasserrohrkessel *m*
e caldera *f* de tubos de agua; caldera *f*
  tubular de agua
n waterpijpketel *m*

2657 water wall
f écran *m* d'eau; chemise *f* d'eau; rideau
  m d'eau; double enveloppe *f* d'eau
d Kühlschirm *m*
e pared *f* de agua
n koelscherm *n*

2658 water washing
  (Centrifuging)

f clairçage m à l'eau
d Waschen n mit Wasser; Decken n mit Wasser; Wasserdecke f; Abdecken n mit Wasser
e lavado m con agua
n dekken n met water; afdekken n met water

2659 weak juice
f jus m pauvre
d zuckerarmer Saft m
e jugo m pobre; guarapo m pobre
n suikerarm sap n

* **weaselsnout** → 1456

2660 webworm *(US)*
(sb; pest)
l Platynota stultana

* **weed catcher** → 2532

2661 weevils; snout beetles *(US)***; billbugs** *(US)*
(sc; pest)
f charançons mpl; scarabées mpl à trompe
d Rüßler mpl; Rüsselkäfer mpl; echte Rüsselkäfer mpl; Schnauzenkäfer mpl
e curculiónidos mpl; gorgojos mpl
n snuitkevers mpl; snuittorren fmpl
l Curculionidae

2662 weighbridge
f bascule f; pont m à bascule
d Waage f; Brückenwaage f
e báscula f; puente-báscula m; puente m pesador
n weegbrug f

* **welted thistle** → 1778

2663 Western potato flea beetle *(US)*
(sb; pest)
l Epitrix subcrinita

2664 Western ragweed *(US)***; perennial ragweed** *(US)*
(sb; weed)
l Ambrosia psilostachya

2665 Western twelve-spotted cucumber beetle *(US)*
(sb; pest)
l Diabrotica soror

2666 Western yellows
(sb; disease)
f "Western yellows"

d "Western yellows"
e "Western yellows"
n "Western yellows"

2667 wet air pump
f pompe f à air humide
d Naßluftpumpe f
e bomba f de aire húmedo; bomba f de aire y de agua
n natte-luchtpomp f

2668 wet cleaning of cane
f nettoyage m humide de la canne
d Naßreinigung f des Zuckerrohrs
e lavado m húmedo de la caña; limpieza f húmeda de la caña
n natte reiniging f van het suikerriet

2669 wet crushing
f pression f humide
d Naßpressung f
e presión f húmeda
n natte persing f

2670 wet-crushing mill
f moulin m de pression humide
d mit Naßpressung arbeitende Mühle f
e molino m de presión húmeda
n met natte persing werkende molen m

2671 wet extraction
f extraction f humide
d Naßextraktion f
e extracción f húmeda
n natte extractie f

2672 wet pulp
(Pulp leaving the diffuser)
f pulpes fpl humides; pulpes fpl vertes
d Naßschnitzel npl; Extraktionsschnitzel npl
e pulpa f húmeda
n natte pulp f

2673 wet rot
(sb; disease)
f pourriture f molle
d Naßfäule f
e podredumbre f húmeda; podredumbre f blanda
n natrot n

2674 wet unloading *(of beets)*
f déchargement m hydraulique
d Naßentladung f
e descarga f húmeda; descarga f por vía húmeda
n leegspuiten n; natte lossing f

2675 **wheat wireworm** *(US)*
      (sb; pest)
      l Agriotes mancus

2676 **white campion; white cockle; evening
      campion**
      (sb; weed)
      f compagnon *m* blanc; lychnide *m* du soir
      d weiße Lichtnelke *f*; weiße Nachtnelke *f*;
      Marienräßchen *n*
      e colleja *f* blanca
      n avondkoekoeksbloem *f*
      l Melandrium album (Mill.) Garcke; Silene
      alba (Mill.) Krause; Lychnis alba

   * **white centrifugal** → **2685**

   * **white charlock** → **2697**

   * **white couch** → **664**

2677 **white grub**
      (sb; pest)
      f ver *m* blanc
      d Engerling *m*
      e gusano *m* blanco; gusano *m* turco; larva
      f del abejorro; gusano *m* arador
      n engerling *m*
      l Melolontha melolontha *(larva)*

2678 **white grub** *(US)*
      (sb; pest)
      l Phyllophaga sp.

   * **white-line dart moth** → **724**

2679 **white melilot; white sweet clover; honey
      clover; Bokhara clover; white
      sweetclover** *(US)*
      (sb; weed)
      f mélilot *m* blanc
      d weißer Steinklee *m*; weißer Honig-
      klee *m*; Bucharaklee *m*; Hanfklee *m*;
      Riesenklee *m*; Pferdeklee *m*
      e meliloto *m* blanco; trébol *m* de Santa
      Maria; trébol *m* dulce; trébol *m* de olor
      blanco
      n witte honingklaver *f*
      l Melilotus albus; Melilotus alba Med.

   * **white molasses** → **2683**

2680 **white mustard**
      (sb; weed)
      f moutarde *f* blanche; sené *m* blanc
      d weißer Senf *m*; englischer Senf *m*
      e mostaza *f* blanca
      n witte mosterd *m*; gele mosterd *m*
      l Sinapis alba

   * **white pigweed** → **2561**

2681 **white rash**
      (sc; fungous disease)
      d Weißfleckigkeit *f*
      e moteado *m* blanco de la hoja
      n witte-vlekkenziekte *f*

2682 **white root rot**
      (sb; disease)
      f pourriture *f* blanche de la racine
      d Weißfäule *f*
      e podredumbre *f* blanca de la raíz
      n witrot *n*

2683 **white run-off; white molasses**
      f égout *m* riche A
      d Weißablauf *m* A
      e miel *f* blanca; derrame *m* blanco
      n witte stroop *f* A

2684 **white sugar**
      f sucre *m* blanc
      d Weißzucker *m*
      e azúcar *m* blanco
      n witte suiker *m*; witsuiker *m*

2685 **white sugar centrifugal; white centrifugal**
      f centrifugeuse *f* à sucre blanc; essoreuse
      f à sucre blanc
      d Weißzuckerzentrifuge *f*
      e centrífuga *f* para azúcar blanco
      n witsuikercentrifuge *f*

2686 **white sugar factory**
      f sucrerie *f* travaillant en blanc; sucrerie *f*
      de blanc; sucrerie *f* en blanc; fabrique *f*
      de sucre blanc
      d Weißzuckerfabrik *f*
      e fábrica *f* de azúcar blanco
      n witsuikerfabriek *f*

2687 **white sugar quotation**
      (eco)
      f cours *m* du sucre blanc; cotation *f* du
      sucre blanc
      d Weißzuckernotierung *f*
      e cotización *f* del azúcar blanco
      n witsuikernotering *f*

2688 **white sugar value; white value**
      (eco)
      f valeur *f* en sucre blanc
      d Weißzuckerwert *m*
      e valor *m* blanco
      n witsuikerwaarde *f*

2689 **white sugar yield**
      f rendement *m* en sucre blanc; rendement

　　m en blanc
　d Weißzuckerausbeute *f*
　e rendimiento *m* de azúcar blanco
　n witsuikeropbrengst *f*

2690 white sugar yield per ha *(in tons)*
　f rendement *m* de sucre blanc par ha
　d Weißzuckerertrag *m* je ha
　e rendimiento *m* de azúcar blanco por ha
　n witsuikeropbrengst *f* per ha

* **white sweet clover** → **2679**

* **white sweetclover *(US)*** → **2679**

* **white value** → **2688**

* **whitlow peppergrass** → **1252**

* **whole seed *(US)*** → **1664**

2691 whole stalk harvester; whole stick harvester; complete cane harvester
　(agr)
　f récolteuse *f* de cannes entières
　d Erntemaschine *f* für ganze Zuckerrohr-stengel
　e máquina *f* para la recogida de cañas enteras
　n oogstmachine *f* voor hele suikerriet-stengels

2692 wild beet; sea beet
　(sb; weed)
　f betterave *f* sauvage; bette *f* maritime
　d wilde Runkelrübe *f*; Wildbete *f*; See-mangold *m*
　e acelga *f* silvestre
　n strandbiet *f*
　l Beta vulgaris ssp. maritima

* **wild buckwheat** → **276**

* **wild bugloss** → **349**

* **wild camomile** → **2694**

2693 wild carrot; lace flower; Queen Anne's lace
　(sb; weed)
　f carotte *f* sauvage
　d Gelbrübe *f*; wilde Möhre *f*; gelbe Rübe *f*; gemeine Mohrrübe *f*
　e dauco *m*; zanahoria *f* silvestre; zanahoria *f* cimarrona
　n peen *f*
　l Daucus carota

2694 wild chamomile; wild camomile; scented mayweed; German chamomile; German camomile
　(sb; weed)
　f matricaire *f* camomille; camomille *f* ordinaire; petite camomille *f*; camomille *f* vraie
　d echte Kamille *f*
　e manzanilla *f* común; manzanilla *f* de Alemania
　n echte kamille *f*; gewone kamille *f*
　l Matricaria chamomilla; Matricaria recutita

2695 wild licorice *(US)*; wild liquorice *(US)*; American licorice *(US)*
　(sb; weed)
　l Glycyrrhiza lepidota

* **wild mallow** → **578**

* **wild morning glory** → **997, 1240**

2696 wild mustard; charlock; chadlock
　(sb; weed)
　f sanve *f*; sené *m*; moutarde *f* des champs; moutarde *f* sauvage; sénevé *m*
　d Acker-Senf *m*; falscher Hederich *m*; wilder Senf *m*; Ackerkohl *m*
　e mostacilla *f*; jaramago *m* amarillo; yerbana *f* amarilla; mostaza *f* de los campos; mostaza *f* silvestre
　n herik *m*; hederik *m*; krodde *f*; wilde mosterd *m*
　l Sinapis arvensis; Brassica arvensis; Sinapis sinapistrum; Brassica kaber var. pinnafitida

* **wild oat** → **2239**

* **wild pansy** → **1002**

2697 wild radish; white charlock; runch; jointed charlock; wild rape
　(sb; weed)
　f ravenelle *f*; raveluche *f*; raifort *m* sauvage; ravenelle *f* sauvage
　d Hederich *m*; echter Hederich *m*; Acker-rettich *m*
　e rábano *m* silvestre; jaramago *m* blanco; yerbana *f* blanca; erviana *f*; rabaniza *f* común; rabanillo *m* blanco; rabanilla *f*; rábano *m* cimarrón; rabaniza *f*
　n knopherik *m*
　l Raphanus raphanistrum

2698 wild turnip; bird rape
　(sb; weed)
　f navette *f* sauvage; chou *m* champêtre

d Rübsen *m*; Sommerrübsen *m*; Rüben-
kohl *m*
e naba *f*
n raapzaad *n*
l Brassica campestris

2699 wilt
(sc; fungous disease)
f flétrissement *m* des touffes
d Zuckerrohrwelke *f*
e marchitez *f*
n verwelkingsziekte *f*

* **wind bent grass** → **2144**

* **wind grass** → **2144**

* **window** → **2141**

2700 windrowing beet harvester
(agr)
f décolleteuse-arracheuse-aligneuse *f*
d Längsschwadköpfroder *m*
e descoronadora-hileradora-arrancadora *f*
de remolacha

2701 windrowing beet topper and lifter
(agr)
f décolleteuse-arracheuse-groupeuse *f*
d Querschadköpfroder *m*; Sammelköpf-
roder *m*
e arrancadora-amontonadora *f* de
remolachas

2702 windrowing harvester
(agr)
f arracheuse-aligneuse *f*
d Rodegerät *n* mit Reihenablage
e arrancadora-hileradora *f*
n rooier *m* met zwadlegger

2703 winter cutworm *(US)*; **cutworm**
(sb; pest)
f ver *m* gris; court vert *m*; noctuelle *f* des
moissons *(larve)*
d Erdraupe *f*; Ackermade *f*; Larve *f* der
Wintersaateule
e rosquilla *f* cortacuellos de la remolacha;
lobillo *m*; cuc *m* dormidor;
malduerme *m*; cachazudo *m*
l Agrotis segetum; Scotia segetum

2704 winter wild oat
(sb; weed)
f folle avoine *f* d'hiver
d Wildhafer *m*; Winter-Flughafer *m*
e avena *f* loca; ballueca *f*; cugula *f*
l Avena ludoviciana

2705 wireworm; skipjack
(sb; pest)
f taupin *m (larve)*; ver *m* fil-de-fer
d Drahtwurm *m*
e gusano *m* de alambre; doradilla *f*
n ritnaald *f*
l Agriotes sp. *(larva)*

2706 wireworm *(US)*
(sb; pest)
l Melanotus cribulosus

2707 wireworm *(US)*
(sb; pest)
l Asaphes mennonius

2708 wireworm
(sb; pest)
l Hypolithus bicolor

2709 wireworm
(sb; pest)
f ver *m* jaune
d Drahtwurm *m*
e gusano *m* alambre
n ritnaald *f*
l Agriotes sp.

2710 wireworms; pithworms
(sc; pest)
f vers *mpl* fil-de-fer
d Drahtwürmer *mpl*; echte Drahtwürmer
mpl
e gusanos *mpl* de alambre
l Elateridae

2711 witchgrass *(US)*; **tumble grass** *(US)*
(sb; weed)
f panic *m* à tiges grêles; panic *m*
capillaire
d Haarhirse *f*
l Panicum capillare

* **witchgrass** → **664**

2712 withdrawal of incondensables
f prise *f* d'incondensables
d Entnahme *f* nichtkondensierbarer Gase;
Abzug *m* nichtkondensierbarer Gase
e toma *f* de incondensables
n ontluchting *f*

2713 withdrawal pipe *(for incondensables)*
f tuyau *m* d'appel
d Abführungsrohr *n*
e tubo *m* de evacuación
n afvoerpijp *f*

**2714 wood sorrel; shamrock; wood-sorrel
oxalis**
(sb; weed)
f oxalis *f* petite oseille; oxalide *f* petite
oseille; surelle *f*; petite oxalide *f*; surelle-
petite-oseille *f*; alléluia *f*
d gemeiner Sauerklee *m*; Waldklee *m*;
Wald-Sauerklee *m*
e trébol *m* ácido; aleluya *f*
n witte klaverzuring *f*
l Oxalis acetosella

2715 woody beet
f betterave *f* ligneuse
d holzige Rübe *f*
e remolacha *f* leñosa
n houtachtige biet *f*

2716 woody tissue
(anat)
f tissu *m* ligneux
d holziges Gewebe *n*
e tejido *m* leñoso
n houtachtig weefsel *n*

2717 working day
f jour *m* de roulaison; jour *m* de travail
d Arbeitstag *m*
e día *m* de trabajo
n werkdag *m*

2718 working pressure *(of boiler)*
f timbre *m*
d Betriebsdruck *m*
e presión *f* de régimen; presión *f* de
servicio; presión *f* de trabajo
n bedrijfsdruk *m*

2719 world sugar consumption
f consommation *f* mondiale de sucre
d Weltzuckerverbrauch *m*
e consumo *m* mundial del azúcar
n wereldsuikerconsumptie *f*;
wereldsuikerverbruik *n*

2720 world sugar contract
(ISA; particularly world contract Nr. 11)
f contrat *m* mondial
d Weltkontrakt *m*
e contrato *m* mundial
n wereldcontract *n*

**2721 World Sugar Research Organization;
WSRO**
f Organisation *f* Mondiale de la Recherche
sur le Sucre
d Weltorganisation *f* für Zuckerforschung
n Wereldorganisatie *f* voor
Suikeronderzoek

2722 worm conveyor
f transporteur *m* à vis d'Archimède
d Schneckenförderer *m*
e transportador *m* de tornillo sin fin
n schroefspiltransporteur *m*

* **wormseed mustard** *(US)* → 2536

* **WSRO** → 2721

X

2723 Xanthomonas gall; bacterial canker
    (sb; disease)
  f gommose *f* bacillaire; galle *f* à
    Xanthomonas
  d Rübentuberkulose *f*; Pockenkrankheit *f*
  e agalla *f* de Xanthomonas
  n Xanthomonas-gal *f*

2724 xylenol
  f xylénol *m*
  d Xylenol *n*
  e xilenol *m*
  n xylenol *n*

Y

2725 yeast
f levure *f*
d Hefe *f*
e levadura *f*
n gist *m*

2726 yellow and blue forget-me-not
(sb; weed)
f myosotis *m* versicolore
d buntes Vergißmeinnicht *n*
l Myosotis versicolor

2727 yellow-bear caterpillar *(US)*
(sb; pest)
l Spilosoma virginica

* **yellow bristlegrass** → 2730

* **yellow clover** → 281

* **yellow dock** → 334, 711

2728 yellow-flowered sweet clover *(US)*;
yellow sweet clover *(US)*; **field melilot;**
common melilot; ribbed melilot
(sb; weed)
f mélilot *m* officinal; mélilot *m* des
champs; couronne *f* royale; mélilot *m*
jaune
d echter Steinklee *m*; echter Ackerhonig-
klee *m*; Steinklee *m*; gelber Steinklee *m*; gebräuchlicher Steinklee *m*;
echter Honigklee *m*
e meliloto *m* oficinal; meliloto *m* de los
campos; trébol *m* oloroso; coronilla *f*
real; meliloto *m* común
n akkerhoningklaver *f*
l Melilotus officinalis (L.) Pallas

2729 yellow flower pepperweed; clasping
pepperweed
(sb; weed)
d durchwachsenblättrige Kresse *f*; durch-
wachsene Kresse *f*
n doorgroeide kruidkers *f*
l Lepidium perfoliatum

2730 yellow foxtail *(US)*; **pigeongrass** *(US)*;
yellow pigeongrass *(US)*; **cat-tail millet;**
yellow bristlegrass
(sb; weed)
f sétaire *f* glauque
d niedrige Borstenhirse *f*; graugrüne
Borstenhirse *f*; gelbe Borstenhirse *f*;
blaugrüne Kolbenhirse *f*; gelbhaariger
Fennich *m*; fuchsgelbes Fennichgras *n*

e almorejo *m*
n zeegroene naaldaar *f*
l Setaria glauca

* **yellow goatsbeard** → 1559

* **yellow goat's-beard** *(US)* → 1559

* **yellow net, beet** ~ → 261

* **yellow nut grass** → 1682

* **yellow nut sedge** *(US)* → 1682

* **yellow ox-eye daisy** → 650

2731 41-yellows
(sb; disease)
f jaunisse *f* 41
d "41-yellows"
e "41-yellows"
n "41-yellows"

2732 yellow spot disease
(sc; fungous disease)
f maladie *f* des taches jaunes
d Gelbfleckenkrankheit *f*
e peca *f* amarilla
n gele-vlekkenziekte *f*

2733 yellow-striped armyworm *(US)*
(sb; pest)
l Prodenia ornithogalli

* **yellow-stripe disease** → 2346

* **yellow sweet clover** *(US)* → 2728

2734 yellow toadflax; common toadflax;
butter-and-eggs
(sb; weed)
f linaire *f* commune; linaire *f* vulgaire
d Frauenflachs *m*; gemeines Leinkraut *n*;
echtes Leinkraut *n*; gelbes Flachs-
kraut *n*; gewöhnliches Leinkraut *n*; Hanfkraut *n*
e linaria *f*; pajarita *f*
n vlasbekje *n*; vlasleeuwebek *m*;
vlasbek *m*
l Linaria vulgaris

* **yellow trefoil** → 281

2735 yellow wilt; beet yellow wilt disease
(sb; disease)
f "Beet Yellow Wilt Disease"
d viröse Gelbwelke *f*
e marchitez *f*; "Beet Yellow Wilt Disease"
n "Yellow Wilt"

* **yellow wood sorrel** → 2585

2736 yield
 f rendement *m*
 d Ausbeute *f*; Ausbeutegrad *m*
 e rendimiento *m*
 n rendement *n*; opbrengst *f*

2737 Yorkshire fog; velvet grass; meadow soft
 grass; common velvet grass
 (sb; weed)
 f houlque *f* laineuse; foin *m* de mouton;
 blanchard *m* velouté; avoine *f* laineuse
 d wolliges Honiggras *n*
 e holco *m* lanoso; holco *m* velloso; pasto
 m lanudo; heno *m* blanco
 n echte witbol *m*; witbol *m*
 l Holcus lanatus

Z

2738 zebra caterpillar *(US)*
   (sb; pest)
   l Ceramica picta

2739 zig-zag
   (A two-roller crusher)
   f Zig-Zag *m*
   d Zick-Zack-Crusher *m*
   e desmenuzadora *f* zig-zag; desmenuzadora
   f en zig-zag
   n zigzag-crusher *m*; zigzag-kneuzer *m*

2740 zinc deficiency
   (sb; disease)
   f carence *f* en zinc
   d Zinkmangel *m*
   e deficiencia *f* de zinc; carencia *f* de zinc
   n zinkgebrek *n*

2741 zipper closed conveyor-belt
   f courroie *f* transporteuse à fermeture
   éclair
   d Förderband *n* mit Reißverschluß
   e correa *f* de transporte con cierre-
   relámpago
   n transportband *m* met ritssluiting

Français

navette sauvage 2698
N bétaïne 271
nécrose terminale sur les
  feuilles du coeur 1665
nématode de la betterave 188
nématode de la tige 352
nématode de la tige et du bulbe
  352
nématode des racines 1996
nématodes des racines 730
nettoyage humide de la canne
  2668
nettoyeur de betteraves 183
nielle des blés 646
nielle des champs 646
niveau du jus 1385
niveleur 1458
nochère 2547
nochère à petit jus 2445
noctuelle 521, 2610
noctuelle antique 520
noctuelle de la betterave 176,
  1453
noctuelle des artichauts 2000
noctuelle des moissons 724,
  2565, 2703
noctuelle du chou 363
noctuelle gamma 1156
noctuelle grise 1156
noctuelle hongroise 1279
noctuelle ypsilon 278
noir animal 314
noir-âtre 1157
noir d'os 314
nombre de divisions d'un
  couteau de coupe-racines 842
nombre de Reynolds 1974
non-saccharose 1678
non-sucre 1680
non-sucre minéral 1318
non-sucre organique 1698
non-sucre résiduel 1966
non-sucres totaux 2525
noria 346
notopèdes 512
N.S. 1680
nucléation 1684

oeil de vache 645
oïdium 1797
O.I.S. 1341
oligosaccharide 1692
omicron nébuleux 363
oplismène 1722
organisation commune du
  marché du sucre 579
Organisation Internationale du
  Sucre 1341

Organisation Mondiale de la
  Recherche sur le Sucre 2721
organisme d'intervention 1343
orge à crinière 1116
orge à épis à crinière 1116
ortie brûlante 2187
ortie dioïque 2283
ortie épineuse 575
ortie ornée 1427
ortie pourpre 1908
ortie romaine 2187
ortie rouge 1910
ortie royale 575
oseille acide 582
oseille à feuilles obtuses 334
oseille commune 582
oseille crépue 711
oseille des brebis 2130
otiorrhynque de la livèche 47
ouverture arrière 765
ouverture arrière réelle 2552
ouverture avant 978
ouverture d'entrée 978
ouverture de sortie 765
ouverture moyenne 1560
ouvrier cuiseur 1723
ouvrier turbineur 456
oxalate 1712
oxalate de potassium 1793
oxalide petite oseille 2714
oxalis droite 2585
oxalis petite oseille 2714
oxalis raide 2585

PAC 565
"Pahala" 1716
pain de sucre 2379
pales agitatrices 1715
pales d'agitation 1715
palpeur de niveau 1457
panic à tiges grêles 2711
panic capillaire 2711
panic pied-de-coq 1722
panic sanguin 1426
panic vert 1195
panier cylindrique 729
panier de centrifuge 450
panier de centrifuge avec fond
  conique en pente forte 451
panier de centrifugeuse à fond
  plat 1082
panier d'essoreuse à fond plat
  1082
panier oscillant 1701
panneau de bagasse 120
panneau de fibre de bagasse
  127
panneau dur 1218
panneau isolant 1322

papier à filtrer 1030
papier de phénolphtaléine 1755
papier filtrant 1030
papier filtre 1030
papier goudronné 274
papier indicateur 1307
pâquerette 731
pâquerette vivace 731
paquet de cannes 378
parée 257
pas-d'âne 559
pas-d'âne commun 559
pas du coupe-cannes 1769
pas d'un couteau de coupe-
  racines 842
passerage à fleurs denses 1800
passerage champêtre 1004
passerage des champs 1004
passerage de Virginie 2626
passerage drave 1252
passerage lépidier 1252
pâte à papier 1727
pâte de sucre moulu 2183
patience crépue 711
patience sauvage 334
patte de lièvre 533
pâturin annuel 69
Pauly 1732
pavot coquelicot 652
pavot sauvage 652
pays à sucre 2357
pays et territoires d'outre-mer
  1711
pays exportateur de sucre 2372
pays importateur de sucre 2378
pays producteur de sucre 2357
pays producteur de sucre de
  betterave 242
pays producteur de sucre de
  canne 407
pays sucrier 2357
pays tiers 2494
PCI 1667
P.C.S. 1201
pectine 1738
pégomyie de la betterave 1545
peigne 561, 2061
peigne à messchaert 1579
peigne de Vénus 2131
pellet 1739
pelletisation 1742
pensée sauvage 1002
pentosane 1745
peptide 1746
perce-oreille 892
percolation du jus 1747
période d'intercampagne 1688
perlite 1748
persicaire douce 1750

Deutsch

Greiferkran 1174
Greiferlader 1175
Greifwinkel 65
Grießzucker 531
Grindkraut 1203
großähriges Liebesgras 2285
große Brennessel 2283
große (europäische) Wühlmaus 2630
große Nessel 2283
großer Ampfer 582
großer Sauerampfer 582
großer Wegerich 1193
große Schildläuse 1548
Grünablauf 1198
grünähriger Fuchsschwanz 2191
Grundmenge 157
Grundpreis 156
Grundquote 158
Grundverordnung 159
grüne Baumwollblattlaus 662
grüne Borstenhirse 1195
grüne Erbsenblattlaus 1734
grüne Futterwanze 574
grüne Gurkenblattlaus 662
grüne Melonenblattlaus 662
grüne Pfirsichblattlaus 1735
grüne Pfirsichlaus 1735
grüner Ablauf 1198
grüner Fennich 1195
grüner Kurs 1197
grüner Pippau 2190
Gummi 1210
Gummikrankheit 1211
Gummosis 1211
Gur 1212
Gurkenmosaik 707
Gurkenmosaikvirus 708
Gurkenspringschwanz 1158
Gürtelschorf 1167
Gurtförderer 146
Gußwürfelverfahren 17
Guter Heinrich 1173
GZT 572

Haarhirse 2711
Haarwurzel 1997
Häckselmaschine für Zuckerrohr 380
Häckselmesser 2176
Häckselrolle 484
Haferdistel 676
Haferwurz 2611
Haferwurzel 2611
Hagelzucker 2019
Hahnenfuß 361
halbautomatische Zentrifuge 2109
halbmagere Rübe 1214

Halbzuckerrübe 1103
Hammer 1215
Hammershredder 1216
Handausdünnung 1217
Handelszucker 564
Handvereinzelung 1217
hanfartige Hundswolle 1305
Hanfhundsgift 1305
Hanfklee 2679
Hanfkraut 2734
Hanfpappel 578
hängende Heizkammer 1090
hängende Schneidmaschine 2436
harter Kristall 1219
hartes Korn 1220
Hartfaserplatte 1218
Harz 1968
Hasenpappel 1537
Hauptausschreibung 1836
Hauptfiltrat 1531
Hauptkalkung 1532
Hauptscheidung 1532
Hauptüberschußgebiet 1533
Hawaii-Quotient 1224
Hawaii-Verhältnis 1224
Hederich 2697
Heerwurm 86, 176
Hefe 2725
Heidekraut-Wanze 965
Heißdampfkühler 786
heiße Digestion 1262
heiße Kalkung 1265
heiße Mazeration 1266
heißes Filtrat 1263
heißes Saccharat 1269
heiße Sulfitation 1270
heiße Vorkalkung 1267
heiße Vorscheidung 1267
Heizdampf 1234
Heizfläche 83, 1235
Heizkammer 366, 1229
Heizkammer mit festen Rohrböden 1075
Heizkammer mit schrägen Rohrböden 1299
Heizkörper 1228
Heizmittel 1231
Heizrohr 1236
Heizschlange 1230
Heizstufe 1233
Heizwert 372
hellgrüne Zwergzikade 1196
Hemmstoff 1313
Herz- und Trockenfäule 315, 882
Hexenkraut 585
Hexose 1242
High-Test-Melasse 1248

Hilfspumpe 2245
hintereinander fahrende Vollerntemaschinen 1104
Hirtentäschelkraut 2132
Hochrinne 23
Hochsteigen 1008
Höchstquote 1557
hochtourige Zentrifuge 1247
Hochvakuum 1249
hohe Ambrosie 581
Hohlköpfigkeit 74
holzige Rübe 2715
holziges Gewebe 2716
Hopfenklee 281
Hopfenluzerne 281
Horizontalschneidmaschine 237
Horizontaltrockner 1257
Hubrad 259
Huflattich 559
Hügel-Vergißmeinnicht 891
Hühnerhirse 1722
Hühnerkraut 568
Hühnertod 283
Hundsgras 533
Hundshirse 268
Hundskohl 70
Hundszahn 268
Hundszahngras 268
Hutzucker 1494
Hydraulikkolben 1281, 1282
hydraulischer Deckel 1280
hydraulischer Druck 1283
Hydrometer 1285
hydrostatischer Druck 1286
Hygrometer 1287

"Icing Sugar" 1288
ICUMSA 1336
Iliaukrankheit 1289
Imbibition 1290
Imbibitionssaft 1291
Imbibitionstrog 1292
Imbibitionswasser 1293
Imbibitionswasserzerstäuber 1294
impfen 2088
Impfkorn 2097
Impfkristall 2094
Impfling 697
Impfung mit Puderzucker 1180
Impfzucker 2101
Indanthrenblau 1304
Indikator 1306
Indikatorpapier 1307
Infusorienerde 797
Inhibitor 1313
Injektionswasser 1315
Inkrustationen 1303
intermediäres Gebiet 1332

Español

AAC 572
ababa 652
ababol 652
abastecimiento de azúcar 2395
abejorro (común) 532
abertura de entrada 978
abertura delantera 978
abertura de salida 765
abertura trasera 765
abertura trasera verdadera 2552
abertura verdadera de salida
    2552
ablandamiento 2197
ablandar 2196
abono nitrogenado 1673
abremanos 649
abrepuños 647, 676
abrojo terrestre 1864
abutilón 2612
ácaro 1918
ácaro rojo 1920
acción bacteriana 108
acedera 711
acedera común 582
acedera menor 2130
acedera obtusifolia 334
acederilla 2130
acelga silvestre 2692
acetato básico de plomo 155
acetato de plomo 1436
acetona 6
aciano 648
acidez del guarapo 9
acidez del jugo 9
acidez del suelo, daños por ~
    11
acidez efectiva 897
acidez final 1049
acidez total 2523
acidificación 7
acidificar 8
ácido carbónico 427
ácido clorhídrico 1284
ácido fosfórico 1759
ácido láctico 1419
ácido málico 1536
ácido orgánico 1697
ácido oxálico 1713
ácido picrámico 1764
ácido pícrico 1765
ácido sulfámico 2402
ácido sulfúrico 2415
aclareadora para remolachas
    246
aclarear 2152
aclareo 2161
aclareo a mano 1217
aclareo con escardillo 1254

acondicionamiento de agua de
    alimentación 982
Acuerdo Internacional
    Azucarero 1339
Acuerdo Internacional del
    Azúcar 1339
Acuerdo Internacional sobre el
    Azúcar 1339
acumulador de vapor 2250
adherencia de la inflorescencia
    1311
adición de cal 1469
aditivo de clarificación 497
aditivo para el silaje 2142
adjudicación 2475
adjudicación complementaria
    2431
adjudicación permanente 1749
adjudicación principal 1836
adsorbente 18
adsorción 19
áfido del algodón 662
afilado 2125
afiladora 2126
afilar 2124
afinabilidad 29
afinación 25
afinado 24
afinar 32
afinidad por el agua 33
agalla de corona 679
agalla del cuello 679
agalla de Xanthomonas 2723
agallas de las raíces
    secundarias 1995
agente de floculación 1094
agente para de(s)colorar 743
agitación 2288
agitador 2287
aglomerado 606
aglomerarse 40
agotabilidad de la melaza 941
agotamiento 942
agotamiento de la melaza 1623
agotamiento de la miel 941
agricultor de caña (de azúcar)
    388
agricultor de remolacha 195
agrilla 582
agropiro común 664
agua complementaria 1534
agua de alimentación 981
agua de alimentación de
    calderas 296
agua de circulación 1644
agua de condensador 602
agua de conducción 1101
agua de constitución 2580
agua de difusión 815

agua de dilución 820
agua de enfriamiento 641
agua de extracción 815
agua de filtros 1041
agua de flotación 1101
agua de imbibición 1293
agua de inyección 1315
agua de las prensas 1826
agua de lavado 2639, 2640, 2641
agua de lavadura 2641
agua de movimiento 1644
agua de prensado de difusión
    810
agua de refrigeración 641
agua fresca 1130
agua higroscópica 2580
agua indeterminada 2580
agua lavadora 2641
agua original 1700
agua para saetines 1101
agua refrigerante 641
aguas condensadas 596
aguas de difusión 169
aguas dulces 2446
aguas endulzadas 2446
aguas para el lavado 2446
aguas residuales 2119
aguas residuales de fábrica de
    sacaratos por el
    procedimiento de Steffen
    2280
aguas residuales de la
    instalación de sacarato 2280
aguas servidas 2119
aguas sucias 2119
aguja 586
aguja de pastor 334, 2131
aire comprimido 591
aire primario 1833
aire secundario 2079
ajuste de salida 768
alacrán cebollero 1632
alcalinidad 52
alcalinidad efectiva 898
alcalinidad final 1050
alcalinidad natural 1660
alcalinidad residual 1961
alcalinización 53, 1483
alcalinización continua 621
alcalinización en caliente 1265
alcalinización en frío 543
alcalinización fraccionada 1117
alcalinización fraccionada y
    doble calefacción 1118
alcalinización insuficiente 1321
alcalinización intermedia 1328
alcalinización intermitente 1333
alcalinización principal 1532
alcalinización secundaria 2082

azúcar cristalizado 695
azúcar crudo 1893
azúcar crudo a granel 355
azúcar de almidón 2247
azúcar de alta calidad 1243
azúcar de baja calidad 38
azúcar de caña 404
azúcar de consumo 611, 1505
azúcar de envase 564
azúcar de la última templa 38
azúcar de melaza 1627
azúcar de pilón 1494
azúcar de primera 94
azúcar de remolacha 239
azúcar de segunda 343
azúcar de segundo producto 38
azúcar de tan alta polarización 2619
azúcar de tercera 701
azúcar de vainilla 2603
azúcar de vainillina 2604
azúcar dextra-rotatorio 793
azúcar dextrógiro 793
azúcar disponible 1900
azúcar doble refinado 853
azúcar embolsado 135
azúcar en cristales 695
azúcar en cuadradillos 705
azúcar en cubos 705
azúcar en pan 1494
azúcar en pancitos 705
azúcar en polvo 436, 1796
azúcar ensacado 135
azúcar en terrones 705
azucarera 2373
azucarera de caña 405
azucarera de remolacha 240
azucarera sin instalación de sacarato 1677
azucarero 2382
azúcares reductores 1927
azúcares totales 2527
azúcar fermentable 989
azúcar granulado 1187
azúcar "icing" 1288
azucarillo 2379
azúcar incristalizable 1676
azúcar invertido 1350
azúcar levógiro 1423
azúcar levo-rotatorio 1423
azúcar líquido 1488
azúcar mascabado 1893
azúcar melado 1627
azúcar moreno 435
azúcar no centrífugo 1675
azúcar preferencial 1813
azúcar preferencial ACP 13
azúcar primera 94
azúcar pulverizado 1796

azúcar quemado 414
azúcar recuperable 1900
azúcar refinado 1934
azúcar refinado líquido 1486
azúcar segunda 343
azúcar sin refinar 435
azúcar soluble 2207
azufre 2411
azular 291
azul de bromofenol 336
azul de indantreno 1304
azulear 291
azulejo 648
azul ultramar 2573
azul ultramarino 2573

babosa Agriolimax reticulatus 1005
babosa Deroceras reticulatum 1005
bacteriosis 111
bagacera 2533
bagacero 767
bagacillo 713
bagazo 119
bagazo comprimido 592
bagazo del primer molino 1070
bagazo del segundo molino 2085
bagazo del último molino 119
bagazo de presión seca 873
bagazo final 119
bagazo fino 713
bagazo fresco 1127
bagazo intermedio 1324
bajo vacío 1518
bala de bagazo 142
balance azucarero 2324
balance de azúcar 2323
balance de la instalación de cocción 299
balance del azúcar invertido 1351
balance del Brix 329
balance de melaza 1621
balance de vapor 2251
balance térmico 2476
balanza de azúcar 2389
balanza de guarapo 1395
balanza de jugo 1395
balsa de decantación 2116
ballueca 2239, 2704
banda de vapor 2609
banda esclerótica de la hoja 148
banda transportadora transversal 2138
baño refrigerante 635
barba cabruna 1559
barbón 1559
barita 153

barra de acoplamiento 2457
barrilla fina 2025
barrilla pinchosa 2025
báscula 2662
báscula de azúcar 2389
báscula de remolachas 177
báscula discontinua 166
báscula para ensacar 138
báscula para ensacar y envasar 136
batería de centrífugas 168
batería de difusión 803
batería de difusores 803
batería de molinos 1601
"beet necrotic yellow vein virus" 215
"Beet Savoy Disease" 2052
"Beet Yellow Net" 261
"Beet Yellow Wilt Disease" 2735
bellorita 731
bentonita 267
betaína 270
bibio de las huertas 1547
bledo 1845, 1914, 2191
bledo blanco 2561
BMYV 211
"BNYVV" 215
bolinas de mieses 2188
bolsa de pastor 2132
bolsa filtrante 1015
bolsillo de filtro 1032
bomba aspiradora de lodo espeso 2490
bomba centrífuga 457
bomba centrífuga con anillo líquido 2654
bomba de agua 2653
bomba de agua caliente 1271
bomba de agua de alimentación 983
bomba de agua fresca 1132
bomba de agua natural 1896
bomba de aguas condensadas 599
bomba de aire 45
bomba de aire húmedo 2667
bomba de aire y de agua 2667
bomba de alimentación 979
bomba de cachaza 1650
bomba de circulación 493
bomba de condensados 599
bomba de cosetas 659
bomba de chorro 1364
bomba de desplazamiento 839
bomba de diafragma 796
bomba de dosificar 844
bomba de émbolo 1768
bomba de filtrado 1043

condensador barométrico de contracorriente 665
condensador barométrico de corriente paralela 535
condensador corto de contracorriente 2134
condensador de contracorriente 667
condensador de corriente paralela 536
condensador de chorro 1362
condensador de chorros múltiples 1653
condensador de eyector 899
condensador de mezcla 1362
condensador de superficie 2433
condensador en contracorriente 667
condensador-eyector 899
condensador independiente 1308
condensador largo 151
condensados 596
conductibilímetro 604
conductivímetro 604
conductor auxiliar de cañas 102
conductor con cadenas de rascadores 432
conductor con cinta de persianas 2170
conductor de azúcar 2355
conductor de bagazo 121
conductor de cañas 379
conductor de chapulín 1189
conductor de gusano 2070
conductor de rascadores 2063
conductor de sacudidas 1189
conductor fijo 1076
conductor intermedio 1325
conductor saltamontes 1189
conductor tembladero 1189
conductor transversal de cañas 102
conducto transportador 633
Confederación Internacional de Remolacheros Europeos 1334
conglomerado 606
conjunto palpador 2518
Consejo Internacional del Azúcar 1340
conservabilidad 2294
consumo de azúcar 2351
consumo de vapor 2257
consumo mundial del azúcar 2719
contador de agua 2651
contenido de azúcar 2353
contenido de azúcares totales 2528

contenido de azúcar invertido 1352
contenido de cenizas 91
contenido de cristales 686
contenido de humedad 1278
contenido de marco 1546
contenido de materia seca 885
contenido de nitrógeno amino 62
contenido de no-azúcar 1681
contenido de no-sacarosa 1679
contenido de sacarosa 2314
contenido de sales de calcio 1476
contenido de sólidos 2202
contenido de sustancias reductoras 1926
contingente arancelario 2469
contracción 626
contracuchillo 671
contraplaca de la caja de cuchillos 1407
contrapresión 105
contrarregla de portacuchillas 1407
contrato a plazo 1149
contrato de almacenamiento 2291
contrato de suministro 764
contrato mundial 2720
control de la instalación de cocción 300
control de la tara 2464
control de molinos 1592
control de tachos 1721
corazoncillo 585
corazón negro 284
cornezuelos 916
coronilla real 2728
correa de transporte 630
correa de transporte con cierre-relámpago 2741
correa de transporte con nervios transversales 631
correa de transporte plana 1083
correa extractora de remolacha 1463
correa transportadora 146, 630
correa transportadora de sección de artesa 2550
corregüela anual 276
corregüela de los caminos 1409
corregüela mayor 1240
corregüela menor 997
corregüela negra 276
corregüela silvestre 997
correhuela anual 276
correhuela de los caminos 1409
correhuela mayor 1240

correhuela menor 997
correhuela negra 276
correhuela silvestre 997
cortacañas 383
cortadora con accionamiento inferior 2578
cortadora de remolachas 235
cortadora-desmochadora-atadora de caña 719
cortadora-desmochadora-hileradora de caña 720
cortadora horizontal de remolachas 237
cortadora suspendida 2436
cortadora vertical de remolachas 870
cortar en pedazos pequeños 2137
cortarraíces con disco horizontal 237
cosecha de las remolachas azucareras 2328
cosecha de remolacha 185
cosechadora de caña 483
cosechadora de remolacha 433, 588
cosechadora divisora de tallos 482
cosechadora integral de remolacha 431, 588
cosechadoras de remolacha de reata 1104
cosechadoras de remolacha en paralelo 1728
cosechadora troceadora 482
cosetas 660
cosetas agotadas 939
cosetas coladas 939
cosetas de cobija 1981
cosetas desazucaradas 939
cosetas frescas 1129
cosetas prensadas 1820
cotización 627, 1881
cotización a la producción 1839
cotización de almacenamiento 2292
cotización del azúcar blanco 2687
cotización diferencial 800
crecer 1205
crecer la masa cocida 351
crecimiento del grano 350
crecimiento de los cristales 687
criadero de semillas 2093
criba de difusor 812
criba de sacudidas 1763, 2123
cribadora 2069
criba mecánica 2067
criba oscilante 2123

inhibidor 1313
insolubilizar 1957
instalación clarificadora de
  aguas residuales 2644
instalación de calderas 297
instalación de cocción 1724
instalación de cuadradillos
  prensados 1818
instalación de difusión 957
instalación de evaporación de
  múltiple efecto 1656
instalación de extracción 957
instalación de melazar 1615
instalación de molienda para
  azúcar 2375
instalación desespumadora 2165
instalación despolvoreadora 889
instalación de toma de pruebas
  de remolacha 230
instalación fija de descarga en
  seco 2249
instalación para apagar cal 1480
instalación para cuadradillos
  706
instalación para el azúcar en
  cubos 706
instalación para la clasificación
  del azúcar 2390
instalación para la escaldadura
  de cosetas 661
Instituto Internacional de
  Investigaciones sobre la
  Remolacha 1337
instrumento de preparación de
  la caña 398
instrumento para control de
  tacho 709
intercambiador de calor 1226
intercambiador de iones 1355
intercambiador iónico 1355
intervención 1342
introducción de jarabe 1323
inversión 1345
inversión de la sacarosa 2318
invertasa 1348
invertina 1348
invertir 1347
inyección de vapor 2263
inyector de vapor 2264
irión 2536
isoglucosa 1358
isopropanol 1359
ISSCT 1338

jaggery 1360
jarabe 2450, 2451
jarabe de afinación 30
jarabe de glucosa 2248
jarabe espeso 2489

jarabe invertido 1248
jarabe rico 2638
jaramago 2460, 2536
jaramago amarillo 2696
jaramago blanco 2697
Java ratio 1361
juego de cuchillas 2114
jugo 1368
jugo absoluto 2
jugo ácido 10
jugo alcalinizado 749
jugo alcalino 50
jugo carbonatado 419
jugo clarificado 499, 1576, 2491
jugo claro 499, 2491
jugo colado 2295
jugo concentrado 2489
jugo crudo 1891
jugo crudo caliente 1268
jugo crudo frío 544
jugo de alta pureza 1244
jugo de bagacillo 1057
jugo de baja pureza 1514
jugo decantado 499
jugo de caña 390
jugo de carbonatación 419
jugo de circulación 489
jugo de decantación 2309
jugo de defecadoras 499
jugo de difusión 805
jugo defecado 749
jugo de imbibición 1291
jugo del último molino 1429
jugo de maceración 1524
jugo de molino 1597
jugo de molino de laboratorio
  1415
jugo denso 2489
jugo de presión 1819
jugo de presión seca 875
jugo de primera carbonatación
  1067
jugo de primera expresión 1069
jugo de pulpa prensada 1860
jugo de remolacha 199
jugo de segunda carbonatación
  2084
jugo desespumado 758
jugo desmineralizado 772
jugo de última expresión 1428
jugo difícil 1946
jugo diluido 819, 1609
jugo encalado 749
jugo estrujado 1819
jugo extraido 950
jugo filtrado 1024
jugo intermedio 1327
jugo medio 1584
jugo mezclado 1609

jugo neutral 1669
jugo normal 2581
jugo original 1699
jugo perdido 1504
jugo pobre 2659
jugo prealcalizado 1810
jugo predefecado 1810
jugo preencalado 1810
jugo primario 1835
jugo purificado 1865
jugo refractario 1946
jugo residual 1963
jugo saturado 2046
jugo secundario 2081
jugo sin diluir 2581
jugo sobrecarbonatado 1705
jugo sulfitado 2406
jugo termoestable 2481
jugo termolábil 2479
jugo turbio 519
juncia 1686
juncia avellanada 1682
juncia redonda 1686
junco de esteras 2199
junco de sapo 2512
junquilla 1686

kieselgur 797
Krajewski 1413

laboratorio de recepción 2466
lactato 1418
lactosa 1420
lado de alimentación 975
lado de escape del vapor 2259
lado de salida 769
ladrillo de bagazo 325
lámina raspadora 2062
lamio púrpura 1908
lampazo menor 1454
langosta emigrante 1587
langosta migratoria 1587
langosta migratoria europea
  1587
lanzadero helicoidal 2225
larva del abejorro 2677
larva de la c-negra 2232
lavado 2633
lavado a vapor 2272
lavado con agua 2658
lavado con aspersores 2634
lavado con vapor 2272
lavado de las espumas 2075
lavado exterior 948
lavado húmedo de la caña 2668
lavadora de ciclón 728
lavadora de remolachas 255
lavador de ácido carbónico 428
lavador de gas 1161

lavador de hoja de remolacha 207
lavador de muestras de remolacha 231
lavador de nabos de remolacha 245
lavador de remolachas 255
lavador de toberas 1365
lavador de tobera tipo "jet" 1365
lavado simple 2151
lavado sistemático 2455
lavar 787, 2632, 2637
lavar la torta 2444
LDP 1496
LDP(R) 1434
LDP(W) 1435
lechada de cal 1472
lechada de sacarato 2029
lechecino 1832
lecherulla 2421
lechetrezna 1449
lechetrezna común 2421
lecho fijo 1074
lechuguilla 653
lechuguilla silvestre 583
lengua de buey 566
lengua de pájaro 1409
lengua de vaca 711
lepra 227
levadura 2725
levulosa 1140
libre circulación 1123
licopsis arvensis 349
licor clarificado 1576
licor de azúcar lavado 1576
licor del equipo Quentin 1875
licor flotante 2427
licor madre 1641
licor standard 2241
liebrecilla 648
ligadura myriogenospora de la hoja 1659
limpiadora cargadora de remolacha 1492
limpiadora centrífuga 678
limpiadora de remolachas 183
limpieza húmeda de la caña 2668
limpiezar con vapor 2267
linaria 2734
linaria amarilla de hoja estrecha 655
lingotes de azúcar 1312
líquido de extracción 951
líquido flotante 2427
lixiviación 1491
lixiviar 949
lobillo 2703

lodo 1651, 2181
lonja del azúcar 2370
losa de azúcar 2167
lupulina 281

llantén blanquecino 1253
llantén común 1193
llantén grande 1193
llantén lanceolado 1979
llantén major 1193
llantén mediano 1253
llantén menor 1979
llave de prueba 1842
llave de vapor 2254
llenadora 1007

maceración 1523
maceración caliente 1266
maceración en baño 1523
macerador de cosetas 658
macerador de cosetas en contracorriente 668
magarza inodora 2059
magarzuela 2286
magma 1526
magma C 525
magma de afinación 27
magma de alta pureza 1245
magma de bajo grado 1508
magma de cristales para semilla 1527
maicillo 1366
malato 1535
mal del esclerocio 2211
malduerme 2703
maltosa 1538
malva 1537
malva común 578
malva de flor pequeña 2186
malva enana 577
mal vinoso 2625
malla 1162, 1577
malla de sostén 104
malla sandwich 2045
mancha alada 2613
mancha café 338
mancha concéntrica 2467
mancha de anillo 1986
mancha de ojo 960
mancha púrpura de la hoja 1911
mancha roja de la hoja 1911
mancha roja de la vaina 1921
manómetro 1824
manzana espinosa 2497
manzanilla común 2694
manzanilla de Alemania 2694
manzanilla de los campos 645
manzanilla fétida 2286

manzanilla hedionda 2286
manzanilla inodora 2059
manzanilla silvestre 645
máquina centrífuga 449
máquina completa 588
máquina cortadora de remolachas 235
máquina cortarraíces 235
máquina de limpieza a chorro de vapor 2266
máquina de llenado de sacos 137
máquina de llenar 1007
máquina de vapor 2258
máquina llenadora de sacos 137
máquina para cortar azúcar 515
máquina para cortar azúcar en cubos 703
máquina para cortar cuadradillos 703
máquina para la recogida de cañas enteras 2691
máquinas combinadas en paralelo 1728
máquinas combinadas en serie 1104
máquina simple 2147
máquinas para el manejo de las cañas 389
marco 1120
marco de remolachas 209
marco filtrante 1025
marchitez 2699, 2735
margarita 731
margarita de los sembrados 650
margarita del trigo 650
margen de fabricación 1838
margen de maniobra 1959
margen de refinación 1940
margen de refinado 1940
mariquita 1421
martillo 1215
masa 1553
masa cocida 1554
masacocida 1554
masa cocida A 58
masa cocida AB 1
masa cocida B 293
masa cocida C 526
masa cocida de afinación 28
masa cocida de alta calidad 58
masa cocida de alta graduación 58
masa cocida de alta pureza 1246
masa cocida de baja calidad 1509
masa cocida de baja pureza 1509, 1515

prensa helicoidal 2071
prensa helicoidal doble 854
prensa para cinta de filtro 149
prensa para la pulpa fina 1060
prensa para la pulpilla 1060
prensa para pulpa 1858
prensa para pulpa seca 1856
prensa por azúcar en cubos y
  tabletas 704
preparación de la cama de
  siembra 2091
preparación de la caña 397
preparación del agua 2645
preparación de la lechada de
  cal 1473
presión absoluta 3
presión de agua 2652
presión de espuma 1137
presión de filtración 1047
presión de régimen 2718
presión de servicio 2718
presión de trabajo 2718
presión de vapor 2268
presión hidráulica 1283
presión hidrostática 1286
presión húmeda 2669
presión seca 872
prima de desnaturalización 775
primera carbonatación 1066
primera centrífuga 1068
primero producto 1072
primeros molinos 1071
procedimiento de Adant 17
procedimiento de difusión 811
procedimiento de extracción
  811
procedimiento de Steffen 2279
producción azucarera 2385
producción de azúcar 2385
producto azucarado 2352
producto de bajo grado 37
producto de fábrica de azúcar
  2374
producto final 908
productor de azúcar 2382
productor de semillas 2098
productos A 81
productos B 320
productos C 673
producto transformado 2531
proteína 1846
protuberancias de las raíces
  secundarias 1995
provisión de la noche 1709
provisión nocturna 1709
prueba al hilo 2301
prueba de soplado 345
pseudomonas 109
pudrición café 337

pudrición cytospora de la vaina
  2128
pudrición de la base del tallo
  154
pudrición de la espiga 87
pudrición de la raíz 1999
pudrición de la vaina 2129
pudrición del tallo 1148
pudrición del tronco 547
pudrición phytophthora de las
  estacas (semilla) 1762
pudrición roja de la vaina 1916
pudrición seca 883
pudrición seca del cogollo 886
puente-báscula 2662
puente-báscula de camión 2551
puente pesador 2662
puente pesador para camiones
  2551
pulgón del algodón 662
pulgón del algodonero 662
pulgón de la remolacha 275
pulgón del melocotonero 1735
pulgón del melón 662
pulgón de los melones 662
pulgón gris del melocotonero
  1735
pulgón negro 275
pulgón verde de la arveja 1734
pulgón verde del duraznero
  1735
pulgón verde del guisante 1734
pulgón verde del melocotonero
  1735
pulguilla 2566
pulguilla de la remolacha 1544
pulpa de melaza 1620
pulpa de papel 1727
pulpa de procedimiento de
  escaldado 1855
pulpa de remolacha 939
pulpa desazucarada 784
pulpa fina 1059
pulpa húmeda 2672
pulpa melazada 1620
pulpa prensada 1820
pulpa saturada de azúcar 2325
pulpa seca 861
pulpa seca melazada 1619
pulpa Steffen 1855
pulpilla 1059
pulverizador de agua de
  imbibición 1294
puntista 1723
punto de cocción 302
punto de ebullición 302
punto de fusión 1575
punto de rocío 790
punto de saturación 2050

pureza 1866
pureza aparente 78
pureza del guarapo 1392
pureza del jugo 1392
pureza en peso 1192
pureza en pol porcentaje Brix
  78
pureza por gravedad 1192
pureza verdadera 2555
purgado 454
purgado doble 852
purgador de vapor 2270
purgado simple 2159
purgar 460
purificación de agua de
  alimentación 984
purificación del guarapo 1390
purificación del jugo 1390

quíntuple efecto 1877
quintuplo efecto 1877

rabanilla 2697
rabanillo blanco 2697
rabaniza 2697
rabaniza común 2697
rábano cimarrón 2697
rábano silvestre 2697
rafinosa 1883
raicilla 1997
raicillas de remolachas 244
raíz 1994
rajadura de la hoja 1445
raleadora para remolachas 246
ralear 2152
raleo 2161
ramularia 1887
ramulariosis 1887
ranúnculo 361
ranúnculo arvense 647
ranúnculo de los campos 647
ranúnculo rastrero 675
ranúnculo silvestre 647
ranurado de un cilindro 1200
ranura messchaert 1578
ranuras de un cilindro 1200
raquitismo de la caña,
  enfermedad del ~ 1890
raspador 2061
raspadores giratorios 1973
raspador Messchaert 1579
raspador superior 2520
raspadura 323
raspa para remolachas 223
rastrillo 1885
rastrillo amontonador 1869
rastrillo amontonador de
  coronas 252
rastrillo de cañas 399

tambor rotativo para apagar la cal 1479
tambor secadero de azúcar 1188
tambor secador 879
tamisador de lechada de cal 1590
tamiz 2064
tamizado 1980
tamizado del guarapo 1398
tamizado del jugo 1398
tamizador de rendijas 2180
tamizar 2140
tamiz de bagacillo 715
tamiz del bagazo 133
tamiz de sacudidas 2123
tamiz giratorio 2010
tamiz para guarapo 1396
tamiz para jugo 1396
tamiz reforzado 104
tamiz rotativo 2010
tamiz vibratorio 2123, 2622
tándem de molinos 1601
tanque alimentador del evaporador 934
tanque de agitación 41
tanque de agua de alimentación 985
tanque de alcalinizar 1484
tanque de alimentación 985, 2432
tanque de alimentación del evaporador 934
tanque de carbonatación 426
tanque de carbonatar 426
tanque de decantación 2118
tanque de depósito 2432
tanque de encalar 1484
tanque de expansión 1080
tanque de flash 1080
tanque de guarapo de circulación 492
tanque de jugo de circulación 492
tanque de medición 1562
tanque mezclador 1616
tanque para refundición 1572
tapa 411
tapa de bayoneta 170
tara 2463
tara de hojas y coronas 2522
tara de tierra 825
tara tanto por ciento 825
tara total 2529
taraxacón 733
Tarifa Aduanera Común 572
tasa verde 1197
técnico azucarero 2396
tecnología azucarera 2397
tecnólogo azucarero 2396

tejido filtrante 1027
tejido leñoso 2716
tejido para tamices 2068
tela de apoyo 104
tela de sostén 104
tela filtrante 1022
tela para filtrar 1022
tela perforada 2179
telaspio 1003
telaspios de los campos 1003
temperatura de difusión 813
temperatura de ebullición 306
temperatura de extracción 813
temperatura de saturación 2051
temperatura final 1056
temperatura inicial 1314
templa 2300
templa de baja calidad 1512
templa de tercer producto 1512
templa final 1054
tensión superficial 2434
terceros países 2494
tercer producto 37
termocompresión 2477
termocompresor 2478
termoestabilidad del guarapo 2480
termoestabilidad del jugo 2480
tiempo de permanencia 1960
tiempo de residencia 1960
tiempo muerto 1688
tierra de diatomeas 797
tierra de infusorios 797
tijereta 892
timol 2504
tiña de la remolacha 214
tiocarbamida 2493
tiourea 2493
tipo de cambio 937
tipo de cambio representativo 1958
tipo de color 558
tipo verde 1197
tipula 674
tipula de las huertas 570
tirado forzado 1110
titración 2510
titrar 2508
titulación 2510
titular 2508
tizón de la hoja 1438
tolueno 2513
tolva 1256
tolva de alimentación 973
tolva de azúcar 2376
tolva de carga 973
tolva de imbibición 1292
tolva de parrilla de raíces 551
tolva de remolachas 198

toma de incondensables 2712
toma de los cilindros 1199
toma de muestras de caña 402
toma de vapor 2605
tomador automático de muestras de guarapo 99
tomador automático de muestras de jugo 99
tomador de muestras de guarapo 1394
tomador de muestras de jugo 1394
tomapruebas 2041
tomtatillos 283
topillo campesino 613
tornabagazo 2533
tornasol 2420
tornillo sin fin 2070
tornillo sin fin de remolachas 232
tornillo transportador 2070
torre de difusión 958
torre de enfriamiento 640
torre de extracción 958
torre de filtros 1039
torre de refrigeración 640
torre de sulfitación 2418
torta 1016
torta de cachaza 1016
torta de filtro 1016
torta de filtro-prensa 1034
torta de sacarato 2028
torta porosa 1786
tortuguilla de la remolacha 517
trabajo con pie de semilla 1108
tráfico de perfeccionamiento activo 1353
trampa de alimentación automática 100
trampa de vapor 2270
transportador 629
transportador de cadena con artesa 2548
transportador de correa 146
transportador de cosetas 656
transportador de las tortas 1017
transportador de rastrillo 2063
transportador de rastrillos 1886
transportador de remolachas 184, 254
transportador de rodillos 1991
transportador de tornillo sin fin 2722
transportador oscilante 2623
transportador transversal 2138
transportar en el canal de flotación 1099
transporte de remolachas 253
tratamiento del agua 2645

Nederlands

voorfabriek 190
voorfixatie 20
voorgekalkt sap 1810
voorkalken 1809
voorkalking 1816
voorkalking volgens Brieghel-
   Müller 1841
voorlegstuk van messenkast
   1407
voorperssap 1069, 1835
voorraadrooier op dwarszwad
   1462
voorscheiden 1809
voorscheiding 1816
voorschot 1731
voorste onderste rol 980
voorverdamper 1811
voorverwarmer 1814
voorwarmer 1814
voorwarmingsruimte 1815
voorwarmingszone 1815
vorming van kristalkernen 1684
vrije suikermarkt 1125
vrijhandelszone 1126
vrij verkeer 1123
vruchtbeginselbrand 1703
vruchtesuiker 1140
vuilklasseerapparaat 826
vuilsap 419
vuilsappomp 421
vuilsapverdeler 420
vuilwater 2119
vulling 470
vulmachine 1007
vulmassa 1554
vulmassa indikken, de ~ 351
vulmassaklonters 145
vulmassapomp 1556
vulmassaschuif 1555
vulmassa van hoge reinheid
   1246
vultrechter 973
vuurkruid 2461

waar temperatuurverschil 2556
wachtbak 2432
wachtkristallisoir 828
wagonipinrichting 2631
wagontipper 2631
ware afvoeropening 2552
ware densiteit 2553
ware diffusie 2554
ware reinheid 2555
ware zuiverheid 2555
warkruid 843
warme digestie 1262
warme imbibitie 1264
warme kalking 1265
warme maceratie 1266

warme sulfitatie 1270
warme voorkalking 1267
warme voorscheiding 1267
warm filtraat 1263
warm ruwsap 1268
warm saccharaat 1269
warmtebalans 2476
warmtedoorgangscoëfficiënt 537
warmtemeter 373
warmteuitwisselingsoppervlak
   1227
warmteverliezen 1237
warmtewisselaar 1226
warmwaterpomp 1271
wasfiltraat 2635
wasmolen voor bietenstaartjes
   245
wassen 787
waswater 2639, 2641
waterafscheider 2655
waterbereiding 2645
waterdruk 2652
waterkanon 1635
waterkoeling 2648
watermeter 2651
waterpijpketel 2656
waterpomp 2653
waterringpomp 1487, 2654
waterstraal 1363
waterstraalcondensor met
   meerdere straalpijpen 1653
weegbrug 2662
weegbrug voor vrachtauto's
   2551
weegschaal voor het opzakken
   138
week kristal 2195
weerhouding 1970
weggras 1409
wereldcontract 2720
Wereldorganisatie voor
   Suikeronderzoek 2721
wereldsuikerconsumptie 2719
wereldsuikerverbruik 2719
werkdag 2717
werken met pied-de-cuite 1108
"Western yellows" 2666
wieltaster 1992, 2012
wikke 2621
wikkelmachine 550
wilde boekweit 276
wilde dragon 2536
wilde goudsbloem 650
wilde haver 2239
wilde kamille 645
wilde latuw 1831
wilde mosterd 2696
wilde sla 1831
winbare saccharose 1899

winbare suiker 1900
winbare witsuiker 1901
winbare witte suiker 1901
windhalm 2144
wisselkoers 937
witbol 2737
witrot 2682
witsuiker 2684
witsuikercentrifuge 2685
witsuikerfabriek 2686
witsuikernotering 2687
witsuikeropbrengst 2689
witsuikeropbrengst per ha 2690
witsuikerwaarde 2688
witte amarant 2561
witte honingklaver 2679
witte klaverzuring 2714
witte krodde 1003
witte mosterd 2680
witte steenraket 1221
witte stroop A 2683
witte suiker 2684
witte-vlekkenziekte 2681
woelrat 2630
wortel 1994
wortelbrand 280
wortelknobbel 679
wortelknobbelaaltje 1996
wortelknobbels 1995
wortelrot 1999
wortelstokziekte 154

Xanthomonas-gal 2723
xylenol 2724

"41-yellows" 2731
"Yellow Wilt" 2735
Yellow Wilt-virus 265

zaadbed 2090
zaadbiet 196
zaaddrager 2089
zaaddroger 2096
zaadselecteerder 2092
zaadveredelingscentrum 2093
zaadvermeerderaar 2098
zaaibed 2090
zaaidichtheid 2095
zachte ooievaarsbek 857
zacht grein 2198
zacht kristal 2195
zakfilter 134
zakken doen, in ~ 117
zakken plaatsen, in ~ 117
zakkenstapelaar 2037
zakkenvul- en afweegmachine
   136
zakkenvulmachine 137
zandafscheider 2043

Latine

321 *Xanthium strumarium*

Perkinsiella saccharicida 2344
Phalaris sp. 2650
Philaenus leucophthalmus 1558
Philaenus spumarius 1558
Philopedon plagiatum 2044
Phlyctaenodes sticticalis 257
Phragmites communis 1930
Phthorimaea ocellatella 214
Phyllophaga sp. 2678
Phyllotreta nemorum 2566
Phyllotreta striolata 2304
Physalis longifolia 1497
Physalis subglabrata 2189
Phytometra gamma 1156
Piesma cinerea 1417
Piesma quadratum 1416
Plantago lanceolata 1979
Plantago major 1193
Plantago media 1253
Platynota stultana 2660
Plusia gamma 1156
Poa annua 69
Polydesmus angustus 1087
Polygonum argyrocoleon 2146
Polygonum aviculare 1409
Polygonum convolvulus 276
Polygonum lapathifolium ssp.
  lapathifolium 353
Polygonum lapathifolium ssp.
  pallidum 1717
Polygonum pensylvanicum 1744
Polygonum persicaria 1750
Polygonum ramosissimum 915
Polygonum sp. 1410
Portulaca oleracea 1868
Prodenia ornithogalli 2733
Pseudaletia unipuncta 86
Psylliodes punctalata 1255

Ranunculus arvensis 647
Ranunculus repens 675
Ranunculus sp. 361
Raphanus raphanistrum 2697
Rhopalosiphonimus staphyleae
  494
Rumex acetosa 582
Rumex acetosella 2130
Rumex crispus 711
Rumex obtusifolius 334

Saccharum officinarum 2339
Salsola kali 2025
Scandix pecten-veneris 2131
Scotia segetum 2703
Scotogramma trifolii 521
Scrobipalpa ocellata 1279
Scutigerella immaculata 1159
Senebiera coronopus 2447
Senecio aureus 1171

Senecio jacobaea 1884
Senecio vulgaris 1203
Setaria faberii 1166
Setaria glauca 2730
Setaria viridis 1195
Sida hederacea 49
Silene alba (Mill.) Krause 2676
Silene dioica (L.) Clairv. 1907
Silene inflata Sm. 288
Silene vulgaris (Moench) Gar.
  288
Silpha bituberosa 2223
Sinapis alba 2680
Sinapis arvensis 2696
Sinapis sinapistrum 2696
Sisymbrium altissimum 2460
Sisymbrium sophia 2461
Sminthurus sp. 2237
Solanum carolinense 1260
Solanum nigrum 283
Solanum rostratum 347
Solanum triflorum 717
Sonchus arvensis 653
Sonchus asper (L.) Hill. 1832
Sonchus oleraceus 583
Sonchus sp. 2214
Sorghum halepense 1366
Spergula arvensis 654
Spilosoma virginica 2727
Spodoptera exigua 176
Stachys arvensis 996
Stachys palustris 1552
Stellaria media (L.) Vill. 568
Syrphus corollae 2449
Syrphus sp. 1276
Systena blanda 1719
Systena taeniata 147

Tanymecus palliatus 258
Taraxacum officinale Web. 733
Tetanops myopaeformis 2335
Tetranychus althaeae 1918
Tetranychus sp. 1919
Tetranychus telarius 1920
Tetranychus urticae 1920
Thlaspi arvense 1003
Thrips anguisticeps 364
Thrips spp. 2503
Thrips tabaci 1693
Thyanta pallido-virens 1917
Tipula oleracea 570
Tipula paludosa 571, 1451
Tipulidae 674
Tragopogon porrifolius 2611
Tragopogon pratensis 1559
Trianthema portulacastrum
  1261
Tribulus terrestris 1864
Triticum repens 664

Tussilago farfara 559

Urtica dioica 2283
Urtica urens 2187

Vaccaria pyramidata 672
Veronica officinalis 584
Veronica sp. 2221
Vicia sp. 2621
Viola tricolor 1002

Xanthium pensylvanicum 569
Xanthium strumarium 1454